D0805913

FOUNDATIONS OF
INFINITESIMAL STOCHASTIC ANALYSIS

STUDIES IN LOGIC

AND

THE FOUNDATIONS OF MATHEMATICS

VOLUME 119

NORTH-HOLLAND
AMSTERDAM • NEW YORK • OXFORD

FOUNDATIONS OF INFINITESIMAL STOCHASTIC ANALYSIS

K. D. STROYAN
Mathematics Department
The University of Iowa
Iowa City, Iowa 52242
U.S.A.

and

José Manuel BAYOD
Facultad de Ciencias
Universidad de Santander
Santander, Spain

1986

NORTH-HOLLAND
AMSTERDAM • NEW YORK • OXFORD

ISBN: 0 444 87927 7

Published by:

ELSEVIER SCIENCE PUBLISHERS B.V.
P.O. Box 1991
1000 BZ Amsterdam
The Netherlands

Sole distributors for the U.S.A. and Canada:

ELSEVIER SCIENCE PUBLISHING COMPANY, INC.
52 Vanderbilt Avenue
New York, N.Y. 10017
U.S.A.

Library of Congress Cataloging-in-Publication Data

Stroyan, K. D.
Foundations of infinitesimal stochastic analysis.

(Studies in logic and the foundations of mathematics ; v. 119)
Bibliography: p.
Includes index.
1. Stochastic analysis. 2. Mathematical analysis, Nonstandard. I. Bayod, José Manuel. II. Title. III. Series.
QA274.2.S79 1986 519.2 85-28540
ISBN 0-444-87927-7

PRINTED IN THE NETHERLANDS

CONTENTS

A SHORT OUTLINE WITH ONLY PROBABILITY MEASURES.

This outline avoids infinite measures and delicate Loeb measurability questions. It is all that is needed for probability.

Section (0.4)

Section (1.1), prove Lemma (1.2.13) in the case $\mu[V] \approx 1$.

Section (1.3), note all functions have finite carrier when $\mu[V] \approx 1$.

Section (1.4), condition (S-MC) is automatic when $\mu[V] \approx 1$.

Section (1.5)

Section (2.1)

Section (2.3), if you skip (2.2), verify the comment after the proof of (2.3.4): $\alpha \circ st^{-1}$ extends μ.

Section (2.4)

Chapter 3

ACKNOWLEDGEMENTS

This project has taken much longer than expected. Our final worry is that we will forget to thank one of the many people who offered us their help during the many years! We appreciate even the smallest suggestions because we know that a sum of infinitesimals can be infinite.

Most of all we thank H. Jerome Keisler for his seminar notes, ideas, examples, criticism, preprints and encouragement. This book would not exist without his help. C. Ward Henson also gave us a great deal of help and an example. Jörg Flum, L. C. Moore, Jr., Robert M. Anderson, and Tom L. Lindstrom generously gave us detailed criticism of parts of early drafts. Douglas N. Hoover, Edwin Perkins, L. L. Helms, Nigel Cutland, J. E. Fenstad and Peter A. Loeb sent us their preprints and discussed the project with us at meetings. K. Jon Barwise, Juan Gatica, Eugene Madison, Robert H. Oehmke, John Birch, Constantin Drossos, Gonzalo Mendieta, David Ross, Vitor Neves, Anna Roque, Lee Panetta and others participated in seminars on various parts of the book. We thank all these people for their help and encouragement. It seems to us that the combined effort of many people is what has made this branch of Robinson's Theory of Infinitesimals blossom.

Bayod thanks the Fulbright Foundation for partial support during the first part of the project and his colleagues at the University of Santander, who, by increasing their work load, allowed him to take a two-semester leave in Iowa. Stroyan gratefully acknowledges support, years ago, of the National Science Foundation of The United States for summer research that appears in parts of the book. Stroyan thanks The University of Iowa for its tolerance and sometimes generous support of his peculiar research interests.

We thank Ada Burns for her superb typing of infinitely many drafts, revisions and corrections of author errors. We also thank Laurie Estrem for excellent typing of part of the next-to-last draft. We thank the staff of North-Holland in advance for the production task they are about to undertake.

The final draft of this book was prepared with the excellent new technical word processor T^3 from TCI Software Research, Inc. and printed by them on an HP LaserJet+ printer.

The series editors, Arjen Sevenster and others at North-Holland have been patient and helpful in making arrangements for this book to appear in *The Studies in Logic* and *The Foundations of Mathematics*. We are delighted that this book will appear in the same series as Abraham Robinson's classic book on infinitesimals.

We dedicate this book to Jerry Keisler for his professional help and to our wives for their emotional support during the project.

to

Jerry,
Carol and Cristina

FOREWORD

Bayod obtained support from the Fulbright Foundation to visit The University of Iowa for the 78-79 academic year in order to learn about Abraham Robinson's Theory of Infinitesimals (so-called "Non-Standard Analysis"). We agreed to focus our seminar on infinitesimal analysis of probability and measures because of exciting work of Keisler and Perkins then in progress and with hopes of further applications. We made careful notes, while, unknown to us, Keisler was doing the same thing with his students. The second draft of this book combined both sets of notes and comprised roughly the present chapters 0 to 4.

Chapter 0 contains all the 'nonstandard stuff' that our reader needs in order to learn about the applications in this book. The reader who is familiar with the basic principles of infinitesimal analysis can go directly to Chapter 1. Chapter 0 tries to give the beginner in infinitesimal analysis the working tools of the trade without proof. We feel that the logical principles such as Leibniz' (transfer) Principle and the Internal Definition Principle, together with Continuity Principles such as Robinson's Sequential Lemma, saturation and comprehension are the things our beginning reader should focus his attention on.

Section (0.1) gives the definition of "all of classical analysis" in the form of a "superstructure." Section (0.2) explains the meaning of Leibniz' Principle and begins to show its usefulness. We believe that our reader can get a working knowledge of these tools of logic by working several of the exercises. Section (0.3) contains more explanation of some basic notions of infinitesimal analysis that are used throughout the book. Section (0.4) contains the important saturation property that we base our measure constructions upon.

The approach to measures in this book was initiated by Peter A. Loeb [1975]. Most of the basic results are due to him, but we have given a more elementary new exposition based on inner and outer internally generated measures. This approach replaces the use of Caratheodory's extension theorem by direct elementary arguments. We have also added some fine points and examples not found in the literature. Section (1.1) deals with probability measures, while section (1.2) treats infinite measures. Since infinite measures cause extra technicalities, we have given a short outline for the reader who is anxious to apply Loeb's construction to probability. (It appears right after the Table of Contents.) C. Ward Henson [1979a,b] discovered the connections between Loeb, Borel and Souslin sets and first proved uniqueness of hyperfinite extension and the unbounded case. The remainder of Chapter 1 explains the relationship between *finite sums and integrals against hyperfinite measures. Robert M. Anderson [1976, 1982] systematically investigated Radon measures, filled in some of the basics on S-integrability and studied product measures. S-integrability is the main ingredient needed to relate sums and integrals.

In Chapter 2 we study the relation between Borel and hyperfinite measures. The basic idea there is to 'pull back with the standard part map'. The case of a completed Borel measure is technically easier to treat, so section (2.1) treats Lebesgue measure independently, while sections (2.2) and (2.3) treat naked sigma algebras and measures. This draws on the works of Anderson and Henson cited above. Anderson and Salim Rashid [1978] and Loeb [1979a] investigated weak standard parts of measures; section (2.4) presents a simple case of those results.

Chapter 3 contains a Fubini-type theorem due to H. Jerome Keisler [1977] as well as a mixed Fubini-type theorem. Anderson's work above showed that a hyperfinite product measure extends the product of hyperfinite measures, while Douglas N. Hoover [1978] showed that that extension is strict. The Fubini theorem holds anyway.

Chapter 4 has a basic treatment of distributions, laws and independence from the point of view of infinitesimal analysis.

Even the "foundations" of stochastic analysis consist of more than measures. Anderson [1976] discussed Brownian motion, including path continuity and Ito's lemma, using Loeb's techniques. Loeb [1975] treated infinite coin tossing and a different approach to the Poisson process. Keisler's [1984] preprint investigated more general processes. (P. Greenwood and R. Hersh [1975] and Edward Nelson [1977] have some infinitesimal analysis of stochastic processes using different techniques.) We discussed these things in the spring of 1979, but only wrote rough notes including some extensions of this work. In the meantime we learned of other fundamental work of Tom L. Lindstrom [1980] and of Hoover and Edwin Perkins [1983]. Our treatment of Chapters 5, 6 and 7 relies most heavily on the work of Keisler, Hoover-Perkins and Lindstrom.

Chapter 5 is devoted to 'path properties' of processes. Our treatment of paths with only jump discontinuities is a little different from Lindstrom-Hoover-Perkins'. We precede that with the less technical case of continuous paths and include a lot of details in both cases. Section (5.4) contains results and extensions of results from Keisler [1984] relating Loeb, Borel and Souslin sets on the product space $[0,1] \times \Omega$. Section (5.5) sketches how one makes the extension of these results to $[0,\infty)$. The infintesimal analysis is very similar to $[0,1]$, but since the classical metrics are rather technical we avoided a complete account.

Chapter 6 gives the basic theory of how events evolve measure-theoretically in time on a hyperfinite scheme. We follow Keisler [1984] again, but extend his results to include "pre-visible measurability" as well as "progressive measurability." A hyperfinite evolution scheme has 'measurability' and 'completeness' properties that make it richer than an arbitrary 'filtered (or adapted) probability space' of the "general theory of processes." We prove that the error "adapted implies progressively measurable" is actually true. We show that previsibility arises from a left filtration. Hence, while we borrowed extensively from Lindstrom [1980] and especially Hoover and Perkins [1983] in later chapters, we adapted their results to Keisler's more specific combinatorial framework. We feel that this would be justified simply because of the concrete "liftings" we obtain. However, Hoover and Keisler [1982] show that this results in no loss of generality in a certain specific logical sense described briefly in the Afterword.

The aim of all our chapters is to give fundamental results needed to apply infinitesimal analysis to the study of stochastic processes. We drew the line of what we call "foundations" at "measure theory" but our reader should not take this seriously. We hope many people will work to extend the foundations of infinitesimal stochastic analysis as well as to give new applications of these methods in solving problems about stochastic processes. We doggedly adhered to a hyperfinite bias. That made certain things 'nice' and should not hinder our reader from learining and using other 'Loeb-space' techniques.

Chapter 7 gives a hyperfinite treatment of semimartingale integrals. We have written it at two levels. Sections (7.1), (7.2), (7.3) and the beginning of (7.7) treat the easily integrated case in complete detail. Section (7.4) shows how the more general local theory at least partly parallels the square integrable case. The remainder of the chapter only outlines the main ideas of the best known contemporary theory. We hope that the statement of results and examples will act as a guide to further study.

The survey article of Cutland [1983b] could be read by a beginner to get an overview of hyperfinite measure theory.

CHAPTER 0: PRELIMINARY CONSTRUCTIONS

(0.0) Motivation with a Finite Probability Experiment

Consider the probability experiment of tossing a *fair coin* n times. We can represent each possible outcome as a sequence of H's and T's . These outcomes can be viewed as the elements of the set:

$$\Omega_1 = \{H, T\}^n.$$

Rolling a *fair die* n times could be viewed as elements of the set:

$$\Omega_2 = \{I, II, III, IV, V, VI\}^n.$$

Coin tossing can be modeled in the die experiment by considering an even outcome as "heads" and an odd outcome as "tails". Sampling n times from an urn with *replacement and complete mixing* after each sample can be modeled on the set

$$\Omega_3 = W^n,$$

where W is a set with m elements representing the balls in the urn. A particular sequence of draws from the urn is represented by a function

$$\upsilon \in \Omega_3$$

$$\upsilon : \{1, \cdots, n\} \to W.$$

This means that $\upsilon(1)$ is the particular ball in the urn W on the first draw, $\upsilon(2)$ is the ball on the second draw, and so on.

If m is even, coin tossing can be modeled inside Ω_3 by considering even outcomes as "heads" and odd outcomes as "tails". This can be coded by any function

$$c: W \to \{H,T\}, \quad \text{with} \quad {}^{\#}[c^{-1}(H)] = {}^{\#}[c^{-1}(T)].$$

(We use ${}^{\#}[\cdot]$ for the finite cardinality function.) The number of sample sequences that have exactly half heads is

$${}^{\#}[\upsilon \in \Omega_3 : {}^{\#}[j : c(\upsilon(j)) = H] = \frac{n}{2}].$$

(There will not be any unless n is even and we already assumed that m is even.) If 6 divides m then the fair die experiment can be modeled inside Ω_3 in a similar way. In order to have lots of divisibility we may as well assume that $n = h!$, and $m = n^n$, for a single integer $h > 1$. (Eventually we will let $h \in {}^*\mathbb{N}$ be an infinite integer.)

We wish to think of the index in the component functions (such as υ above) as a time, so that it is more convenient to represent our random experiment a slightly different way. This means that, when we make h (and hence n and m) larger, more sampling takes place in the same elapsed time.

We first take a set of times

$$\mathbb{T} = \{\Delta t, 2\Delta t, 3\Delta t, \ldots, n\Delta t\},$$

where $\Delta t = \frac{1}{n}$. Then we take a finite set W with m elements and define our space of sample sequences to be the set of all functions from $\mathbb{T}$ into W

(0.0.1) $$\Omega = \{\omega \mid \omega: \mathbb{T} \to W\} = W^{\mathbb{T}}.$$

Now if $\omega \in \Omega$, then $\omega(\frac{1}{3})$ is the ball selected at the time $t = \frac{1}{3}$ for the particular sequence of samples represented by ω.

The idea of a *mixed urn* mentioned above means that each

ball in W is equally likely to be drawn at each time of sampling. Thus the probability of each individual sample sequence $\omega \in \Omega$, is the uniform amount

$$\Delta P(\omega) = \frac{1}{m^n} = \frac{1}{{}^{\#}[\Omega]} . \tag{0.0.2}$$

The set $\mathcal{E}$ of all subsets of Ω is called the *algebra of events* of the experiment. For example, the event consisting of those sample sequences ω that contain exactly half heads is the set $\Lambda_h \in \mathcal{E}$ given by

$$\Lambda_h = \{\omega \in \Omega \mid {}^{\#}[t : c(\omega(t)) = H] = \frac{n}{2}\}.$$

The *probability of an event* $\Lambda \in \mathcal{E}$ is given by the finite sum of the individual probabilities from Λ:

$$\begin{aligned} P[\Lambda] &= \Sigma[\Delta P(\lambda) : \lambda \in \Lambda] \\ &= \frac{{}^{\#}[\Lambda]}{{}^{\#}[\Omega]}. \end{aligned} \tag{0.0.3}$$

For example, the probability of the event Λ_h above is given by the binomial coefficient $\binom{n}{n/2}$ over the number of ways that one can assign heads or tails to a sample along $\mathbb{T}$,

$$P[\Lambda_h] = \binom{n}{\frac{n}{2}} \frac{1}{2^n} . \tag{0.0.4}$$

We could attach values to heads or tails corresponding to winning or losing a step in a game, for example,

$$\Delta B(\omega_t) = \begin{cases} +\sqrt{\Delta t}, & \text{if } c(\omega_t) = H \\ -\sqrt{\Delta t}, & \text{if } c(\omega_t) = T \end{cases} .$$

In this case the running total of signed winnings is the *stochastic process* (a random walk of step size $\sqrt{\Delta t}$)

(0.0.5) $$\begin{aligned} B(t,\omega) &= \Sigma[\Delta B(\omega_s) : s \in \mathbb{T} , \quad \Delta t \leq s \leq t] \\ &= \Delta B(\omega_{\Delta t}) + \Delta B(\omega_{2\Delta t}) + \cdots + \Delta B(\omega_t). \end{aligned}$$

We have written very much about a simple example, because we want to point out what sorts of mathematical entities are involved. First, we have an n element set $\mathbb{T}$ and an m element set W. Next, we have the set Ω of functions from $\mathbb{T}$ into W. Then we have a function P defined on the set $\mathcal{E}$ of all subsets of Ω in terms of the (constant uniform) function ΔP and the summation function $\Sigma[\cdot : \cdot]$. Moreover, we associate certain simple combinatorial formulas with P when computing an event defined in terms of the function c from W into $\{H,T\}$. Finally, we have a function $B : \mathbb{T}\times\Omega \longrightarrow \mathbb{R}$ given by the summation function and another function, ΔB. We want to let Δt be an infinitesimal δt and use the same kinds of formulas to analyse Brownian motion. Our infinitesimal analysis will require more than just extending numbers to include infinitesimals, but will also require extensions of functions, functions of functions (including summation), sets of these, and combinatorial formulas relating to them all.

Abraham Robinson's contemporary **Theory of Infinitesimals** allows us to enlarge "all of classical analysis" to include infinitesimals and infinite numbers as well as *certain* functions, sets, sets of functions, and so on. This extension procedure satisfies a precise transfer principle akin to Leibniz' old idea that what holds for ordinary numbers, curves, etc., also holds for the ideal extensions. Robinson's formulation of the principle uses a formal language for precision, but in practice we need only care with quantifiers and some training in the limitations of the formal transfer.

LEIBNIZ' TRANSFER PRINCIPLE (heuristic form)

A property Φ *is true in classical analysis if and only if its* **transform* $^*\Phi$ *is true in infinitesimal analysis.*

The precise formulation of the transfer principle is in section (0.2). Appendix 2 sketches some set-theoretic constructions needed for the formulation. The aim of this chapter is to show the reader how to use Leibniz' Principle— proofs can be found in the references given in the appendices. For now you should think of * as a mapping defined on all the objects of classical analysis (in an informal sense).
For example, $^*\mathbb{R}$ is a set extending the set of real numbers, $\mathbb{R}$ in the following sense. The mapping * is defined on each fixed number $r \in \mathbb{R}$, for example, ^{*0}, ^{*1}, $^*\sqrt{2}$, $^*\pi$. The restriction of * to the set $\mathbb{R}$ maps bijectively onto the set $^\sigma\mathbb{R} = \{r \in {}^*\mathbb{R} : (\exists\ s \in \mathbb{R})[r = {}^*s]\}$, because * preserves the property of two numbers being unequal to each other. Nevertheless, $^\sigma\mathbb{R}$ is a proper subset of the set $^*\mathbb{R}$. The binary functions $+$ and $\cdot$ and the relation $<$ on $\mathbb{R}$ extend to binary functions $^*+$ and $^*\cdot$ and a relation $^*<$ on $^*\mathbb{R}$.

The tuple $({}^*\mathbb{R}, {}^*0, {}^*1, {}^*+, {}^*\cdot, {}^*<)$ is a real closed ordered field. The extensions of fixed numbers retain their basic properties (such as being an identity), so ${}^\sigma\mathbb{R}$ is an isomorphic copy of $\mathbb{R}$ in ${}^*\mathbb{R}$. For simplicity of notation we usually drop the σ on ${}^\sigma\mathbb{R}$ and the $*$'s on the field operations and order.

Transcendental functions like sine and cosine have natural extensions via $*$ which satisfy the same first order properties as the originals. For example, ${}^*\sin(x)$ takes on all the values $y \in {}^*\mathbb{R}$ satisfying $-1 \leq y \leq +1$ while still satisfying the trig identities (with $*$'s on the functions) for all x in ${}^*\mathbb{R}$. Summation extends to the same type of operation and in general $*$ preserves types or levels of sets. However, above the level of numbers there are restrictions on the extensions. The general rule for deciding how an extended quantity acts is to examine the transform under the $*$ map of properties of the classical quantity. We call the resulting properties "*-transforms."

***-TRANSFORMS** (heuristic form)

You just put $*$'s *on everything!*

A statement in classical analysis has a *-transform given by extending each fixed set, function, or point used in the statement via the $*$ map. The quantifiers in such a statement must have specific bounds. 'Specific bounds' means that "for all positive $\epsilon \in \mathbb{R}$" is O.K., but "for all ϵ" is not, "for every element A in the power set of $\mathbb{N}$" is O.K., but "for every subset of $\mathbb{N}$" is not. The *-transform of a property like continuity is referred to as *continuity.

Not all the properties of the extension of classical analysis arise by transfer. Otherwise the extension would be isomorphic to the original and thus not contain infinitesimals. The properties of "internal" sets and functions arise by transfer, whereas "external" sets and functions are not subject to transferred properties. This important distinction is characteristic of contemporary infinitesimal analysis and is the main thing this chapter tries to uncover.

(0.1) Superstructures

We want an object big enough to include "all of classical analysis" and since this might later include some probability space we begin with a fixed but arbitrary ground set X_0 containing $\mathbb{R}$. For example, if we were given a space S we could take $X_0 = S \cup \mathbb{R}$. The elements of X_0 must be *individuals* (or set-theoretic "atoms" or "urelements"); they have no elements. If $x \in X_0$, then $y \in x$ is not defined. (In particular, we shall think of numbers $r \in \mathbb{R}$ as "points" and not as sets of sequences of rationals or as Dedekind cuts, etc.)

Let $\mathfrak{P}(A)$ denote the power set of all subsets of a set A. We form a sequence of sets inductively as follows:

$$X_1 = \mathfrak{P}(X_0)$$

inductively for $n \in \mathbb{N}$

$$X_{n+1} = \mathfrak{P}\Big(\bigcup_{k=0}^{n} X_k \Big).$$

(0.1.1) **DEFINITION:**

The set $\mathfrak{X} = [X_n \mid n \in \mathbb{N}]$ *is called the superstructure over the ground set* X_0. *The elements of* X_0 *are called individuals while the elements of* $\mathfrak{X} \setminus X_0$ *are called entities.*

Notice that $X_1 \subseteq X_2 \subseteq \cdots$ and $X_0 \cap X_n = \emptyset$, for $n \geq 1$.

We call $\mathfrak{X}$ a structure because we can think of analysis as coded in the theory of $\in$ and $=$ restricted to the entities and individuals. Strictly speaking the superstructure is a triple $(\mathcal{X}, \in, =)$.

The following examples illustrate why a superstructure is big enough to be considered a model of "all of classical analysis."

(a) $\mathbb{R}^2 \in \mathcal{X}$: For $x, y \in \mathbb{R}$, the ordered pair $(x,y) = \{\{x\},\{x,y\}\} \in X_2$, hence $\mathbb{R}^2 \in X_3 \subseteq \mathcal{X}$.

(b) $\sin(\cdot) \in \mathcal{X}$: Any real valued function with domain, $D \subseteq \mathbb{R}$, may be viewed as a set of ordered pairs of real numbers, hence, $\sin(\cdot) \in X_3 \subseteq \mathcal{X}$.

(c) $C[0,1] \in \mathcal{X}$: Every function defined on $[0,1]$ belongs to X_3, so $C[0,1] \in X_4 \subseteq \mathcal{X}$. These remarks apply to many other classical function spaces such as ℓ^∞, the space of all bounded sequences, which may be viewed as a set of functions from $\mathbb{N}$ into $\mathbb{R}$.

(d) $\|\cdot\|_\infty \in \mathcal{X}$: The uniform norm on $C[0,1]$ is a function from an element of X_4 into $\mathbb{R}$, so it belongs to X_6.

(e) $\tau \in \mathcal{X}$: The topology $\tau = \{U \subseteq C[0,1] \mid U \text{ is } \|\cdot\|_\infty\text{-open}\}$ belongs to X_5.

Notice that a sequence $(x_n \mid n \in \mathbb{N})$ defined in such a way that for each n, $x_n \in X_{m+1} \setminus X_n$, is not an element of $\mathcal{X}$. The "von Neuman ordinals",

$$V = \{\emptyset, \{\emptyset\}, \{\emptyset,\{\emptyset\}\}, \{\emptyset,\{\emptyset\},\{\emptyset,\{\emptyset\}\}\},\cdots\}$$

(which can be used as a model for the natural numbers) is **not** an entity, $V \notin \mathfrak{X}$. (We have taken $\mathbb{R}$, and thus $\mathbb{N}$, as a set of atoms, so $V \neq \mathbb{N}$.) However, because we have some flexibility in how we choose to code objects from analysis and because we may take X_0 to be any set of atoms containing $\mathbb{R}$, we feel that it is justified to call $(\mathfrak{X}, \in, =)$ a model of classical analysis.

The rest of this section locates the pieces of the construction of section (0.0) in the superstructure $\mathfrak{X}$. We will take *transforms of these constructions in the next section.

Arithmetic operations can be viewed as functions, hence are elements of $\mathfrak{X}$. Partially defined functions also belong to $\mathfrak{X}$. Define the functions $f_j : \mathbb{R}\times\mathbb{R} \to \mathbb{R}$ by

$$f_1(r,s) = r+s, \quad f_2(r,s) = r\cdot s.$$

Define the function $f_3 : \mathbb{R}\times[\mathbb{R}\setminus\{0\}] \to \mathbb{R}$ by

$$f_3(r,s) = \frac{r}{s}.$$

Define the functions on $\mathbb{N}$ or $\mathbb{N}\times\mathbb{N}$ by

$f_4(h) = h!$, $f_5(h,k) = h^k$, $f_6(h,k) = \binom{k}{h}$, binomial coefficient, as well as,

$$f_7(n) = \binom{n}{\frac{n}{2}}\frac{1}{2^n}, \quad \text{for } n \text{ even}.$$

(0.1.2) **EXERCISE:**

Find p *so that* $f_1, \ldots, f_7 \in X_p$. *Express* f_7 *in terms of* f_2, f_5 and f_6.

Simple set-valued functions are also involved in the constructions of section (0.0). For example, $F_1 : [\mathbb{N}\setminus\{0,1\}] \longrightarrow \mathbb{B}(\mathbb{R})$ defined by

$$F_1(n) = \mathbb{T} = \{t \in \mathbb{R} \mid (\exists\, k \in \mathbb{N})[1 \leq k \leq n!\ \&\ t = \frac{k}{n!}]\}$$

F_2: $[\mathbb{N}\setminus\{0\}] \to \mathcal{B}(\mathbb{N})$ defined by

$$F_2(m) = W = \mathbb{N}[1,m] = \{k \in \mathbb{N} \mid 1 \le k \le m\}$$

(0.1.3) **EXERCISE:**

Find p *so that* $f_1, \cdots, f_7, F_1, F_2 \in X_p$.

We also need some restricted forms of set functions which do not belong to $\mathfrak{X}$ as elements in their unrestricted forms. For each $p > 0$, define functions $G_1^p : X_p \to X_{p+1}$ and $H_1^p : X_p \times X_p \to X_{p+2}$ by the following restrictions

$$G_1^p(Y) = \mathcal{B}(Y), \text{ the power set of all subsets of } Y,$$

and

$$H_1^p(Y,Z) = Z^Y, \text{ the set of all functions from } Y \text{ into } Z.$$

(0.1.4) **EXERCISE:**

Find q *so that* $G_1^p, H_1^p \in X_q$.

If $p' > p$, *show that* $G_1^{p'}$ *extends* G_1^p *as a function, that is,* $G_1^{p'}(Y) = G_1^p(Y)$ *if* $Y \in X_p$. *Also,* $H_1^{p'}$ *extends* H_1^p *as a function.*

For each $p \ge 0$, the set of all finite subsets of $X_p \cup X_0$, Fin_p, satisfies $\mathrm{Fin}_p \subseteq X_{p+1}$, so $\mathrm{Fin}_p \in X_{p+2}$. However, the function $f: \mathbb{N} \to \mathfrak{X}$ given by $f(p) = \mathrm{Fin}_p$ is not an element of $\mathfrak{X}$. The finite cardinality function counting the number of elements of a set can be restricted to any of these sets in order to make it an entity. For each $p \ge 0$, define $G_2^p : \mathrm{Fin}_p \to \mathbb{N}$ by

$$G_2^p(Y) = {}^{\#}[Y].$$

The summation function can also be defined in terms of a restricted entity. Let Fct_q denote the set of all real-valued functions whose domain satisfies $D \subseteq X_q$. Summation can be defined by the function

$$H_2^q(f,Y) = \Sigma[f(y) \mid y \in Y],$$

where H_2^q is defined on the set of pairs (f,Y) with $f \in Fct_q$ and Y a finite subset of the domain of f.

(0.1.5) **EXERCISE:**

For each $p,q > 0$ *find* r *so that* $G_2^p, H_2^q \in X_r$.

(0.2) Results from Logic:

Leibniz' Principle, the Internal Definition Principle

This section gives the main results from logic needed to apply Robinson's Theory of Infinitesimals to analysis along with some examples directed at our basic random sampling scheme. Further examples are given in Appendix 1. A more complete treatment can be found in the references given in the appendices. The main thing one must consider carefully in applying $*$-transforms is the bounds on the quantifiers of the statement being transformed.

(0.2.1) DEFINITION:

A statement in classical analysis is said to have a standard bounded formalization if it can be expressed in terms of individuals and entities of $\mathfrak{X}$ *using the relations* $\in$ *and* $=$ *in a way so that the quantifiers occur only in the bounded forms:*

$$(\forall\ x)[x \in A \Rightarrow \cdots] \quad or \quad (\exists\ y)[y \in B \ \& \cdots],$$

where A *and* B *are entities,* $A, B \in \mathfrak{X}$.

The bounds on the quantifiers, "for all $x \in A$" or "there exists $y \in B$" are sometimes abbreviated

$$(\forall\ x \in A)[\cdots] \quad \text{or} \quad (\exists\ y \in B)[\cdots].$$

The statement that a particular entity in $\mathfrak{X}$ has finite cardinality has a standard bounded formalization. In fact, for any fixed $p \geq 0$, we may characterize the elements of the set Fin_p (from the last section) as follows:

$(\forall\ Y \in X_{p+1})\{[Y \in Fin_p]$

$\Leftrightarrow (\exists\ f \in Fct_p)(\exists\ n \in \mathbb{N})$["f maps Y 1-1 and onto $\mathbb{N}[1,n]$"]

$\Leftrightarrow (\forall\ f \in Fct_p)$["(f maps Y onto Y) $\Leftrightarrow$ (f maps Y 1-1 to Y)"] $\}$

In other words, an entity of X_p is finite if it can be mapped bijectively onto an initial segment of $\mathbb{N}$ or if it has the property that functions from Y to Y are surjective if and only if they are injective. The statements in quotation marks are abbreviations for easily formalized statements. The quantifier bounds, Fin_p and Fct_p play an important role in the *-transform of this statement.

We shall assume that there is another superstructure $(\mathcal{Y}, \in, =)$ over a ground set Y_0 along with an injection

$$^* : \mathcal{X} \to \mathcal{Y}$$

satisfying Leibniz' Principle (0.2.3) below. The principle is stated in terms of *-transforms of statements about $\mathcal{X}$, so first we give the definition of *-transform and an example.

(0.2.2) **DEFINITION:**

*If Φ is a statement with a standard bounded formalization, the *-transform of Φ is the statement $^*\Phi$ obtained by applying the * map to each entity and individual in Φ.*

The *-transform of being a finite entity at the pth level is:

$(\forall\ Y \in {}^*X_{p+1})\{[Y \in {}^*Fin_p]$

$\Leftrightarrow (\exists\ f \in {}^*Fct_p)(\exists\ n \in {}^*\mathbb{N})[$"f maps Y 1-1 and onto ${}^*\mathbb{N}[1,n]$"$]$

$\Leftrightarrow (\forall\ f \in {}^*Fct_p)[$"(f maps Y onto Y) $\Leftrightarrow$ (f maps Y 1-1 to Y)"$]\ \}$

We shall see that the $*$'s on the quantifier bounds cause restrictions on the informal meaning of *finiteness. Infinite sets can be *finite because this property is only tested with functions from ${}^*Fct_p$. On the other hand, *finite sets can be treated as if they were finite as long as the operations are "internal." We give the transfer principle first and then discuss the meaning of "internal."

(0.2.3) LEIBNIZ' TRANSFER PRINCIPLE:

The mapping $^* : \mathfrak{X} \to \mathfrak{Y}$ *is an injection of the superstructure* $\mathfrak{X}$ *over* X_0 *into the superstructure* $\mathfrak{Y}$ *over the individuals* Y_0, *where* $Y_0 = {}^*X_0$, *satisfying:*

(a) *Any sentence about* $(\mathfrak{X}, \in, =)$ *with a standard bounded formalization holds in* $\mathfrak{X}$ *if and only if its* $*$*-transform holds in* $(\mathfrak{Y}, \in, =)$.

(b) * *is polyenlarging, in particular,* ${}^*\mathfrak{X}$ *is* $card^+(\mathfrak{X})$ *saturated.*

Part b of the transfer principle will be explained further in section (0.4). A special case of part (b) is that the natural extension of every infinite set is strictly larger than the set of (extended) standard elements of the set. In particular, the standard real numbers, embedded in ${}^*\mathbb{R}$ as ${}^\sigma\mathbb{R} = \{{}^*r \mid r \in \mathbb{R}\}$, form a proper subset of ${}^*\mathbb{R}$. We call the elements of ${}^*\mathbb{R}$ *hyperreal* numbers. The Dedekind numbers, $\mathbb{R}$, are a maximal archimedean ordered field (in a given set theory), so we see that Leibniz' Principle asserts the existence of infinitesimal numbers $\delta \in {}^*\mathbb{R}$ such that ${}^*0 < \delta < {}^*r$ for every positive $r \in \mathbb{R}$.

The nonstandard extension map acts at many levels (or "types") of set theory and this can be quite confusing at first. At the ground level we have ^{*0} in ${}^*\mathbb{R}$ as an additive identity, ^{*1} as a multiplicative identity, ${}^*1+{}^*1 = {}^*2$, and so on, by part (a) of Leibniz' Principle. Since the extensions of the fixed numbers behave the same way as the originals, we usually do not write $*$'s on them. Hence, $\pi \in {}^*\mathbb{R}$ is the *area of the *unit *circle, 0 is the additive identity of ${}^*\mathbb{R}$ and 1 is the multiplicative identity. Writing no $*$'s on standard numbers may be thought of as an identification

$$\mathbb{R} \equiv {}^\sigma\mathbb{R} = \{{}^*r : r \in \mathbb{R}\} \subset {}^*\mathbb{R}.$$

Hence, we may drop the sigma on the $\mathbb{R}$, but there is a danger in this. The set ${}^\sigma\mathbb{R}$ is an *external* set; no *transferred* statement about $\mathbb{R}$ can refer to all the standard numbers in ${}^*\mathbb{R}$ and no more. This is explained fully below.

(0.2.4) DEFINITION:

If $X \in \mathfrak{X} \backslash X_0$ *is an entity, the discrete standard part of* X *is the set*

$$^{\sigma}X = \{^{*}x : x \in X\}.$$

If $X \in X_1$ we usually drop the σ and identify $X \equiv {}^{\sigma}X = \{^{*}x : x \in X\}$, for example, $\mathbb{N} \subset {}^{*}\mathbb{N}$ means ${}^{\sigma}\mathbb{N} \subset {}^{*}\mathbb{N}$. In fact, this is quite reasonable at the *first* level of the superstructure, but σ is useful above that level. Whenever we feel there might be confusion, we will pedantically include all the $*$'s and σ's.

(0.2.5) **DEFINITION:**

a) *An element* $Y \in \mathcal{Y}$ *is called (an extended) standard entity or individual if there is an* $X \in \mathcal{X}$ *such that* $Y = {}^{*}X$.

b) *An element* $Y \in \mathcal{Y}$ *is called an internal entity (or individual) if there is an* $X \in \mathcal{X}$ *such that* $Y \in {}^{*}X$.

c) *An element* $Y \in \mathcal{Y}$ *is called an external entity if* Y *is not internal.*

Every hyperreal number $r \in {}^{*}\mathbb{R}$ is internal because it belongs to ${}^{*}\mathbb{R}$. Similarly, every individual of $\mathcal{Y}$, $y \in Y_0 = {}^{*}X_0$, is internal. The extended set ${}^{*}\mathbb{R}$ is "standard", while it is shown in Appendix 1 that $\mathbb{R}$ viewed as ${}^{\sigma}\mathbb{R} \subset {}^{*}\mathbb{R}$ is external! In fact, ${}^{\sigma}X$ is *external for every infinite entity* $X \in \mathcal{X}$. Extended standard sets are internal because if $X \in X_p$ for some p, then ${}^{*}X \in {}^{*}X_p$ by Leibniz Principle.

(0.2.6) **EXERCISE:**

Show that for any entity, $A \in \mathfrak{X}\backslash X_0$, ${}^\sigma\mathfrak{P}(A)$ *is the set of standard subsets of* A, ${}^*\mathfrak{P}(A)$ *is the set of internal subsets of* A, *and* $\mathfrak{P}({}^*A)\backslash{}^*\mathfrak{P}(A)$ *is the set of external subsets of* A.

Let us consider a special case of entities of $\mathfrak{Y}$ which are subsets of the *natural numbers, ${}^*\mathbb{N}$. If

$$A \in {}^\sigma\mathfrak{P}(\mathbb{N}),$$

then there is a $B \in \mathfrak{P}(\mathbb{N})$ such that $A = {}^*B$, that is, A is an extended standard entity. For example, A might be equal to the set of *prime numbers. In that case, by Leibniz Principle,

$$A = \{p \in {}^*\mathbb{N} \mid (\forall a,b \in {}^*\mathbb{N})[1 < a \le b < p \Rightarrow a \cdot b \ne p]\} = {}^*B.$$

Part (b) of Leibniz' Principle asserts that ${}^*B\backslash{}^\sigma B$, the set of nonstandard prime numbers, is nonempty. There are unlimited (or infinite) primes!

Next we consider an internal subset of ${}^*\mathbb{N}$ which is not the extension of any standard set. We use the function $F_2 : \mathbb{N} \to \mathfrak{P}(\mathbb{N})$ given by $F_2(m) = \{k \in \mathbb{N} \mid 1 \le k \le m\}$. We know that

$$(\forall m \in \mathbb{N})[[F_2(m) \in \mathfrak{P}(\mathbb{N})] \;\&\; (\forall k \in \mathbb{N})[k \in F_2(m) \Leftrightarrow 1 \le k \le m]]$$

so by $*$-transform

$$(\forall m \in {}^*\mathbb{N})[[{}^*F_2(m) \in {}^*\mathfrak{P}(\mathbb{N})] \;\&$$
$$(\forall k \in {}^*\mathbb{N})[k \in {}^*F_2(m) \Leftrightarrow 1 \le k \le m]].$$

For an infinite integer $m \in {}^*\mathbb{N}\backslash\mathbb{N}$, ${}^*F_2(m)$ is internal but not standard, that is, ${}^*F_2(m) \in {}^*\mathfrak{P}(\mathbb{N})\backslash{}^\sigma\mathfrak{P}(\mathbb{N})$.

The fact that ${}^\sigma\mathbb{N}$ is external means that

$${}^\sigma\mathbb{N} \in \mathfrak{P}({}^*\mathbb{N})\backslash{}^*[\mathfrak{P}(\mathbb{N})].$$

Hence we have the strict inclusions

$$\mathfrak{P}({}^*\mathbb{N}) \supset {}^*[\mathfrak{P}(\mathbb{N})] \supset {}^\sigma[\mathfrak{P}(\mathbb{N})].$$

(0.2.7) EXERCISE:

Let $h \in {}^*\mathbb{N}\backslash\mathbb{N}$ *be an unlimited* **natural number. Use the extended functions* ${}^*f_4$, ${}^*f_3$ *and* ${}^*F_1$ *of section* (0.1) *to show that the set*

$$\mathbb{T} = \{t \in {}^*\mathbb{R} \mid (\exists k \in {}^*\mathbb{N})[t = \frac{k}{h!} \;\&\; 1 \le k \le h!]\}$$

is an element of ${}^*[\mathfrak{P}(\mathbb{R})]$ *and hence is an internal set.*

Also, use f_1, f_5 *and* F_2 *to show that* $W = \{j \in {}^*\mathbb{N} \mid 1 \le j \le m\} \in {}^*[\mathfrak{P}(\mathbb{N})]$ *where* $m = n^n$ *and* $n = h!$

Informally, we might denote $\mathbb{T}$ by

$$\mathbb{T} = \{\delta t, 2\delta t, \cdots, 1\}, \quad \text{where} \quad \delta t = \frac{1}{h!},$$

and

$$W = \{1, 2, \cdots, m\}.$$

The set of all *internal* sample sequences taken from urns with m elements along the $\mathbb{T}$-time axis is the internal set

$$\Omega = W^{\mathbb{T}} = {}^*H_1^1(\mathbb{T}, W),$$

where H_1^1 is the function of section (0.1) and $\mathbb{T}$ and W are taken from the previous exercise. The set Ω is not only internal, but *finite and contains m^n elements (where $n = h!$). Here is what we mean. "There are m^n functions from an n-element set into an m-element set,"

or

$$(\forall X \in Fin_p)(\forall Y \in Fin_p)[G_2^p(H_1^p(X,Y)) = f_5(G_2^p(Y), G_2^p(X))].$$

(0.2.8) EXERCISE:

Verify that $\mathbb{T}, W \in {}^*Fin_1$, *so that* ${}^\#[\Omega] = m^n$, *and there is an internal function mapping* Ω *bijectively onto the set* $\{k \in {}^*\mathbb{N} : 1 \leq k \leq m^n\}$. *(See the *-transform of "finite" at the beginning of the section.)*

The power set function $X \xrightarrow{\mathfrak{B}} \mathfrak{B}(X)$ is <u>not</u> officially an entity. However, because of exercise (0.1.4), we may write ${}^*\mathfrak{B}(\mathbb{T})$ to mean ${}^*G_1^1(\mathbb{T})$ and, in general, if Y is any internal entity, $Y \in {}^*X_p$, then the internal set of all internal subsets of Y is denoted ${}^*\mathfrak{B}(Y)$, but it is officially,

$$ {}^*\mathfrak{B}(Y) = {}^*G_1^p(Y), $$

the extended function ${}^*G_1^p$ applied at a, possibly nonstandard, but certainly internal set Y.

(0.2.9) EXERCISE:

Show that the internal set of internal events,

$$ \mathcal{E} = {}^*\mathfrak{B}(\Omega) $$

is a **finite set and compute its* **cardinality,* ${}^\#[\mathcal{E}]$.

ANDERSON'S INFINITESIMAL RANDOM WALK

Finally, use of the *summation function, H_2^p, of section (0.1) makes it clear that the extended formula,

$$ B(t,\omega) = \Sigma[\beta(\omega_s)\sqrt{\delta t} \mid s \in \mathbb{T} \ \& \ \delta t \leq s \leq t], \tag{0.2.10} $$

defines an internal function called Anderson random walk once we have given an internal function $\beta : W \to \{-1,+1\}$ with the property that ${}^{\#}[\beta^{-1}(\{-1\})] = {}^{\#}[\beta^{-1}(\{+1\})] = \frac{m}{2}$. This can be done explicitly in several ways. Here are two independent ways.

(0.2.11) EXERCISE:

a) *Show that the definitions*

$$\beta_1(w) = (-1)^w$$

and

$$\beta_2(w) = \begin{cases} -1, & \text{if} \quad w \leq m/2 \\ +1, & \text{if} \quad w > m/2 \end{cases}$$

both define internal functions from $W = \{w \in {}^*\mathbb{N} : 1 \leq w \leq m\}$ *with the same number of* $+1$'s *and* -1's, *where* $m = n^n$, $n = h!$ *and* h *is an unlimited* **natural number*, $h \in {}^*\mathbb{N}\setminus\mathbb{N}$.

b) *For* $(i,j) = (1,2)$ *and* $(2,1)$ *and for* $k = +1$ *and* -1, *show that*

$$\frac{{}^{\#}[\beta_i^{-1}(+1) \cap \beta_j^{-1}(k)]}{{}^{\#}[\beta_j^{-1}(k)]} = \frac{1}{2}.$$

There is a very useful logical principle that one may use in situations like the previous exercises. We used explicit functions for the exercises to help the beginner. Here is the more general rule.

(0.2.12) THE INTERNAL DEFINITION PRINCIPLE:

A formula of infinitesimal analysis is said to be a bounded internal formula if it can be expressed in terms of $\in$, $=$ *and internal entities of* $\mathcal{Y}$ *using bounded quantifiers. An entity* $A \in \mathcal{Y}$ *is internal if and only if* $A = \{y \in B : \Phi(y)\}$, *where* B *is an internal entity and* $\Phi(y)$ *is a bounded internal formula with* y *as the only free variable of* $\Phi(y)$.

For example,

$$W = \{k \in {}^*\mathbb{N} \mid 1 \le k \le m\}$$

is internal because it is described with standard constants ${}^*\mathbb{N}$, $\le$, 1 and the internal constant m. The formulas for β_1 and β_2 above are internal, so the exercise (0.2.9) follows from this principle. However, the reader may solve the exercise by using extensions of appropriately defined standard functions ${}^*K_1$, ${}^*K_2$ applied at the nonstandard point m. The general moral of this functional approach is that if $b = {}^*f(a)$, even for nonstandard a, then b is internal. (This is trivial, if confusing, because $f \subseteq D\times R$, so ${}^*f \subseteq {}^*D\times{}^*R$, and $b \in {}^*R$, making b internal.)

(0.3) Basic Infinitesimals

At times the notation of infinitesimal analysis can be cumbersome. We remarked in section (0.2) that we drop the *'s on standard individuals and drop the $^\sigma$'s on familiar sets in X_1 such as $^\sigma\mathbb{N}$. Of course, we also drop the *'s on the arithmetic operations $+$, $\cdot$, $\leq$ as well as $h!$, m^n, etc. Whenever we have a real valued function f, the natural extension $^*f$ is a function extension, that is, $^*f(^*a) = {}^*[f(a)]$. Because of this we drop the * on f. Hence, $\sin(\frac{e}{\pi})$ may mean either the standard sine at Euler's number over pi or its natural extension, whereas, if δ is nonstandard, $\sin(\delta)$ really means $(^*\sin)(\delta)$. If we fear a possibility of confusion, we will use all the *'s and $^\sigma$'s.

(0.3.1) **DEFINITION:**

A number $r \in {}^*\mathbb{R}$ *is limited (or "finite") if* $|r| < m$ *for some standard natural number* $m \in {}^\sigma\mathbb{N}$. *Otherwise, it is unlimited (or "infinite"). We denote the limited scalars by* $\mathcal{O}$ *(big oh).*

A number $\delta \in {}^*\mathbb{R}$ *is infinitesimal if* $|\delta| < \frac{1}{m}$ *for every standard natural number* $m \in {}^\sigma\mathbb{N}$. *We denote the infinitesimal scalars by* o *(small oh). Also, for* $a,b \in {}^*\mathbb{R}$, *if* $a-b$ *is infinitesimal we write* $a \approx b$. *We write* $a \lesssim b$ *if* $a < b$ *or* $a \approx b$, *whereas, we write* $a << b$ *if* $a < b$ *but* $a \not\approx b$.

In Appendix 1 we show that $^*\mathbb{N}$ "looks like" $\mathbb{N}$ followed by infinitely many new numbers, that is, if $n \in {}^*\mathbb{N}\setminus{}^\sigma\mathbb{N}$, then $n > m$ for every $m \in \mathbb{N} = {}^\sigma\mathbb{N}$. In other words, all the nonstandard *natural numbers are "infinite". Since that term may cause confusion with "*finiteness" we also to call these numbers "unlimited."

It follows from algebra that a proper ordered field extension of $\mathbb{R}$ must be non-archimedean. Archimedes' axiom just says, "There are no infinitesimals," $(\forall\ \epsilon \in \mathbb{R})[\epsilon > 0 \Rightarrow (\exists\ m \in \mathbb{N})[\epsilon < \frac{1}{m}]]$. Thus, Leibniz' Principle asserts that there are infinitesimals in $^*\mathbb{R}$. (At the same time, the *archimedean axiom holds!) It also asserts that there are infinitely many distinct infinitesimals, for $\epsilon > 0$ in $\mathbb{R}$,

$$\epsilon > \cdots > \sqrt{\delta} > \delta > \delta^2 > \delta^m > \delta^n > \cdots > 0$$

for $m \in \mathbb{N}$, $n \in {}^*\mathbb{N}\setminus\mathbb{N}$. Leibniz' Principle says that around each standard real, there are infinitely many distinct points an infinitesimal distance away. This is because we may transfer the statement that for all $x \in \mathbb{R}$, $x \neq 0$ implies $x+r \neq r$, so $r+\delta \neq r+\delta^2 \neq r \cdots$. Of course, there are unlimited *real numbers such as $\frac{1}{\delta}$, but the limited numbers are just the standard numbers with a "monad" of infinitesimals clustered around each one. Here is a fancy way to express this idea.

(0.3.2) PROPOSITION:

The limited numbers, $\mathcal{O}$, are a totally ordered ring with maximal order ideal o. The quotient field $\mathcal{O}/\approx$ is isomorphic to $\mathbb{R}$ and the homomorphism $o \subset \mathcal{O} \xrightarrow{st} \mathbb{R}$ is called the standard part.

In other words, every limited $r \in {}^*\mathbb{R}$ has a standard $s \in \mathbb{R}$ infinitely nearby, $st(r) = s$ or $r = s+\iota$ for some infinitesimal ι. The fact that the infinitesimals form a maximal order ideal means that if a number has magnitude less than an infinitesimal, then it is also infinitesimal, only infinitesimals have unlimited reciprocals, and a limited number times an infinitesimal is infinitesimal ('moderate size times very small is very small'). The proof of this assertion can be found in one of the references such as Stroyan & Luxemburg [1976, (4.4.4)-(4.4.7)].

For our work in measure theory it is very convenient to define the extended standard part into the standard two point compactification of $\mathbb{R}$, $[-\infty,+\infty]$.

(0.3.3) **DEFINITION:**

The map $\mathbf{st} : {}^*\mathbb{R} \longrightarrow {}^\sigma[-\infty,\infty]$ *is given by*

$$\mathbf{st}(r) = \begin{cases} st(r), & \textit{if } r \textit{ is limited, } r \in \mathcal{O} \\ +\infty, & \textit{if } r > 0 \textit{ and } r \textit{ is unlimited} \\ -\infty, & \textit{if } r < 0 \textit{ and } r \textit{ is unlimited} \end{cases}$$

The love knot symbols $\pm\infty$ represent standard objects, not nonstandard numbers.

The standard part map extends to the finite dimensional coordinate spaces ${}^*R^d$ by

$$st((x_1,\cdots,x_d)) = (st(x_1),\cdots,st(x_d)).$$

Similarly, we write $x \approx y$ for $x,y \in {}^*\mathbb{R}^d$ if $x_j \approx y_j$ for $1 \leq j \leq d$.

The notion of infinitesimal is external. The interplay

between internal notions, to which transfer applies, and external notions is the "art" of Robinson's theory. The following is a useful "continuity principle" or "permanence principle" that says internal sequences cannot stop being infinitesimal at the boundary between limited and unlimited indices. We will use this frequently in stochastic analysis in conjunction with the saturation principles of section (0.4).

(0.3.4) **ROBINSON'S SEQUENTIAL LEMMA:**

If a_n *is an internal sequence and* $a_m \approx 0$ *for all standard* $m \in {}^{\sigma}\mathbb{N}$, *then there is an unlimited* $n \in {}^{*}\mathbb{N}$ *such that*

$$a_k \approx 0 \quad \textit{for all} \quad k \in {}^{*}\mathbb{N} \quad \textit{such that} \quad 0 \leq k \leq n.$$

PROOF:

The set

$$I = \{n \in {}^{*}\mathbb{N} : (\forall\ k \in {}^{*}\mathbb{N})[k \leq n \Rightarrow k \in \mathrm{dom}(a)\ \&\ |a_k| < \tfrac{1}{k}]\}$$

is internal by the Internal Definition Principle. It contains the external set ${}^{\sigma}\mathbb{N}$ by hypothesis and thus contains a nonstandard $n \in {}^{*}\mathbb{N}\setminus{}^{\sigma}\mathbb{N}$. When k is infinite and $k \leq n$, $|a_k| < \frac{1}{k} \approx 0$.

POISSON PROCESSES:

Now we will construct two independent approximate Poisson processes. Let $a_1, a_2 \in \mathbb{R}$ be standard positive numbers. Recall that $\delta t = \frac{1}{h!}$ and W has $m = h!^{h!}$ elements. Define an internal function $\alpha_1 : W \longrightarrow \{0,1\}$ by

$$\alpha_1(w) = \begin{cases} 1, & \text{if } w \leq [m a_1 \delta t] \\ 0, & \text{otherwise,} \end{cases}$$

where $[\cdot]$ denotes the *greatest integer function. We see that $ma\delta t-1 \leq {}^{\#}[\alpha_1^{-1}(1)] \leq ma\delta t$, so $\frac{{}^{\#}[\alpha_1^{-1}(1)]}{{}^{\#}W} = b_1\delta t$, with $b_1 \approx a_1$. Define an internal function $\alpha_2 : W \longrightarrow \{0,1\}$ by

$$\alpha_2(w) = \begin{cases} 1, & \text{if} \quad w \leq [ma_1a_2\delta t^2] \\ 1, & \text{if} \quad [ma_1\delta t] < w \leq [ma_1\delta t]+[(m-ma_1\delta t)a_2\delta t] \\ 0, & \text{otherwise.} \end{cases}$$

(0.3.5) **EXERCISE:**

a) *Show that* $\frac{{}^{\#}[\alpha_2^{-1}(1)]}{{}^{\#}[W]} = b_2\delta t$ *where* $b_2 \approx a_2$.

b) *For* $(i,j) = (1,2)$ *and* $(2,1)$ *and for* $k = 1$ *and* 0 *show that*

$$\frac{{}^{\#}[\alpha_i^{-1}(1) \cap \alpha_j^{-1}(k)]}{{}^{\#}[\alpha_j^{-1}(k)]} = b_i\delta t, \quad \textit{where} \quad b_i \approx a_i.$$

Now suppose that $(\alpha,a)=(\alpha_j,a_j)$ for either $j = 1$ or $j = 2$. We define an internal stochastic process $J : \mathbb{T}\times\Omega \longrightarrow {}^*\mathbb{N}$ by

(0.3.6) $$J(t,\omega) = \Sigma[\alpha(\omega_s) : s \in \mathbb{T},\ \delta t \leq s \leq t].$$

This process increases by unit jumps as t increases. It is close to a Poisson process with rate a. In terms of the distribution, we have the following with $b \approx a$,

$$(0.3.7)\quad P[\{\omega : J(t,\omega) \le k\}] = \sum_{h=0}^{k} \begin{bmatrix} \frac{t}{\delta t} \\ h \end{bmatrix} p^h (1-p)^{[\frac{t}{\delta t} - h]}, \quad p = b\delta t$$

$$= \sum_{h=0}^{k} \begin{bmatrix} \frac{t}{\delta t} \\ h \end{bmatrix} \frac{(b\delta t)^h}{(1-b\delta t)^h} (1-b\delta t)^{\frac{t}{\delta t}}$$

$$= (1-b\delta t)^{\frac{t}{\delta t}} \sum_{h=0}^{k} \frac{(b)^h}{h!} \frac{\prod_{j=0}^{h-1} (t-j\delta t)}{(1-b\delta t)^h}$$

$$\approx (1-b\delta t)^{\frac{t}{\delta t}} \sum_{h=0}^{k} \frac{a^h}{h!}$$

$$\approx e^{-at} \sum_{h=0}^{k} \frac{a^h}{h!}.$$

One may verify that

$$(1-b\delta t)^{\frac{t}{\delta t}} \approx e^{-at}, \text{ for } b \approx a,$$

by using the series expansion for e^x and applying the *binomial theorem and Robinson's Sequential Lemma to the left-hand-side. See Stroyan & Luxemburg [1976] for this and other elementary properties of e^x using infinitesimals.

Next, we want to calculate the distribution of Anderson's infinitesimal random walk (0.2.9). For this computation we need another elementary property of e^x.

(0.3.8) **STIRLING'S FORMULA:**

For $n \in {}^{*}\mathbb{N}$,

$$e^{\frac{1}{12n+1}} < \frac{n!\ e^{n}}{\sqrt{2\pi n}\ n^{n}} < e^{\frac{1}{12n}}.$$

The proof of this formula can be found in Feller [1968, vol. 1, p. 54].

We also need a very simple form of the integral on the line. The treatment of Lebesgue integrals in Chapter 2 below includes this case, so we present the special case without much explanation.

(0.3.9) **"CAUCHY" INTEGRALS:**

The integral of a continuous standard function $f: [a,b] \longrightarrow \mathbb{R}$ *satisfies:*

$$\int_a^b f(x)dx \approx \sum[f(t)dt : a \leq t \leq b,\ \text{step } \delta t]$$

for any positive infinitesimal δt.

We compute the distribution of $B(t,\omega)$ using the *finite sums. This fact about integrals simply tells us that we have computed the classical normal distribution at the end.

(0.3.10) **EXERCISE:**

Suppose that $f : [a,b] \longrightarrow \mathbb{R}$ *and* $F : [a,b] \longrightarrow \mathbb{R}$ *are standard functions such that whenever* $a \leq x \leq b$ *in* ${}^{*}\mathbb{R}$ *and* δx *is a positive infinitesimal then* $F(x+\delta x)-F(x) = [f(x)+\iota(x,\delta x)]\delta x$ *where* $\iota \approx 0$. *Prove that*

$$\int_a^b f(x)dx \approx \sum[f(x)\delta x : a \leq x \leq b, \text{ step } \delta x] \approx F(b)-F(a)$$ *by using the* **linearity of* **finite summation, the* **triangle inequality, and the estimate*

$$|\sum[\iota(x,\delta x)\cdot\delta x : a \leq x \leq b, \text{ step } \delta x]| \leq \max[|\iota(x,\delta x)| : a \leq x \leq b, \text{ step } \delta x]\cdot(b-a).$$

Observe that the **finite maximum is attained by transfer.*

Of course this exercise is one half of the Fundamental Theorem of Integral Calculus. The usual continuity hypothesis on $f(x)$ is contained in the assumption that $\frac{\delta F}{\delta x} \approx f(x)$ at nonstandard x's between a and b. Such x's certainly appear in the summation from a to b in steps of $\delta x\cdots$, see Barwise [1977, Ch. A.6].

For convenience in our discussion of $B(t,\omega)$, we define some parameters. Let $\delta t = \frac{1}{2n_1}$ and whenever t in $\mathbb{T}$ is an even multiple of δt, $2n_t = \frac{t}{\delta t}$. We first estimate the probability η_t of "exactly half heads" after $n = n_t$ tosses, or equivalently, of $\{\omega : B(t,\omega) = 0\}$.

$$\eta_t = P[B(t,\omega) = 0]$$

$$= \frac{1}{2^{2n}} \binom{2n}{n} = \frac{(2n)!}{(2^{2n})(n!)(n!)},$$

where $n = n_t$. Using Stirling's formula for t such that n_t

is infinite, we obtain:

$$\eta_t = \frac{(2n^{2n})\ \sqrt{2\pi 2n}\ (e^n)^2}{(2^{2n})(n^n)^2(\sqrt{2\pi n})^2(e^{2n})} \cdot (1+\iota_t)$$

$$= \frac{1}{\sqrt{\pi n_t}}\ (1+\iota_t), \quad \text{where} \quad \iota_t \approx 0.$$

Notice that even for infinitesimal $t > \sqrt{\delta t}$ (with n_t defined), $n_t > \frac{1}{\sqrt{\delta t}}$ is infinite. This is all that is needed to make ι_t infinitesimal. Changing back to the time variables, we have

$$\text{(0.3.11)} \qquad \eta_t = \frac{2}{\sqrt{2\pi t}}\ \sqrt{\delta t}\ (1+\iota_t), \quad \iota_t \approx 0.$$

The only values $B(t,\omega)$ can assume are positive and negative integer multiples of $2\sqrt{\delta t}$, so we let $\delta x = 2\sqrt{\delta t}$ and consider the probabilities of $\{\omega : B(t,\omega) = x\}$ for $|x| = k\delta x = 2k\sqrt{\delta t}$. We shall only be interested in these when x is limited (in $\mathcal{O}$) in which case $\frac{k}{n_1} \approx 0$ (since $\sqrt{k/(2n_1)}$ is limited, $\sqrt{2/n_1}$ is infinitesimal and limited times infinitesimal is infinitesimal). Notice that $\{\omega : B(t,\omega) = x\}$ and $\{\omega : B(t,\omega) = -x\}$ have the same *number of elements because an ω is in one by having k more than half "heads," so interchanging "heads" and "tails" internally maps one on the other. Let k be as above and $n = n_t$, where $t \not\approx 0$, so $\frac{k}{n_t} = 2\frac{k}{t}\ \sqrt{\delta t}\ \sqrt{\delta t} \approx 0.$

$$P[B(t,\omega) = x] = \frac{1}{2^{2n}} \begin{bmatrix} 2n \\ n+k \end{bmatrix}$$

$$= \frac{1}{2^{2n}} \begin{bmatrix} 2n \\ n \end{bmatrix} \frac{\prod_{h=0}^{k-1} [n-h]}{\prod_{h=1}^{k} [n+h]}$$

$$= \eta_t \frac{n}{n+k} \prod_{h=1}^{k-1} \left\{ \frac{[1 - \frac{h}{n}]}{[1 + \frac{h}{n}]} \right\}$$

$$= \eta_t \frac{n}{n+k} \prod_{h=1}^{k-1} \left[1 - \frac{2h}{n} + \frac{2h}{n} \frac{h}{(n+h)} \right],$$

by formulas extended from algebra,

$$= \eta_t \frac{1}{(1+\frac{k}{n})} \exp \left\{ \sum_{h=1}^{k-1} \log \left[1 - \frac{2h}{n}(1+\iota_h) \right] \right\}, \quad \iota_h \approx 0,$$

since $h < k$ and $\frac{k}{n_t} \approx 0$, because $t \not\approx 0$, $(n = n_t)$

$$= \eta_t \frac{1}{(1+\frac{k}{n})} \exp \left\{ \sum_{h=1}^{k-1} \left[-\frac{2h}{n}(1+\vartheta_h) \right] \right\}, \quad \vartheta_h \approx 0$$

since log is differentiable at 1,

$$= \eta_t \frac{1}{(1+\frac{k}{n})} \exp \left\{ \frac{-k(k-1)}{n} + \vartheta \frac{k^2}{n} \right\}, \quad \vartheta \approx 0,$$

by the formula for Σh and estimates like (0.3.10) for ϑ,

$$= \eta_t e^{-\frac{k^2}{n_t}} (1+\iota_t), \quad \text{where} \quad \iota_t \approx 0.$$

Notice that $\frac{k^2}{n_t} = \left[\frac{k\sqrt{\delta t}}{t}^2\right]$ is finite and $\frac{k}{n_t} \approx 0$, so $(n=n_t)$

$$\frac{1}{(1+\frac{k}{n})} e^{\frac{k}{n}} e^{\frac{\vartheta k^2}{n}} = (1+\iota_t) \quad \text{with} \quad \iota_t \approx 0.$$

Finally we substitute (0.3.11), $2\sqrt{\delta t} = \delta x$ and $k^2/n = x^2/(2t)$ to obtain:

(0.3.12) **DE MOIVRE'S LIMIT THEOREM:**

If $|x|$ *is finite (or limited in* $\mathcal{O}$*) and* t *(as above) is not infinitesimal, then*

$$P[x < B(t,\omega) \le x+\delta x] = \frac{1}{\sqrt{2\pi t}} e^{-\frac{x^2}{2t}} \delta x(1+\iota), \quad \iota \approx 0.$$

Hence, if $a,b \in \mathcal{O}$, *and* $t \not\approx 0$, *then*

$$P[a < B(t,\omega) \le b] \approx \frac{1}{\sqrt{2\pi t}} \int_a^b e^{-\frac{\xi^2}{2t}} d\xi,$$

$B(t)$ *is nearly normally distributed with mean* 0 *and variance* t.

PROOF:

Aside from the substitution mentioned before the statement,

all that remains is to associate the sum below with the integral. The sum over possible values of $B(t)$,

$$\frac{1}{\sqrt{2\pi t}} \sum_x \left[e^{-\frac{x^2}{2t}} \delta x(1+\iota) : a < x \le b, \text{ step } \delta x \right]$$

is, by (0.3.9) and an estimate like (0.3.10), infinitely close to the integral

$$\frac{1}{\sqrt{2\pi t}} \int_a^b e^{-\frac{x^2}{2t}} dx.$$

This proves the result.

The reader may wish to look ahead to the proof of continuity of the paths of $B(t)$. The preliminary results in Appendix 1 give the infinitesimal formulation of continuity while (5.2.3) shows that for ω off a set $\Lambda = \bigcup_m \bigcap_n \Omega_n^m$, $B(\cdot,\omega)$ is S-continuous. The sets Ω_n^m are internal and of smaller and smaller *probability, but the set Λ is external. The measure of chapter 1 allows us to conveniently discuss such "Loeb sets" and their probabilities.

Results of section (5.3) show that we can take "standard parts" of the paths of $J(t,\omega)$ from (0.3.6) in the appropriate sense. What is required there is that two or more jumps do not occur in infinitesimal time. Again, this happens on an external set which can be approximated by bigger and bigger internal sets of probability finitely less than 1. It is convenient to have a probability defined on the external set so that we can say that J has jumps of size 1 with probability 1.

(0.4) Saturation & Comprehension

Part (b) of Leibniz' Transfer Principle (0.2.3) requires that the * mapping is "polyenlarging." This section explains what that term means. All things considered, we feel that a polyenlargement is the simplest framework for infinitesimal analysis. There are interesting Loeb measure questions in various special nonstandard models, but we shall leave such questions to the specialists. (Some of these require additional axioms for set theory. For example, see Ross [1984].)

The bounded forms of Nelson's axioms for Internal Set Theory hold in a polyenlargement, but not in an enlargement. Moreover, polyenlargements have a stronger "saturation property" than Nelson's Idealization Principle.

(0.4.1) **DEFINITION:**

> A *superstructure extension* $^*: \mathfrak{X} \longrightarrow \mathfrak{Y}$ *is polyenlarging if it is given as a direct limit of* κ *successive enlargements, where* κ *is a regular cardinal satisfying* $\kappa > \mathrm{card}(\mathfrak{X})$.

The image of a polyenlarging extension is called a polyenlargement. Successive enlargements are described in more detail below. First we give the main properties which make polyenlargements useful.

We say that a family of sets $\mathcal{F}$ has the finite intersection property if every finite subfamily of $\mathcal{F}$ has nonempty intersection.

(0.4.2) THE SATURATION PRINCIPLE:

Let $^*: \mathcal{X} \to \mathcal{Y}$ *be a polyenlarging extension. Suppose that* $\mathcal{F}$ *is a family of internal subsets of an internal set* Y *which has the finite intersection property and* $\mathrm{card}(\mathcal{F}) \leq \mathrm{card}(\mathcal{X})$. *Then* $\cap[F : F \in \mathcal{F}] \neq \emptyset$.

The family $\mathcal{F}$ is usually external; only its elements, $F \in \mathcal{F}$, need be internal. If every $F \in \mathcal{F}$ is an internal subset of the internal set $Y \in \mathcal{Y}$, then $Y \in {}^*X_{m+1}$ for some m. The sets $F \in \mathcal{F}$ are subsets of the standard set ${}^*X = {}^*X_m \cup {}^*X_0$.

One of the primary ways that we will use saturation is this. Suppose that $\mathcal{F} = \{A_m \mid m \in {}^\sigma\mathbb{N}\}$ is a countable sequence of internal sets and that $\mathcal{F}$ has the finite intersection property. Then for each $m \in {}^\sigma\mathbb{N}$, $B_m = \bigcap_{j=0}^{m} A_j$ is a nonempty internal set. The sequence of B_m's decreases. If we assign internal probabilities to these sets, then the sequence of standard parts, $\mathrm{st}\, P[B_m] = b_m$, decreases. Saturation says that the external intersection, $\cap[B_m : m \in {}^\sigma\mathbb{N}] \neq \emptyset$, hence we may assign it the probability $\inf[b_m : m \in {}^\sigma\mathbb{N}]$. This is the whole secret of Loeb's measure extension.

Another way to view saturation is as a kind of "completeness." Consider the gap in ${}^*\mathbb{R}$ where we might like to put the "biggest infinitesimal" (no such number ζ exists because $\zeta+\zeta > \zeta$). No countable nested sequence of intervals $[\delta_m, \frac{1}{m}]$, for $\delta_m \approx 0$, will have empty intersection; there will always be an infinitesimal $\eta > \delta_m$ for all $m \in {}^\sigma\mathbb{N}$.

An extremely useful consequence of saturation is a function extension property called "comprehension." The model can comprehend "small" infinite sets by *extending* them (we can even make the extension *finite).

(0.4.3) **THE COMPREHENSION PRINCIPLE:**

Let $f \in \mathcal{Y}$ *be an external function and suppose that the domain of* f, *satisfies* card(dom(f)) $\leq$ *card*($\mathcal{X}$). *Let* D *and* R *be internal entities such that* $D \supseteq$ dom(f) *and* $R \supseteq$ rng(f). *Then* f *has an internal extension* $F : D \to R$, *that is, for each* $x \in$ dom(f), $F(x) = f(x)$.

The main way we will use comprehension is in the countable case. For example, if $\{A_m \mid m \in {}^\sigma\mathbb{N}\}$ is a countable family of internal subsets of an internal set $\mathbb{V}$, then we may take $D = {}^*\mathbb{N}$, $R = {}^*\mathbb{B}(\mathbb{V})$ and find an internal sequence $A: {}^*\mathbb{N} \longrightarrow {}^*\mathbb{B}(\mathbb{V})$ such that $A(m) = A_m$ for standard m.

Many people's first inclination is to treat infinite integers $n \in {}^*\mathbb{N} \setminus {}^\sigma\mathbb{N}$ like countable ordinals. Another temptation is to consider *finite sums as some sort of countable series. These are misleading analogies. The initial segments ${}^*\mathbb{N}[1,n]$ are all externally uncountable sets when n is infinite. The set $\mathbb{T} = \{t \in {}^*\mathbb{R} \mid (\exists\ k \in {}^*\mathbb{N})[t = \frac{k}{n}\ \&\ 1 \leq k \leq n]\}$ is an internal one-to-one image of the initial segment of ${}^*\mathbb{N}$ above. The set $\mathbb{T}$ is S-dense in [0,1], so the external map st shows that the cardinality of $\mathbb{T}$ (and hence ${}^*\mathbb{N}[1,n]$) is at least the continuum. (In certain nonenlargement nonstandard models the cardinalilty can be exactly the continuum.)

Another important property of polyenlargements is the "internal homogeneity" property of Henson's Lemma given below.

One consequence of this property is the following result which we will find useful in construction of nonmeasurable sets. The result says internal sets are homogeneous in size.

(0.4.4) PROPOSITION:

Let $^*\mathfrak{X}$ be a polyenlargement. All infinite internal sets, including unlimited *finite sets, have the same external cardinality.

The common cardinality of the internal sets will be gigantic by non-set-theorists' standards. Internal cardinalities, in particular, *finite ones, "seem smaller to the model" because those cardinalities can only be tested with internal functions. For example, there is no internal functior from $^*\mathbb{N}[1,m]$ onto $^*\mathbb{N}[1,m+1]$, while there is an external bijection.

(0.4.5) EXERCISE:

a) *Use saturation to prove that* $^*X \setminus {}^{\sigma}X \neq \emptyset$ *for every infinite* $X \in \mathfrak{X}$. *(HINT: Let* $\mathcal{F} = \{^*X \setminus \{^*x\} \mid x \in X\}$. *Why is* $^*X \setminus \{^*x\}$ *internal?)*

b) *Prove that there are positive infinitesimals by considering the family* $\mathcal{F} = \{^*(0,\epsilon) \mid \epsilon \in \mathbb{R}_+\}$.

Now we give the detailed definition of a direct limit of successive enlargements. We begin with a single enlargement. Let $\mathfrak{X}^1$ be a superstructure over a set of individuals X_0^1 [$\mathfrak{X}^1 = \cup X_m^1$, where $X_{m+1}^1 = \mathcal{B}(\bigcup_{k=0}^{m} X_k^1)$, as in section 0.1.] Let $\mathfrak{X}^2$

be a superstructure over a set of individuals X_0^2. We say that an injection $i_1^2 : \mathfrak{X}^1 \longrightarrow \mathfrak{X}^2$ is a *superstructure extension* if $i_1^2(X_0^1) = X_0^2$ and i_1^2 satisfies part (a) of Leibniz' Transfer Principle (i_1^2 is an elementary extension of $\mathfrak{X}^1$ for the language of $\in$ and $=$ with a constant for each element of $\mathfrak{X}^1$). If i_1^2 is a superstructure extension, then an entity $Y \in \mathfrak{X}^2$ is i_1^2-*internal* provided there exists $X \in \mathfrak{X}^1$ such that $Y \in i_1^2(X)$. An internal entity $Y \in i_1^2(X_m^1)$ is called i_1^2-*finite* if every i_1^2-internal injection of Y into Y is onto.

(0.4.6) **DEFINITION:**

A superstructure extension $i_1^2 : \mathfrak{X}^1 \longrightarrow \mathfrak{X}^2$ *is an enlargement of* $\mathfrak{X}^1$ *if for every entity* $X \in \mathfrak{X}^1$ *there is an* i_1^2-*finite* $Y \in \mathfrak{X}^2$ *such that* $\{i_1^2(x) \mid x \in X\} \subseteq Y \subseteq i_1^2(X)$.

Enlargements may be constructed by forming "adequate" ultrapowers as defined in Appendix 2.

(0.4.7) EXERCISE:

If $i_1^2 : \mathfrak{X}^1 \longrightarrow \mathfrak{X}^2$ *is enlarging for* $\mathfrak{X}^1$ *and* $i_2^3 : \mathfrak{X}^2 \longrightarrow \mathfrak{X}^3$ *is enlarging for* $\mathfrak{X}^2$, *then* $i_2^3 \circ i_1^2$ *is enlarging for* $\mathfrak{X}^1$.

SUCCESSIVE ENLARGEMENTS:

Let κ be an infinite cardinal. A direct limit of κ successive enlargements of $\mathfrak{X}$ is given inductively as follows. Let $\mathfrak{X}^0 = \mathfrak{X}$, be our original superstructure. Let $i_0^1 : \mathfrak{X}^0 \longrightarrow \mathfrak{X}^1$ be an enlargement. If λ is a cardinal less than or equal to

κ, and the family of enlargements $i_\alpha^\gamma : \mathfrak{X}^\alpha \longrightarrow \mathfrak{X}^\gamma$ is defined and satisfies $i_\alpha^\gamma = i_\beta^\gamma \circ i_\alpha^\beta$, whenever $\alpha < \beta < \gamma$, we proceed inductively in two cases.

<u>Case 1</u>: If $\lambda = \beta + 1$, let $i_\beta^{\beta+1}: \mathfrak{X}^\beta \to \mathfrak{X}^{\beta+1}$ be an enlargement and define $i_\alpha^\lambda = i_\beta^{\beta+1} \circ i_\alpha^\beta$, for $\alpha < \beta$.

<u>Case 2</u>: If λ is a limit ordinal note that the usual algebraic definition cannot be used because the limit mappings go into a superstructure over a set of individuals. First take a set of individuals X_0^λ equivalent to the union $\cup[X_0^\alpha : \alpha < \lambda]$ under the usual identification, $x \equiv y$ if $i_\alpha^\beta(x) = y$, where $x \in X_0^\alpha$ and $y \in X_0^\beta$. Then define i_α^λ by induction on superstructure levels, $i_\alpha^\lambda(A) = \{i_\beta^\lambda(x) \mid x \in i_\alpha^\beta(A)\}$.

This completes the detailed definition of direct limit of successive enlargements. A set $A \in \mathfrak{X}^\kappa$ is *α-standard* if $A = i_\alpha^\kappa(B)$, for some $B \in \mathfrak{X}^\alpha$ and *α-internal* if A is an element of some α-standard set. The terms *standard* and *internal* refer to 0-standard and 0-internal.

A polyenlargement $^* : \mathfrak{X} \to \mathfrak{Y}$ is the direct limit of κ successive enlargements where κ is regular and satisfies $\kappa \geq \text{card}^+(\mathfrak{X})$, the first cardinal greater than $\text{card}(\mathfrak{X})$, and $^* = i_0^\kappa$. Successor cardinals are always regular. (See the proof in Chang-Keisler[1977,A.25,p.505].) Notice that **finite* sets are i_0^κ-finite in the sense above.

PROOF OF THE SATURATION PRINCIPLE:

Let $\mathcal{F}$ be a family of internal subsets of $^*X$ satisfying $\text{card}(\mathcal{F}) \leq \text{card}(\mathfrak{X})$ and having the finite intersection property. Since successor cardinals are regular, there is an $\alpha < \text{card}^+(\mathfrak{X})$

and a family $\mathcal{F}^\alpha \subseteq X^\alpha_m$ such that $\mathcal{F} = \{i^\kappa_\alpha(F_\alpha) \mid F_\alpha \in \mathcal{F}^\alpha\} \subseteq i^\kappa_\alpha(\mathcal{F}^\alpha)$. Since i^κ_α is enlarging, $\mathcal{F}$ is contained in a *finite set $\mathcal{A}$, $\mathcal{F} \subseteq \mathcal{A} \subseteq i^\kappa_\alpha(\mathcal{F}^\alpha)$. The last set has the *finite intersection property by transfer of that property from $\mathfrak{X}^\alpha$ and therefore $\cap \mathcal{F} \supseteq \cap \mathcal{A} \neq \emptyset$.

PROOF OF THE COMPREHENSION PRINCIPLE:

Use the notation of (0.4.3). For each $x \in \text{dom}(f)$ define the set $A_x = \{F \mid F : D \to R$ *is internal* & $F(x) = f(x)\}$. Any $F \in \cap[A_x : x \in \text{dom}(f)]$ satisfies the assertion.

PROOF OF (0.4.4):

Let A and B be infinite internal sets $A = i^\kappa_\alpha(A^\alpha)$, $B = i^\kappa_\alpha(B^\alpha)$, for $A^\alpha, B^\alpha \in \mathfrak{X}^\alpha$. Let $A^{\alpha+m} = i^{\alpha+m}_\alpha(A^\alpha)$, $B^{\alpha+m} = i^{\alpha+m}_\alpha(B^\alpha)$. We know that, "for any finite subset F of A^α there is an injection $f : F \to B^\alpha$." The transfer of this statement to $\mathfrak{X}^{\alpha+1}$ says that every *finite set has such a map. The enlarging property of $i^{\alpha+1}_\alpha$ says that A^α is embedded in a *finite $F^{\alpha+1} \subseteq A^{\alpha+1}$. Let $f_{\alpha+1} : F^{\alpha+1} \to B^{\alpha+1}$ be an injection asising by transfer. The composition $f_{\alpha+1} \circ i^{\alpha+1}_\alpha : A^\alpha \to B^{\alpha+1}$ is an injection. Next, if G is a finite subset of $B^{\alpha+1}$ there is an injection extending $f^{-1}_{\alpha+1}$ and also defined on G. Embedd $B^{\alpha+1}$ in a *finite subset and transfer this property to obtain an injection $g_{\alpha+2} : G^{\alpha+2} \to A^{\alpha+2}$ extending $i^{\alpha+2}_{\alpha+1} \circ f^{-1}_{\alpha+1}$. The map $g_{\alpha+2} \circ i^{\alpha+2}_{\alpha+1}$ injects $B^{\alpha+1}$ into $A^{\alpha+2}$. Continue this procedure back and forth thru a countable number of steps so that the injection $f_{\alpha+m+1}$ extends the injection $i^{\alpha+m+1}_{\alpha+m} \circ g^{-1}_{\alpha+m}$. The limit mapping is a bijection of $A^{\alpha+\omega}$ onto $B^{\alpha+\omega}$.

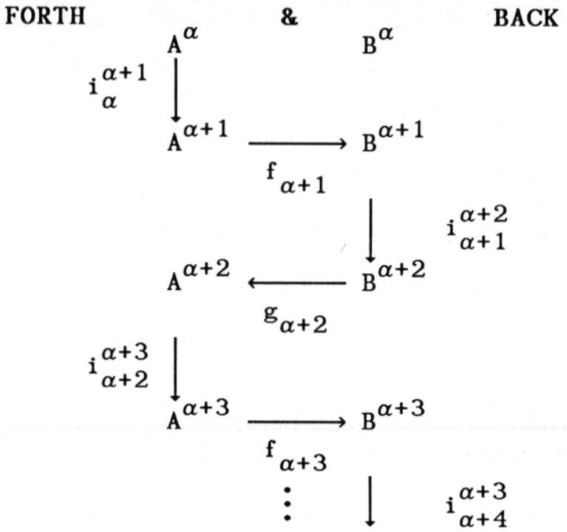

FORTH & BACK

$$
\begin{array}{ccc}
A^{\alpha} & & B^{\alpha} \\
i_{\alpha}^{\alpha+1} \downarrow & & \\
A^{\alpha+1} & \xrightarrow{f_{\alpha+1}} & B^{\alpha+1} \\
 & & \downarrow i_{\alpha+1}^{\alpha+2} \\
A^{\alpha+2} & \xleftarrow{g_{\alpha+2}} & B^{\alpha+2} \\
i_{\alpha+2}^{\alpha+3} \downarrow & & \\
A^{\alpha+3} & \xrightarrow{f_{\alpha+3}} & B^{\alpha+3} \\
 & \vdots & \downarrow i_{\alpha+4}^{\alpha+3}
\end{array}
$$

(0.4.8) **HENSON'S LEMMA:**

For each first order language L with less than card$^{+}(\mathfrak{X})$ constants and relations, if **A** *and* **B** *are elementarily equivalent structures for L whose domains and relations are internal entities of a polyenlargement ${}^{*}\mathfrak{X}$, then* **A** *and* **B** *are isomorphic.*

PROOF:

Let $\mathbf{A} = (A, R_k)$ and $\mathbf{B} = (B, S_k)$ denote the internal domains and relations of the L-structures. Since card$^{+}(\mathfrak{X})$ is regular, there exist $\alpha <$ card$^{+}(\mathfrak{X})$, A^{α}, R^{α}_{k}, B^{α}, S^{α}_{k} in $\mathfrak{X}^{\alpha}$ such that $A = i_{\alpha}^{\kappa}(A^{\alpha})$, etc. Since ${}^{*}\mathfrak{X}$ is an enlargement of $\mathfrak{X}^{\alpha}$, the L-structures $(A^{\alpha}, R^{\alpha}_{k})$ and $(B^{\alpha}, S^{\alpha}_{k})$ are elementarily equivalent.

We may use the enlargement property of $i_{\alpha}^{\alpha+1}: \mathfrak{X}^{\alpha} \rightarrow \mathfrak{X}^{\alpha+1}$ to construct an elementary L-monomorphism from $(A^{\alpha}, R^{\alpha}_{k})$ into $(i_{\alpha}^{\alpha+1}(B^{\alpha}), i_{\alpha}^{\alpha+1}(S^{\alpha}_{k}))$ and continue back and forth as in the proof of (0.4.4) above. See Henson[1974] for details.

CHAPTER 1: *FINITE & HYPERFINITE MEASURES

This chapter is an elementary treatment of Loeb's [1975] construction of measures and related work as described in the foreword. *Finite sets were defined in section (0.2) and the *finite summation operation may be defined by the natural extension of the function $^*H_2^q$ from exercise (0.1.5). We illustrated many uses of *finite sums in (0.2.9) and (0.3.9-12). The standard part function, st, is defined in section (0.3).

According to "Littlewood's Principles," Lebesgue integration just amounts to three basic facts: 1) Lebesgue sets are almost intervals. 2) Lebesgue functions are almost continuous. 3) Convergence is almost uniform. The principles analogous to (1) and (2) for hyperfinite measure theory allow us to replace measurable sets and functions by formally finite ones. Saturation corresponds to the third principle.

(1.1) Limited Hyperfinite Measures

A *finite *weight function* is an internal function $\delta\mu : \mathbb{V} \longrightarrow {}^*[0,\infty)$, where the domain $\mathbb{V}$ is an internal set. The (positive) **finite measure* associated with $\delta\mu$ is the set function $\mu : {}^*\mathfrak{B}(\mathbb{V}) \longrightarrow {}^*[0,\infty)$ defined on all internal subsets of $\mathbb{V}$, $A \in {}^*\mathfrak{B}(\mathbb{V})$, by *summation of the weights,

$$\mu[A] = \Sigma[\delta\mu(a) : a \in A].$$

We say that μ is a *limited (positive)* **finite measure* if

$\mu[\mathbb{V}] \in \mathcal{O}$. The *outer measure* associated with μ is the set function $\overline{\mu} : \mathfrak{B}(\mathbb{V}) \longrightarrow [0,\infty)$ defined by

$$\overline{\mu}[U] = \inf[\mathrm{st}\ \mu[A] : U \subseteq A \in {}^*\mathfrak{B}(\mathbb{V})].$$

The *inner measure* associated with μ is the set function $\underline{\mu} : \mathfrak{B}(\mathbb{V}) \longrightarrow [0,\infty)$ defined by

$$\underline{\mu}[U] = \sup[\mathrm{st}\ \mu[B] : {}^*\mathfrak{B}(\mathbb{V}) \quad B \subseteq U].$$

The inner and outer measures are defined on both internal and external subsets of $\mathbb{V}$, while μ is only defined on internal subsets of $\mathbb{V}$. Since $\mathbb{V}$ is *finite, the weight function $\delta\mu$ may be defined from μ by $\delta\mu(v) = \mu[\{v\}]$. One may generalize our treatment to situations where μ is not given by a weight function (see EXERCISE (1.1.8)).

(1.1.1) **DEFINITIONS:**

The sigma algebra generated by the internal subsets of a **finite set* $\mathbb{V}$ *is called the Loeb algebra,*

$$\Sigma({}^*\mathfrak{B}(\mathbb{V})) = \mathrm{Loeb}(\mathbb{V}).$$

A set $M \subseteq \mathbb{V}$ *is called* **μ-measurable** *provided* $\underline{\mu}[M] = \overline{\mu}[M]$. *The collection of* μ*-measurable sets is denoted*

$$\mathbf{Meas}(\mu).$$

The limited hyperfinite measure μ *associated to the limited* **finite measure* μ *is the set function*

μ : Meas(μ) $\longrightarrow \mathbb{R}$ *given by*

$$\mu[M] = \overline{\mu}[M] = \underline{\mu}[M].$$

Theorem (1.1.6) below justifies this terminology. The intermediate lemmas simply prove various parts of the result.

The *finite measure μ can be manipulated with formal combinatorics, but there are two things "wrong" with μ. First, μ takes values in $^{*}\mathbb{R}$ instead of $\mathbb{R}$. Since μ is limited we can solve this problem by taking standard parts, st μ : $^{*}\mathfrak{B}(\mathbb{V}) \longrightarrow \mathbb{R}$. This brings up the second problem: the class of internal subsets of $\mathbb{V}$, $^{*}\mathfrak{B}(\mathbb{V})$, is not a sigma algebra. Suppose that $(A_m : m \in {}^{\sigma}\mathbb{N})$ is a properly decreasing sequence of internal subsets of $\mathbb{V}$. If $A_\infty \subseteq \cap[A_m : m \in {}^{\sigma}\mathbb{N}]$ is internal, then the countable sequence $(A_m \backslash A_\infty : m \in {}^{\sigma}\mathbb{N})$ has the finite intersection property and cannot have an empty total intersection by the saturation property (0.4.2). Therefore we cannot have $A_\infty = \cap[A_m : m \in {}^{\sigma}\mathbb{N}]$ internal and so countable intersections of internal sets are not internal unless they reduce to finite intersections.

(1.1.2) **REMARK:**

Since μ and st are monotone, $\underline{\mu}[U] \leq \overline{\mu}[U]$ and if $U_1 \subseteq U_2$, $\overline{\mu}[U_1] \leq \overline{\mu}[U_2]$ and $\underline{\mu}[U_1] \leq \underline{\mu}[U_2]$. Thus if $\mu(M) = 0$ and $N \subset M$, N is measurable.

(1.1.3) **LEMMA:**

$M \subseteq \mathbb{W}$ is measurable if and only if for every positive standard ϵ, there exist internal sets A, B such that $B \subseteq M \subseteq A$ and $\mu[A \setminus B] < \epsilon$.

PROOF:

If M is measurable, for every such ϵ there exist internal sets $A \supseteq M \supseteq B$ such that $\mu[B] + \frac{\epsilon}{2} > \underline{\mu}(M) = \overline{\mu}(M) > \mu[A] - \frac{\epsilon}{2}$; then $\mu[A \setminus B] < \epsilon$. Conversely, assume that for each standard $\epsilon > 0$ there are internal A, B such that $B \subseteq M \subseteq A$ and $\mu[A \setminus B] < \epsilon$; then $\overline{\mu}(M) - \underline{\mu}(M) < \epsilon$.

(1.1.4) **LEMMA:**

If M and N are measurable, so are $M \setminus N$, $M \cap N$, $M \cup N$.

PROOF:

We begin with $M = \mathbb{W}$ showing that $\mathbb{W} \setminus N$ is measurable. Fix a standard $\epsilon > 0$ and internal A, B so that $B \subseteq N \subseteq A$ and $\mu[A \setminus B] < \epsilon$. Now the complements satisfy $A^c \subseteq N^c \subseteq B^c$ and $B^c \setminus A^c = A \cap B^c = A \setminus B$, so $\mu[B^c \setminus A^c] < \epsilon$ and N^c is measurable.

For the intersection, let $\epsilon > 0$ be a given standard number and choose internal sets A, B, C, D so that $A \supseteq M \supseteq B$, $C \supseteq N \supseteq D$, $\mu[A \setminus B] < \frac{\epsilon}{2}$, $\mu[C \setminus D] < \frac{\epsilon}{2}$. Then $E = A \cap C$ and $F = B \cap D$ satisfy $E \supseteq M \cap N \supseteq F$ and $\mu[E \setminus F] \leq \mu[A \setminus B] + \mu[C \setminus D] < \epsilon$.

The rest of the proof follows from set algebra, $M \setminus N = M \cap N^c$, $M \cup N = (M^c \cap N^c)^c$.

(1.1.5) **LEMMA:**

If (M_k) *is a countable sequence of measurable sets, then* $\cup M_k$ *is measurable, and if the* M_k *are disjoint,* $\mu[\bigcup_{k=1}^{\infty} M_k] = \sum_{k=1}^{\infty} \mu[M_k]$, *the convergent series.*

PROOF:

Without any loss of generality, we can always assume that the M_k are disjoint to prove that their union M is measurable.

Given $\epsilon \in {}^{\sigma}\mathbb{R}^+$, choose internal sets $A_k \supseteq M_k \supseteq B_k$ with $\mu[A_k \backslash B_k] < \frac{\epsilon}{2^{k+1}}$. Extend the sequence (A_k, B_k) to an internal sequence by the Countable Comprehension Principle (0.4.3). Use the Internal Definition Principle to pick an infinite n_1 such that

$$B_k \subseteq A_k \text{ and } \mu[A_k \backslash B_k] < \frac{\epsilon}{2^{k+1}} \text{ for } 1 \leq k \leq n_1.$$

For each infinite $n \in {}^*\mathbb{N}$, $n \leq n_1$,

$$\overline{\mu}[M] \lesssim \mu[\bigcup_1^n A_k] = \mu[\bigcup_1^n (A_k \backslash B_k) \cup \bigcup_1^n B_k] \leq \sum_1^n \mu[A_k \backslash B_k] + \sum_1^n \mu[B_k]$$
$$< \frac{\epsilon}{2} + \sum_1^n \mu[B_k];$$

hence $\overline{\mu}[M] < \epsilon + \sum_1^n \mu[B_k]$.

Again the Internal Definition Principle tells us that there is a finite $m \in {}^{\sigma}\mathbb{N}$ such that $\overline{\mu}[M] < \epsilon + \sum_1^m \mu[B_k]$. But $\sum_1^m \mu[B_k]$

$= \mu[\bigcup_1^m B_k] \lesssim \underline{\mu}[M]$, so that $\overline{\mu}[M] \leq \epsilon + \underline{\mu}[M]$, and ϵ was any standard positive number. Moreover,

$$\mu[M] - \epsilon < \sum_1^m \mu[B_k] \lesssim \sum_1^m \mu[M_k]$$

so that

$$\mu[M] - \sum_{k=1}^m \mu[M_k] < \epsilon.$$

Since the terms $\mu[M_k]$ are all nonnegative, this proves that $\sum_{k=1}^{\infty} \mu[M_k] = \mu[M]$ and this completes the proof of countable additivity of μ.

(1.1.6) THEOREM:

If $\mathbb{V}$ *is a* **finite set and* $\delta\mu : \mathbb{V} \longrightarrow {}^*[0,\infty)$ *is an internal weight function that sums to a limited* **finite measure,* $\mu[A] = \Sigma[\delta\mu(a) : a \in A]$, *for internal* A, *then the limited hyperfinite measure space* $(\mathbb{V}, \mathrm{Meas}(\mu), \mu)$ *is a complete countably additive finite positive measure space. Moreover,* μ *is the unique countably additive extension of* $\mathrm{st}\,\mu[\cdot]$ *to* $\mathrm{Loeb}(\mathbb{V})$ *and* $\mathrm{Meas}(\mu)$ *is the* μ*-completion of* $\mathrm{Loeb}(\mathbb{V})$.

PROOF:

The remarks and lemmas preceding the theorem show that μ is a complete countably additive measure. It is trivial that internal sets are μ-measurable, so $\mathrm{Loeb}(\mathbb{V}) \subseteq \mathrm{Meas}(\mu)$. The uniqueness and completion remarks follow from the approximation lemma since we can find increasing and decreasing chains of internal sets $B_k \subseteq M \subseteq A_k$ with $\mu[A_k \setminus B_k] < \frac{1}{k}$ forcing $\mu[M] = \text{S-}\lim_k \mu[A_k] = \text{S-}\lim_k \mu[B_k]$.

Because of saturation, we can approximate measurable sets by internal sets up to an error of μ-measure zero. The error is in the sense of the *symmetric set difference*,

$$M \triangledown N = (M \setminus N) \cup (N \setminus M).$$

This result is a special case of Lemma (1.2.13) below. We cannot assert that there is an internal set $A \supseteq M$ such that $\mu[A] \approx \mu[M]$. The set $o(r)$ in EXERCISE (1.1.13) has measure zero, but every internal superset has noninfinitesimal measure.

(1.1.7) **LEMMA:** (Sets are almost finite.)

A set M *is measurable with respect to a limited hyperfinite measure* μ *if and only if it differs from an internal set by a set of measure zero, that is, there is an internal* A *such that* $\mu[M \triangledown A] = 0$.

The proof of (1.1.7) is left as an exercise in case our reader is only interested in probability measures and plans to skip section (1.2). In that case, solve the exercise by observing that the proof of (1.2.13) works here with minor rephrasing.

(1.1.8) EXERCISE:

a) *Let* $\mathbb{V}$ *be an internal set and let* $\mathcal{A}$ *be an arbitrary algebra of internal subsets of* $\mathbb{V}$. *Suppose that a monotone finitely additive function* $\mu : \mathcal{A} \longrightarrow {}^*[0,\infty)$ *with limited* $\mu[\mathbb{V}] \in \mathcal{O}$ *is given. Define inner and outer measures on* $\mathbb{V}$ *by*

$$\overline{\mu}[U] = \inf[\mathrm{st}\ \mu[A] : U \subseteq A \in \mathcal{A}]$$

and

$$\underline{\mu}[U] = \sup[\mathrm{st}\ \mu[B] : U \supseteq B \in \mathcal{A}],$$

and μ*-measurability by agreement of the inner and outer measures. Show that Theorem* (1.1.6) *still holds where* Loeb($\mathcal{A}$) *is the smallest sigma algebra containing* $\mathcal{A}$

b) *Suppose in addition to part* (a) *that the algebra and set function* $\mu : \mathcal{A} \longrightarrow {}^*[0,\infty)$ *are internal. Then* LEMMA (1.1.7) *holds, where* $A \in \mathcal{A}$.

c) *Suppose in addition to parts* (a) *and* (b) *that the algebra* $\mathcal{A}$ *is* **finite. Define an equivalence relation on* $\mathbb{V}$ *by* $u \equiv v$ *if whenever* $u \in A \in \mathcal{A}$, *then* $v \in A$. *The set* $\overline{\mathrm{V}} = (\mathbb{V}/\equiv)$ *is* **finite and the uniform hyperfinite measure over* $\overline{\mathrm{V}}$ *is isomorphic to the extension of part* (b). (*Note*: $[v] = \cap[A: v \in A]$ *belongs to* $\mathcal{A}$ *since it is a* **finite intersection of an internal algebra.*)

Since μ may be external, one must extend more than (A_k, B_k) in the proof of the analogue to Lemma (1.1.5). However, the sequences of numbers have additive extensions out to some infinite n_1.

The next proposition is the famous Caratheodory trick that we could have used in the next section.

(1.1.9) **PROPOSITION:**

Let $\mu[\mathbb{V}]$ *be a limited* **finite measure. A set* $M \subseteq \mathbb{V}$ *is* μ*-measurable if and only if for each* $U \subseteq \mathbb{V}$,

$$\overline{\mu}[U] = \overline{\mu}[U \cap M] + \overline{\mu}[U \setminus M].$$

PROOF:

Fix a standard $\epsilon > 0$. Suppose M is measurable, so that we may pick internal A, B with $B \subseteq M \subseteq A$ and $\mu[A \setminus B] < \frac{\epsilon}{2}$. Now take an arbitrary $U \subseteq \mathbb{V}$ and let $C \supseteq U$ be an internal set so that $\mu[C] < \overline{\mu}[U] + \frac{\epsilon}{2}$. We see that

$$\begin{aligned}\overline{\mu}[U \cap M] + \overline{\mu}[U \setminus M] &\lesssim \mu[C \cap A] + \mu[C \setminus B] \\ &\leq \mu[C] + \mu[A \setminus B] < \overline{\mu}[U] + \epsilon.\end{aligned}$$

Thus, measurability implies the condition, since the opposite inequality always follows from the monotone property of $\overline{\mu}$.

If the equation holds for M and all U, take $U = \mathbb{V}$, so $\overline{\mu}[\mathbb{V}] = \overline{\mu}[M] + \overline{\mu}[\mathbb{V} \setminus M] = \overline{\mu}[M] + \overline{\mu}[\mathbb{V}] - \underline{\mu}[M]$ and $\overline{\mu}[M] = \underline{\mu}[M]$.

(1.1.10) **DEFINITION:**

We say that a **finite measure* μ *and its hyperfinite extension* μ *are non-atomic provided* $\delta\mu(v) \approx 0$ *for every* $v \in \mathbb{V}$.

(1.1.11) **EXAMPLE:**

One of the most important types of example is a uniform *finite measure. This is when the *cardinality $^\#[\mathbb{V}] = n$ and $\delta\mu(v) = \frac{1}{n}$ for all v in $\mathbb{V}$. As long as n is infinite μ is non-atomic. One example of this is the discrete time axis from 0 to 1, $\mathbb{T} = \{t \in {}^*\mathbb{R} : 0 < t \leq 1,\ t = k\delta t,\ k \in {}^*\mathbb{N}\}$, where $\delta t = \frac{1}{n}$ for some infinite n in $^*\mathbb{N}$. In this case $\delta\mu(t) = \delta t$ for all t. Another uniform probability is the space $\Omega = W^{\mathbb{T}}$ of Chapter 0. In this case $^\#[\Omega] = m^n$ and $\delta\mu(\omega) = \delta P(\omega) = \frac{1}{m^n}$.

(1.1.12) **EXAMPLE** (A nonmeasurable set):

Let n be an infinite natural number from $^*\mathbb{N}$ so that the *external* cardinality of $\{k \in {}^*\mathbb{N} : 1 \leq k \leq n\}$ is minimal. [Recall that in polyenlargement models all *finite sets have the same external cardinality.] Let $\mathbb{V} = \{k \in {}^*\mathbb{N} : 1 \leq k \leq \log_2(n)\}$ and let $\mathcal{A}$ be the collection of all infinite *finite subsets of $\mathbb{V}$. Since $\mathcal{A} \subseteq {}^*\mathcal{B}(\mathbb{V})$, $\text{card}(\mathcal{A}) = \text{card}(\mathbb{V}) = \text{card}(A)$ for any $A \in \mathcal{A}$. Let γ be the first ordinal with the cardinality of $\mathcal{A}$ and index the elements, $\mathcal{A} = \{A_\alpha : \alpha < \gamma\}$ We define sets $B_\alpha = \{x_\delta, y_\delta : \delta < \alpha\}$ inductively. Let $B_0 = \emptyset$. For $\beta < \gamma$, since $\text{card}(B_\beta) < \gamma$ and $\text{card}(A_\beta) = \gamma$, we may choose $x_\beta \neq y_\beta$ both from $A_\beta \setminus B_\beta$. For successor ordinals let $B_{\beta+1} = B_\beta \cup \{x_\beta, y_\beta\}$ for such a pair of points. For limit α let $B_\alpha = \bigcup_{\beta<\alpha} B_\beta$. Let $U = \{x_\alpha : \alpha < \gamma\}$ and observe that $U \cap A$ and $A \setminus U$ are nonempty for each A in $\mathcal{A}$.

The only internal subsets of either U or $\mathbb{V} \setminus U$ must be finite, so $\underline{\mu}[U] = \underline{\mu}[\mathbb{V} \setminus U] = 0$. If μ is a non-atomic probability measure then, since $\overline{\mu}[\mathbb{V}] = 1$, $\overline{\mu}[U] \neq \underline{\mu}[U]$.

This shows that the set U *is non-μ-measurable for every non-atomic hyperfinite probability measure* μ.

(1.1.13) **EXAMPLE:**

In this example we let $\mathbb{V} = \mathbb{T} = \{t \in {}^*\mathbb{R} : t = k\delta t,\ k \in {}^*\mathbb{N},\ 1 \leq k \leq n\}$ where $\delta t = \frac{1}{n}$ for some infinite n in ${}^*\mathbb{N}$. We may think of this as an infinitesimally discrete time axis. We let $\mu(t) = \delta t$ for all t in $\mathbb{T}$. If r in $\mathbb{R}$ is a standard number, $0 < r < 1$, let $o(r) = \{t \in \mathbb{T} : t \approx r\}$ denote the "instant" of times such that $st(t) = r$. Prove that $o(r)$ is δt-measurable.

If r in ${}^*\mathbb{R}$ satisfies $0 \leq r \leq 1$, then the internal map $[r] = \max[t \in \mathbb{T} : t \leq r]$ satisfies $0 \leq r-[r] \leq \delta t$. *The external set of 'positive half instants'*

$$H = \{t \in \mathbb{T} : st\,(t) \leq t\}$$

is non-measurable. Suppose I is internal and $I \supseteq H$. For any fixed standard r in $\mathbb{R}$ satisfying $0 \leq r \leq 1$, show that there is a finite m in ${}^\sigma\mathbb{N}$ such that

$$I \supseteq \{t \in \mathbb{T} : [r] < t \leq [r] + \frac{1}{m}\} = I(r,m).$$

Show that $\mu[(I(r,m)] = \frac{1}{m}$. Use that to prove that the outer measure of H is 1, for example, by a sort of covering argument.

Suppose J is internal and $J \subseteq H$. Replace I above by the complement of J, reverse the inequalities and show that

$\mu[J] \approx 0$, so that the inner measure of H is zero.

(1.1.14) **EXAMPLE:**

The example of a nonmeasurable set worked out in the last example for a specific space $(\mathbb{T}, \text{Meas}(\mu), \mu)$ can be generalized to any non-atomic hyperfinite probability space $(\mathbb{V}, \text{Meas}(\alpha), \alpha)$ as follows:

Ignore those points of $\mathbb{V}$ where α takes the value zero, and for a fixed *enumeration of the elements in $\mathbb{V}$, $\mathbb{V} = \{v_1, \cdots, v_n\}$, n infinite, consider the function

$$g : \mathbb{V} \longrightarrow {}^*[0,1], \quad g(v_i) = \delta\mu(v_1) + \cdots + \delta\mu(v_i), \quad i = 1, \cdots, n.$$

Such a function g carries $\mathbb{V}$ internally and bijectively onto an S-dense subset $g(\mathbb{V})$ of $^*[0,1]$. (Verify this.)

Define $\delta\alpha(g(v_i)) = \delta\mu(v_i)$, $i = 1, \cdots, n$, and you have an "isomorphic copy" of $(\mathbb{V}, \mu)$ in the form $(\mathbb{T}, \alpha)$, where $\mathbb{T}$ is S-dense in $^*[0,1]$ and α is non-atomic, with $\alpha[\mathbb{T}] = 1$.

Now we use the same argument as in (1.1.13) to prove that under these wider conditions the set

$$H = \{t \in \mathbb{T} : st(t) \leq t\}$$

is not μ-measurable, and conclude that $g^{-1}(H)$ is not α-measurable.

(1.1.15) REMARK:

Panetta [1978] has shown that no well ordering of $\mathbb{T}$ can be measurable with respect to the uniform *finite measure on $\mathbb{T}^2$.

(1.1.16) EXERCISE:

Show that the Caratheodory trick of (1.1.9) *also works for the inner measure, so that measurability could be defined with either the outer measure or the inner measure by itself. A set* M *is* μ*-measurable if and only if for every* $U \subseteq \mathbb{V}$,

$$\underline{\mu}[U] = \underline{\mu}[U \cap M] + \underline{\mu}[U \backslash M].$$

Notice that $\underline{\mu}$ *is superadditive,* $\underline{\mu}[U_1 \cup U_2] \geq \underline{\mu}[U_1] + \underline{\mu}[U_2]$, *when* U_1 *and* U_2 *are disjoint.*

(1.2) Unlimited Hyperfinite Measures

Let $\mathbb{V}$ be a *finite set and $\delta\mu : \mathbb{V} \longrightarrow {}^*[0,\infty)$ be an internal function. Every internal set $A \in {}^*\mathbb{B}(\mathbb{V})$ has an associated *finite measure

$$\mu[A] = \Sigma[\delta\mu(a) : a \in A].$$

In this section we assume that $\mu : {}^*\mathbb{B}(\mathbb{V}) \longrightarrow {}^*\mathbb{R}$ is an *unlimited* **finite measure*, that is $\mu[\mathbb{V}] \notin \mathcal{O}$. This may seem like a small change, since infinite measures like Lebesgue's are easy to work with. Unfortunately, this is a false first impression. The hyperfinite extension of an unlimited *measure is not sigma finite. This section is much more complicated than the last. For purposes of basic probability, including the later chapters of this book, you may skip this section to avoid these technicalities.

Let $\mathrm{st} : {}^*\mathbb{R} \longrightarrow {}^\sigma[-\infty,\infty]$ denote the extended standard part, so $\mathrm{st}\,\mu[\mathbb{V}] = +\infty$. The *outer measure* associated with μ is given by

$$\overline{\mu}[U] = \inf\{\mathrm{st}\,\mu[A] \mid U \subseteq A \in {}^*\mathbb{B}(\mathbb{V})\}.$$

It is understood that $\overline{\mu}[U] = +\infty$ if $\mu[A]$ is unlimited whenever an internal A contains U. The *inner measure* associated with μ is given by

$$\underline{\mu}[U] = \sup\{\mathrm{st}\,\mu[B] \mid U \supseteq B \in {}^*\mathbb{B}(\mathbb{V})\}.$$

It is understood that $\underline{\mu}[U] = +\infty$ if either the set of values is unbounded or if there is an internal B contained in U with $\mu[B]$ unlimited.

(1.2.1) DEFINITIONS:

An internal set $L \in {}^*\mathbb{B}(V)$ *is called* μ*-limited if* $\mu[L] \in \mathcal{O}$. *Any set* $I \subseteq \mathbb{V}$ *is called* μ*-integrable if* $\underline{\mu}[I] = \overline{\mu}[I] < \infty$. *We denote the class of* μ*-integrable sets by* Intg(μ). *A set* $M \subseteq \mathbb{V}$ *is called* μ*-measurable if* $M \cap I$ *is integrable for every integrable* I. *We denote the class of* μ*-measurable sets by* Meas(μ). *A set* $U \subseteq \mathbb{V}$ *is a* μ*-unique extension set if* $\underline{\mu}[U] = \overline{\mu}[U]$.

We can not define sets to be μ-measurable by the condition $\underline{\mu}[U] = \overline{\mu}[U]$, because these sets do not form a sigma algebra (when μ is unlimited). Extensions of st μ by $\overline{\mu}$ and by $\underline{\mu}$ both yield countably additive measures on the complete sigma algebra Meas(μ). They are not the same and each has an unpleasant feature (caused by non-sigma-finiteness). However, the extension of st μ is unique on the Loeb algebra generated by the internal subsets of $\mathbb{V}$, Loeb($\mathbb{V}$) = $\Sigma({}^*\mathbb{B}(\mathbb{V}))$. We have chosen the larger extension, $\overline{\mu}$, for the hyperfinite measure so that integrable sets are contained in limited ones.

(1.2.2) DEFINITION:

Let μ *be an unlimited* **finite measure on* $\mathbb{V}$. *The unlimited hyperfinite measure* μ *associated with* μ *is the outer measure function* $\overline{\mu}$ *restricted to the* μ*-measurable sets,* $\mu[M] = \overline{\mu}[M]$, *for* $M \in \text{Meas}(\mu)$.

Now we shall justify this terminology, show that μ is the unique extension of $\text{st}\,\mu$ to $\text{Loeb}(\mathbb{V})$, and show that $\underline{\mu}$ restricted to $\text{Meas}(\mu)$ is a different measure extension of $\text{st}\,\mu$ beyond $\text{Loeb}(\mathbb{V})$. Of course, $\underline{\mu}[U] \leq \overline{\mu}[U]$ for all sets, but there can be measurable sets H with $\underline{\mu}[H] = 0$ and $\overline{\mu}[H] = \infty$.

(1.2.3) REMARKS:

The inner and outer measures are monotone, $U_1 \subseteq U_2$ implies $\underline{\mu}[U_1] \leq \underline{\mu}[U_2]$ and $\overline{\mu}[U_1] \leq \overline{\mu}[U_2]$. The outer measure is subadditive, $\overline{\mu}[U_1 \cup U_2] \leq \overline{\mu}[U_1] + \overline{\mu}[U_2]$, while the inner measure is superadditive, $\overline{\mu}[U_1 \cup U_2] \geq \overline{\mu}[U_1] + \overline{\mu}[U_2]$, when U_1 and U_2 are disjoint.

(1.2.4) LEMMA:

The sets $\text{Meas}(\mu)$ *and* $\text{Intg}(\mu)$ *are closed under finite intersection.*

PROOF:

Let $F,G \in \text{Intg}(\mu)$. Then there is an internal $V' \subset \mathbb{V}$ with finite μ-measure $\mu[V']$ such that $V' \supseteq F \cup G$. Thus, F,G are measurable in the new space, V', with limited μ-measure and we can apply the results in the preceding paragraph: (1.1.4).

Suppose $M, N \in \text{Meas}(\mu)$, and take any $F \in \text{Intg}(\mu)$. Then $N \cap F \in \text{Intg}(\mu)$, so $M \cap (N \cap F) \in \text{Intg}(\mu)$.

The pattern of the argument in the last proof, referring to the preceding paragraph applied to some limited superset, will be used systematically in the sequel. The basic fact is that the outer and inner measures of a set U suffer no change if we consider only the internal subsets of a fixed internal set that contains U.

(1.2.5) **PROPOSITION:**

All internal sets of finite measure are integrable, and all internal sets are measurable.

PROOF:

It is obvious that if A is internal, then $\overline{\mu}(A) = \underline{\mu}(A)$, so that A is integrable if $\mu[A]$ is limited.

Now, if A is internal and $F \in \text{Intg}(\mu)$, $\overline{\mu}(A \cap F) < +\infty$, so that there is an internal $B \supseteq A \cap F$ with $\mu[B]$ limited. The sets $A \cap B$ and F are measurable for the limited measure obtained by restricting μ to B, hence $A \cap B \cap F = A \cap F$ is measurable.

(1.2.6) **PROPOSITION:**

(i) A set is integrable if and only if it is measurable and has finite measure.

(ii) A set is integrable if and only if it is measurable and is included in some limited set.

(iii) *A set is measurable if and only if its intersection with any integrable set is measurable.*

PROOF:

(i) If $F \in \text{Intg}(\mu)$ then $F \in \text{Meas}(\mu)$ according to Lemma (1.2.4). Conversely, assume that $M \in \text{Meas}(\mu)$ and $\mu[M] < +\infty$. Since $\overline{\mu}[M] < +\infty$, there is an internal $A \supseteq M$ with $\mu[A]$ finite, $A \in \text{Intg}(\mu)$. By definition of $\text{Meas}(\mu)$ $M = M \cap A \in \text{Intg}(\mu)$.

(ii) follows from finiteness of $\overline{\mu}$.

(iii) In one direction, it follows from $\text{Intg}(\mu) \subseteq \text{Meas}(\mu)$. Conversely, if for each integrable F, $M \cap F \in \text{Meas}(\mu)$, applying (ii) we have $M \cap F \in \text{Intg}(\mu)$.

(1.2.7) LEMMA:

$\text{Meas}(\mu)$ *and* $\text{Intg}(\mu)$ *are closed under countable intersections.*

PROOF:

Let $(F_n)_{n \in {}^{\sigma}\mathbb{N}} \subseteq \text{Intg}(\mu)$ and define $F_n' = F_1 \cap F_n$ for each $n \in {}^{\sigma}\mathbb{N}$. Then if V' is an internal set of finite measure containing F_1, $\cap F_n = \cap F_n'$ can be regarded as a countable intersection of measurable sets in the space with finite measure V'. Apply (1.1.4) and (1.1.5).

Suppose now $(M_n)_{n \in {}^{\sigma}\mathbb{N}} \subseteq \text{Meas}(\mu)$ and take any $F \in \text{Intg}(\mu)$. Then $\cap M_n \cap F = \cap(M_n \cap F)$, and the result follows from the first part.

(1.2.8) **LEMMA:**

$Meas(\mu)$ *and* $Intg(\mu)$ *are closed under differences. And* $Meas(\mu)$ *is closed under complementation.*

PROOF:

Since $\mathbb{V}$ is measurable, the second part is an immediate consequence of $Meas(\mu)$ being closed under differences.

Also, the fact that $Meas(\mu)$ is closed under differences follows from the same property of $Intg(\mu)$: if M,N are measurable, then for each integrable set F, the set $(M \backslash N) \cap F = (M \cap F) \backslash (N \cap F)$ is integrable.

Thus, it only remains to be proved that $F,G \in Intg(\mu)$ implies $F \backslash G \in Intg(\mu)$. As usual, this follows from the results in section (1.1): embed F in an internal V' of finite measure and apply (1.1.4) to prove that $F \backslash G = F \cap (V' \backslash G) \in Intg(\mu)$.

(1.2.9) **LEMMA:**

(i) $Meas(\mu)$ *is closed under countable unions.*

(ii) $Intg(\mu)$ *is closed under dominated countable unions.*

PROOF:

(i) Let $(M_n)_{n\in{}^\sigma\mathbb{N}} \subseteq Meas(\mu)$. Then $(\cup M_n)^c = \cap M_n^c \in Meas(\mu)$ according to the last two lemmas. Again by (1.2.8), $\cup M_n \in Meas(\mu)$.

(ii) Suppose $(F_n)_{n\in{}^\sigma\mathbb{N}} \subseteq Intg(\mu)$ and $\cup F_n \subseteq F \in Intg(\mu)$. then by (i) and Proposition (1.2.6)(ii), $\cup F_n \in Intg(\mu)$.

(1.2.10) **LEMMA:**

If $(M_n : n \in {}^{\sigma}\mathbb{N})$ *is a disjoint countable sequence of measurable sets, then*

a) $\overline{\mu}[\cup M_n] = \sum_{n=0}^{\infty} \overline{\mu}[M_n]$

and

b) $\underline{\mu}[\cup M_n] = \sum_{n=0}^{\infty} \underline{\mu}[M_n]$.

PROOF:

In case some M_n is not integrable, $\cup M_n$ is not either (by Proposition (1.2.6)(ii)), hence both sides of the equation are $+\infty$ for the outer measure. If all M_n and $M = \cup M_n$ are integrable, the property is proved using (1.1.5) by embedding in some limited superset. This case is the same for both the inner and outer measure.

To complete (a) we only need to show that if $M \notin \mathrm{Intg}(\mu)$ and $M_n \in \mathrm{Intg}(\mu)$ for all $n \in {}^{\sigma}\mathbb{N}$, then $\sum_n \overline{\mu}[M_n] = +\infty$. Observe that since it is a series of nonnegative terms, it is enough to prove that for some infinite n, the nth partial sum is infinite. For each $n \in {}^{\sigma}\mathbb{N}$, there is an internal $A_n \supseteq M_n$ such that $\mu[A_n] < \frac{1}{2^n} + \overline{\mu}[M_n]$. Extend $(A_n)_{n \in {}^{\sigma}\mathbb{N}}$ to an internal $(A_n)_{n \in {}^{*}\mathbb{N}}$. Let S_n be the *extension of the externally defined but standard series $\sum_{k=0}^{n} \overline{\mu}[M_k] = S_n$. The set

$$\{n \in {}^{*}\mathbb{N} \mid \sum_{1}^{n} \mu[A_k] < 1 + S_n\}$$

is internal and contains ${}^{\sigma}\mathbb{N}$, so that there is an infinite n

in it. For that n, the set $\bigcup_1^n A_k$ is internal and contains $\bigcup_n M_n$, therefore $\mu[\bigcup_1^n A_k]$ is infinite. But $\mu[\bigcup_1^n A_k] \leq \sum_1^n \mu[A_k] < 1 + S_n$, hence S_n is infinite.

Any case for the inner measure where each M_n is integrable may be treated by choosing internal sets $B_n \subseteq M_n$ with $\mu[B_n] > \underline{\mu}[M_n] - \frac{\epsilon}{2^n}$. When m is finite, $\underline{\mu}[\cup M_n] \geq \mathrm{st} \sum_{n=0}^m \mu[B_n] > \sum_{n=0}^m \underline{\mu}[M_n] - 2\epsilon = \sum_{n=0}^m \bar{\mu}[M_n] - 2\epsilon$. Since we know $\sum_{n=0}^m \bar{\mu}[M_n] \longrightarrow \bar{\mu}[\cup M_n]$, we see that $\underline{\mu}[\cup M_n] \geq \bar{\mu}[\cup M_n]$ in this case.

The final case for the inner measure is when some M_n is not integrable. Of course $\cup M_n$ is also not integrable, but the question is: What do internal subsets of $\cup M_n$ measure up to? We claim there are only two cases. Either there is an increasing sequence B_k of limited sets, $B_k \subseteq \cup M_n$, with $\underline{\mu}[B_k] \longrightarrow \underline{\mu}[\cup M_n]$ or there is one point $a \in \cup M_n$ with $\delta\mu_2(a)$ unlimited. We prove this claim after we show why it suffices.

This condition establishes the countable additivity of the inner measure trivially in the second case of a point with unlimited weight. In the first case,

$$\underline{\mu}[B_k] = \sum_{n=0}^{\infty} \underline{\mu}[M_n \cap B_k] \leq \sum_{n=0}^{\infty} \underline{\mu}[M_k],$$

so the series is infinite. For each standard m, the superadditivity of $\underline{\mu}$ shows that

$$\sum_{n=0}^{m} \underline{\mu}[M_n] \leq \underline{\mu}[\bigcup_{n=0}^{m} M_n] \leq \underline{\mu}[\cup M_n] \leq \sum_{n=0}^{\infty} \underline{\mu}[M_n] = \infty.$$

Now we establish our claim about the weights for μ. If no sequence B_k exists, then UM_n must contain an unlimited internal set B, yet there is a limited $b \in \mathcal{O}$ such that no limited $A \subseteq UM_n$ has $\mu[A] > b$. Order the weights $\{\delta\mu(v) : v \in B\}$ as an increasing internal sequence $\{\delta_1, \delta_2, \cdots, \delta_p\}$. The weight δ_q is unlimited where $q = \min\{j : \sum_{i=1}^{j} \delta_j > b\}$.

This completes all the cases of our proof.

(1.2.11) **DEFINITIONS:**

A sigma algebra $\mathfrak{B}$ *is said to be* μ*-saturated if whenever a set* U *has the property that* $U \cap F \in \mathfrak{B}$ *for every* μ*-integrable* F, *then* $U \in \mathfrak{B}$.

A measure space is called semifinite if each measurable set of infinite measure contains integrable subsets of arbitrarily large finite measure.

Clearly, we have defined $\text{Meas}(\mu)$ so that it is saturated and complete (every subset of a set of measure zero is measurable).

An infinite measure space is called sigma finite if it is a countable union of sets of finite measure. Hence a (infinite) sigma finite measure is semifinite. Non-atomic unlimited hyperfinite measures (as well as hyperfinite measures with unlimited weights $\delta\mu(v) \notin \mathcal{O}$) are not even semifinite, but the inner measure usually is.

(1.2.12) THEOREM:

a) *An unlimited hyperfinite measure space* $(\mathbb{V}, \mathrm{Meas}(\mu), \mu)$ *is a complete saturated countably additive infinite measure space.*

b) *The inner measure,* $\underline{\mu}$, *restricted to* $\mathrm{Meas}(\mu)$ *is also a complete countably additive infinite measure space.*

c) *The inner and outer measures agree on the Loeb algebra,* $\mathrm{Loeb}(\mathbb{V})$, *so the extension of* $\mathbf{st}\ \mu$ *from the internal sets to* $\mathrm{Loeb}(\mathbb{V})$ *is unique.*

d) *An unlimited inner measure* $\underline{\mu}$ *is never sigma-finite, but if the weights are all limited,* $\delta\mu(v) \in \mathcal{O}$, *then* $\underline{\mu}$ *is semifinite.*

e) *If the weight function for* μ *has only infinitesimal values,* $\delta\mu[v] \approx 0$, *then there is a measurable set* $H \in \mathrm{Meas}(\mu)$ *such that* $\underline{\mu}[H] = 0$ *and* $\overline{\mu}[H] = \infty$. *The extension of* $\mathbf{st}\ \mu$ *to* $\mathrm{Meas}(\mu)$ *is not unique.*

f) *Non-atomic unlimited hyperfinite (outer) measures are not semifinite, and thus are also not sigma finite.*

PROOF:

The proof of (a) and (b) is contained in the previous lemmas.

To prove (c), observe that the collection $\{M \in \mathrm{Meas}(\mu) \mid \underline{\mu}[M] = \overline{\mu}[M]\}$ is a sigma algebra which contains all the internal sets. Therefore it contains $\mathrm{Loeb}(\mathbb{V})$. If ν is a monotone set function that extends $\mathbf{st}\ \mu$, then $\underline{\mu}[M] \leq \nu[M] \leq \overline{\mu}[M]$. Thus any measure on $\mathrm{Loeb}(\mathbb{V})$ that agrees with μ on internal sets must equal μ on Loeb($\mathbb{V}$).

Now we prove part (d). We show that $\underline{\mu}$ is never sigma finite by showing that any countable union of measurable sets of finite inner measure always omits an infinite amount of $\mathbb{V}$. Suppose $\mathbb{V} \supseteq \cup I_m$ with each $\underline{\mu}[I_m] \leq \infty$. We may assume that the sequence is increasing, $I_m \subseteq I_{m+1}$ (by taking finite unions). Next, choose an increasing sequence of internal sets $B_m \subseteq I_m$ with $\mu[B_m] \geq \underline{\mu}[I_m] - \frac{1}{m}$. We know that $\mu[\mathbb{V}\backslash B_m] > m$, since $\mu[\mathbb{V}]$ is unlimited and $\mu[B_m] \lesssim \underline{\mu}[I_m] < \infty$. Extend B_m to an increasing internal sequence using countable comprehension. The internal set of indices n such that $\mu[\mathbb{V}\backslash B_n] > n$ must contain an infinite n, thus $\underline{\mu}[\mathbb{V}\backslash \cup I_m] \geq \underline{\mu}[\mathbb{V}\backslash B_n] = \infty$.

To prove the rest of (d) suppose that a set M has $\underline{\mu}[M] = \infty$ and $B \subseteq M$ is unlimited. Order the weights $\{\delta\mu(v) \mid v \in B\}$ in increasing order $\{\delta_1, \delta_2, \cdots, \delta_p\}$ corresponding to points $\{b_1, \cdots, b_p\} = B$. Define a sequence $B_1 = \{b_i \mid i \leq \min[j : \sum_{h=1}^{j} \delta_h > 1]\}$ and $B_{k+1} = \{b_i \mid i \leq \min[j : \sum_{h=1}^{j} \delta_h > \mu[B_k] + 1]\}$. We have $k \leq \underline{\mu}[B_k] < \infty$.

The second part of (e) and part (f) follow from the first because if H contains an integrable set I, then $\underline{\mu}[I]$ may be approximated by internal subsets of I. Since $\underline{\mu}[H] = 0$, H cannot contain integrable sets of positive standard measure.

Now we complete the proof of the theorem by constructing H. Let $\mathcal{I}_\alpha$ be the collection of all internal subsets of $\mathbb{V}$ of limited μ-measure constructed at or before the stage $\mathfrak{X}_\alpha$ in the direct limit defining our polyenlargement. Since $\mathbb{V}$ has unlimited measure the complements of sets of $\mathcal{I}_\alpha$, $\mathbb{V}\backslash I$, have the finite intersection property. Hence there exist points in $\mathbb{V} \cap \mathfrak{X}_{\alpha+1}$ which are not in any $I \in \mathcal{I}_\alpha$. Thus we may select a family $H = \{x_\alpha \mid \alpha < \text{card}^+(\mathfrak{X})\}$ with $x_\alpha \in \mathbb{V}\backslash\cap\mathcal{I}_\alpha$.

Clearly, H has $\text{card}(H) = \text{card}^+(\mathfrak{X})$, but for any μ-limited internal set A, $\text{card}(H \cap A) < \text{card}^+(\mathfrak{X})$, because $A \in \mathcal{I}_\alpha$ for some $\alpha < \text{card}^+(\mathfrak{X})$. We know that any set I of finite outer measure is contained in a μ-limited set, so $\text{card}(H \cap I) < \text{card}^+(\mathfrak{X})$ also. This means that $\overline{\mu}(H) = \infty$, because $\text{card}(H) = \text{card}^+(\mathfrak{X})$.

Let I be a μ-integrable set. We know that $\text{card}(H \cap I) < \text{card}^+(\mathfrak{X})$ by the above. We will show that $\overline{\mu}[H \cap I] = 0$ by finding an internal $A \supseteq H \cap I$ with $\mu[A] < \epsilon$ where ϵ is an arbitrary standard positive number. Since $\underline{\mu}[H \cap I] \leq \overline{\mu}[H \cap I] = 0$, this makes H μ-measurable. Fix an arbitrary ϵ and consider the family of internal sets $\{\mathcal{A}_\epsilon(a) \mid a \in H \cap I\}$, where

$$\mathcal{A}_\epsilon(a) = \{A \in {}^*\mathcal{B}(\mathbb{V}) \mid a \in A \;\&\; \mu[A] < \epsilon\}.$$

Given any finite set $a_1,\cdots,a_m \in H \cap I$, the set $\{a_1,\cdots,a_m\}$ belongs to each $\mathcal{A}_\epsilon(a_i)$ since μ is non-atomic. In other words, $\bigcap_{i=1}^{m} \mathcal{A}_\epsilon(a_i) \neq \emptyset$. By $\text{card}^+(\mathfrak{X})$-saturation $\cap[\mathcal{A}_\epsilon(a) \mid a \in H \cap I] \neq \emptyset$, so there is a single A such that $H \cap I \subseteq A$ and $\mu[A] < \epsilon$.

Finally, we show that $\underline{\mu}[H] = 0$. If $B \subseteq H$ is a limited subset of H, then $\text{card}(B) < \text{card}^+(\mathfrak{X})$. Therefore, by (0.4.4), B is finite and $\mu[B] \approx 0$. An unlimited set B cannot be a subset of H because we may write $B = \{b_1, b_2, \cdots, b_n\}$ for an internal sequence and define the subset $B' = \{b_j \in B \mid j \leq \min_j [k : \sum_{i=1}^{k} \mu(b_i) > 1]\}$. By this definition, $2 > \mu[B'] > \frac{1}{2}$. If $B' \subseteq H$, then B' is finite, so $\mu[B'] \approx 0$. This completes the proof.

The next result is the "Littlewood Principle" that says, 'sets are almost finite'.

(1.2.13) THE SET LIFTING LEMMA:

A set F is μ-integrable if and only if there is a μ-limited internal A such that the symmetric difference $A \triangledown F = (A\backslash F) \cup (F\backslash A)$ satisfies $\mu[A \triangledown F] = 0$.

PROOF:

The set F is integrable if and only if for standard m, there exist internal sets $B_m \subseteq B_{m+1} \subseteq F \subseteq A_{m+1} \subseteq A_m$ with $\mu[A_m \backslash B_m] < \frac{1}{m}$ and $\mu[A_m] < \mu[F] + \frac{1}{m} < \infty$. Extend the sequence (A_m, B_m) to an internal sequence and select an infinite n so that $m \leq n$ implies $B_m \subseteq B_{m+1} \subseteq A_{m+1} \subseteq A_m$ and $\mu[A_m \backslash B_m] \leq \frac{1}{m}$. The sets A_n, B_n satisfy $\mu[F] = \mu[\cup B_m] \leq \mu[B_n] \leq \mu[A_n] \leq \mu[\cap A_m] = \mu[F]$, and $\cup[B_m \mid m \in {}^{\sigma}\mathbb{N}] \subseteq F \subseteq \cap[A_m \mid m \in {}^{\sigma}\mathbb{N}]$, so that the symmetric difference has measure zero for either A_n and F or B_n and F. The converse of the second part follows from completeness of μ and the measurability of internal sets.

(1.2.14) THEOREM:

A set M *is* μ*-measurable if and only if it is measurable in the sense of Caratheodory: for every* $T \subseteq \mathbb{V}$, $\bar{\mu}(T) = \bar{\mu}(T \cap M) + \bar{\mu}(T \setminus M)$.

PROOF:

Assume that Caratheodory's condition holds for $M \subseteq \mathbb{V}$ and let F be an integrable set. Take an internal $V' \supseteq F$ of finite measure; then for each subset of V', the outer measure $\bar{\mu}$ agrees with the outer measure $\bar{\mu}_{V'}$, so that

$$\bar{\mu}_{V'}[T] = \bar{\mu}[T] = \bar{\mu}[T \cap M] + \bar{\mu}[T\setminus M]$$
$$= \bar{\mu}_{V'}[T \cap (M \cap V')] + \bar{\mu}_{V'}[T\setminus(M \cap V')],$$

for every $T \subseteq V'$, i.e., $M \cap V'$ satisfies the Caratheodory condition in V'. Therefore, by (1.1.9), $M \cap V'$ is measurable in V', and then it is integrable as a subset of $\mathbb{V}$. Hence $M \cap F = (M \cap V') \cap F$ is also integrable.

Suppose now $M \in \mathrm{Meas}(\mu)$. For each $T \subseteq \mathbb{V}$ of finite outer measure (in case $\bar{\mu}[T] = +\infty$ the Caratheodory property follows from Remark (1.2.3)), there is an integrable $F \supseteq T$ with $\mu[F] = \bar{\mu}[T]$. But $M \cap F$ is integrable, hence it is Caratheodory-measurable (apply Lemma (1.1.9)), so

$$\bar{\mu}[T] = \bar{\mu}[T \cap M \cap F] + \bar{\mu}[T\setminus(M \cap F)] = \bar{\mu}[T \cap M] + \bar{\mu}[T\setminus M].$$

The next exercise shows why we cannot define a set to be measurable simply if $\underline{\mu}[U] = \bar{\mu}[U]$. The function $\mathrm{st}\,\mu$ has a unique extension to such sets (if the extension is continuous), but these sets do not form an algebra. To show this we need a non-measurable set. Example (1.1.12) yields a non-measurable subset of a μ-limited set when μ is non-atomic.

(1.2.15) EXERCISE:

Let μ *be an unlimited hyperfinite measure with a nonmeasurable set* $T \subseteq \mathbb{V}$ *and an integrable* F *such that* $\underline{\mu}[T \cap F] < \overline{\mu}[T \cap F]$. *Let* $U = \mathbb{V} \setminus (T \cap F)$.

a) *Show that* $\underline{\mu}[U] = \overline{\mu}[U]$, *but* $\underline{\mu}[U^c] < \overline{\mu}[U^c]$.

b) *Show that* $\underline{\mu}[U] = \overline{\mu}[U]$ *and* $\underline{\mu}[F] = \overline{\mu}[F]$, *but* $\underline{\mu}[U \cap F] < \overline{\mu}[U \cap F]$.

c) *Show that* U *is not measurable.*

Another approach one might try to extend an internal measure is to take the sigma algebra generated by the integrable sets. These sets form a sigma algebra with the property that $\underline{\mu}[M] = \overline{\mu}[M]$ for each M, however, the internal sets are not all included.

(1.2.16) EXERCISE:

Let μ *be an unlimited hyperfinite measure whose weight function takes only limited values,* $\delta\mu(v) \in \mathcal{O}$. *Show that one can divide* $\mathbb{V}$ *roughly in half by two internal sets* $A \cup B = \mathbb{V}$ *with neither* A *nor* B *in the sigma algebra generated by the integrable sets.*

(1.3) Almost and Nearly Sure Events, Measurable and Internal Functions

Recall that a real-valued function, f, is measurable with respect to a sigma algebra, Σ, if $f^{-1}(B) \in \Sigma$ for every Borel set $B \subseteq \mathbb{R}$. It is sufficient to show that $f^{-1}(-\infty, r) = \{v \mid f(v) < r\} = \{f < r\} \in \Sigma$ for every r in $\mathbb{R}$. Moreover, we may allow f to also take the extended standard values $+\infty$ and $-\infty$, still f is meaurable if and only if $f^{-1}[-\infty, r) = \{f < r\} \in \Sigma$ for each r in $\mathbb{R}$.

(1.3.1) DEFINITION:

Let $f : \mathbb{V} \longrightarrow {}^*\mathbb{R}$ *be an internal function. The projection of* f *is the extended-real-valued function:*

$$\tilde{f}(v) = \mathrm{st}(f(v)).$$

(1.3.2) THE FUNCTION PROJECTION LEMMA:

Let $\mathbb{V}$ *be a* **finite set. The projection* $\tilde{f} : \mathbb{V} \longrightarrow [-\infty, \infty]$ *of an internal function* $f : \mathbb{V} \longrightarrow {}^*\mathbb{R}$ *is Loeb measurable and since* $\mathrm{Loeb}(\mathbb{V}) \subseteq \mathrm{Meas}(\mu)$, f *is* μ*-measurable for any hyperfinite measure* μ *on* $\mathbb{V}$.

PROOF:

$$\{\tilde{f} < r\} = \cup[\{f < r-1/m\} : m \in {}^{\sigma}\mathbb{N}]$$

$$\{\tilde{f} \geq r\} = \cap[\{f \geq r-1/m\} : m \in {}^{\sigma}\mathbb{N}], \quad r \in \mathbb{R}$$

$$\{\tilde{f} \leq r\} = \cap[\{f \leq r+1/m\} : m \in {}^{\sigma}\mathbb{N}]$$

(1.3.3) DEFINITION:

If $g: \mathbb{V} \longrightarrow [-\infty,+\infty]$ *is a function,* $f : \mathbb{V} \longrightarrow {}^*\mathbb{R}$ *is internal and if* $\tilde{f} = g$, *then* f *is called a (uniform) lifting of* g.

$$\begin{array}{ccc} \mathbb{V} & \xrightarrow{f} & {}^*\mathbb{R} \\ \| & & \downarrow st \\ \mathbb{V} & \xrightarrow[g]{} & [-\infty,+\infty] \end{array}$$

Naturally, f is a lifting of its own projection $\tilde{f}$. The proof of (1.3.2) above shows one implication of our next result.

(1.3.4) THE UNIFORM LIFTING THEOREM:

A function $g : \mathbb{V} \longrightarrow [-\infty,\infty]$ *has a uniform lifting if and only if for each rational* r *in* $\mathbb{R}$ *the sets* $\{v : g(v) \le r\}$ *and* $\{v : g(v) \ge r\}$ *are countable intersections of internal sets.*

PROOF (of converse):

We shall construct a sequence of internal approximating functions by first making partitions with g. For a rational r in $\mathbb{Q}$, let

$$\{g \ge r\} = \{v : g(v) \ge r\} = A(r,\infty) = \cap A(r,n)$$

and

$$\{g \le r\} = \{v : g(v) \le r\} = B(r,\infty) = \cap B(r,n),$$

where $A(r,n), B(r,n)$ are sequences of internal sets. We also assume $A(r,n) \supseteq A(r,n+1)$ and $B(r,n) \supseteq B(r,n+1)$ because $\bigcap_{k=1}^{n} C'(r,k) = C(r,n)$ is internal and decreasing when C' is just internal. Let $\{r_1, r_2, \cdots\}$ enumerate $\mathbb{Q}$. For each m, let $\{s_1,\cdots,s_m\} = \{r_1,\cdots,r_m\}$ with $s_1 < s_2 < \cdots < s_m$. The following sets partition $\mathbb{V}$:

$$\{g < s_1\} = B(s_1,\infty) \cap A^c(s_1,\infty)$$
$$= B(s_1,\infty)\cap A^c(s_1,\infty)\cap B(s_2,\infty)\cap A^c(s_2,\infty)\cap\cdots\cap B(s_m,\infty)\cap A^c(s_m,\infty)$$

$$\{g = s_1\} = B(s_1,\infty) \cap A(s_1,\infty)$$
$$= B(s_1,\infty) \cap A(s_1,\infty) \cap B(s_2,\infty) \cap A^c(s_2,\infty) \cap \cdots \cap A^c(s_m,\infty)$$

$$\{s_i < g < s_j\} = [A(s_i,\infty)\backslash B(s_i,\infty)] \cap [B(s_j,\infty)\backslash A(s_j,\infty)]$$
$$= [\bigcap_{h\leq i} B^c(s_h,\infty)] \cap [\bigcap_{h\leq i} A(s_i,\infty)] \cap [\bigcap_{k\geq j} B(s_k,\infty)] \cap [\bigcap_{k\geq j} A^c(s_k,\infty)],$$
$$\text{for } 1 \leq i < j \leq m,$$

$$\{g = s_i\} =$$
$$[\bigcap_{h<i} B^c(s_h,\infty)] \cap [\bigcap_{h\leq i} A(s_h,\infty)] \cap [\bigcap_{k\geq i} B(s_k,\infty)] \cap [\bigcap_{k>i} A^c(s_k,\infty)],$$
$$\text{for } 1 \leq i \leq m,$$

$$\{g > s_m\} = [\bigcap_k B^c(s_k,\infty)] \cap [\bigcap_k A(s_i,\infty)].$$

Notice that each of these sets may be written as a 2m-fold intersection with each set $B(s_k,\infty)$ or $A(s_k,\infty)$ represented once with either itself or its complement in the intersection. We may code this with a function $\epsilon : \{1,\cdots,2m\} \longrightarrow \{\text{blank},c\}$ so each set above is

$$Q(\epsilon) = [\bigcap_k B^{\epsilon(2k-1)}(s_k,\infty)] \cap [\bigcap_k A^{\epsilon(2k)}(s_k,\infty)]$$

for an appropriate function ϵ with the blank or complement values as described. We shall call this partition $Q(m)$ and refer to the sets above in terms of the corresponding ϵ function.

Our next step is the following claim: For each m there exists $n \geq m$ so that for each nonempty $Q(\epsilon)$ in $Q(m)$, the corresponding set

$$P(\epsilon,n) = [\bigcap_k B^{\epsilon(2k-1)}(s_k,n) \cap [\bigcap_k A^{\epsilon(2k)}(s_k,n)]$$

satisfies

$$P(\epsilon,n) \cap Q(\epsilon) \neq \emptyset.$$

Suppose $Q(\epsilon) = \{s_i < g < s_j\}$ contains v, so

$$\begin{aligned} v &\in [A(s_i,\infty)\backslash B(s_i,\infty)] \cap [B(s_j,\infty)\backslash A(s_j,\infty)] \\ &= [A(s_i,\infty)\backslash \bigcap_n B(s_i,n)] \cap B(s_j,\infty)\backslash \bigcap_n A(s_j,n)] \end{aligned}$$

and there is an n so that $v \notin B(s_i,n)$ and $v \notin A(s_j,n)$. Since A's and B's decrease we may take $n(i,j) = \max(m,n)$. Since $A(s_i,n) \supseteq A(s_i,\infty)$ and $B(s_j,n) \supseteq B(s_j,\infty)$ we see that $P(\epsilon,n(i,j))$ contains v. A similar argument works for the full expression of each of the $Q(\epsilon)$ sets in $\mathcal{Q}(m)$ and we let n be the maximum of all the $n(i,j)$'s, etc. This n fulfills the claim.

For this n, let $\mathcal{R}(m)$ be the partition generated by the sets $\{A(s_k,n);B(s_k,n) : 1 \leq k \leq m\}$, which, by the way, may be written as the collection of $2m$-fold intersections

$$[\bigcap_k A^{\vartheta(2k)}(s_k,n)] \cap [\bigcap_k B^{\vartheta(2k-1)}(s_k,n)]$$

for the set of all functions $\vartheta : \{1,\cdots,2m\} \longrightarrow \{blank, c\}$. Define a function by summing over the ϵ-functions describing nonempty $Q(\epsilon)$'s:

$$f_m(v) = \Sigma[g(v_{P(\epsilon,n)})I_{P(\epsilon,n)}(v) : Q(\epsilon) \neq \emptyset]$$

where $I_{P(\epsilon,n)}$ is the indicator function and $v_{P(\epsilon,n)}$ is a sample of points taken from $P(\epsilon,n) \cap Q(\epsilon)$ for all the nonempty $Q(\epsilon)$'s. Each f_m is internal, because $R(m)$ is and finite sequences $\{v_P\}$ are. This function enjoys the following property: "for each $j,k \leq m$,

$$\{f_m \leq r_j\} \subseteq B(r_j,n) \subseteq B(r_j,m)$$

and

$$\{f_m \geq r_k\} \subseteq A(r_k,n) \subseteq A(r_k,m)."$$

Use comprehension to extend $\{(f_m, B(r_1,m)\cdots A(r_m,m))\}$ to an internal sequence with decreasing sets and satisfying the property in quotes above up to an infinite n. Let $f = f_n$. Then for each $r < s$ in $\mathbb{Q}$, $\{r \leq f \leq s\} \subseteq \{r \leq g \leq s\}$, so $st[f(v)] = g(v)$ as claimed in the theorem.

(1.3.5) **DEFINITION:**

Let $(\mathbb{V}, Meas(\mu), \mu)$ *be a hyperfinite measure space, We say that the property* $v \in \mathbb{W} \subseteq \mathbb{V}$ *holds* μ almost surely (μ-a.s.) *or* μ-almost everywhere (a.e.) *if* $\mu[\mathbb{V}\setminus\mathbb{W}] = 0$. *We say that* $v \in \mathbb{W}$ *holds* μ nearly surely (μ-n.s.) *if there is an internal set* $U \subseteq \mathbb{V}$ *such that* $U \supseteq \mathbb{V}\setminus\mathbb{W}$ *and* $\mu[U] \approx 0$.

Clearly, if $v \in \mathbb{W}$ n.s., then $\mu[\mathbb{V}\setminus\mathbb{W}] = 0$, so "nearly surely implies almost surely." The converse fails.

(1.3.6) **EXERCISE:**

Let $\mathbb{T}$ *and* μ *be as in* (1.1.13). *Show that* $t \notin o(1/2)$ *a.s. while if* U *is internal and* $o(1/2) \subseteq U$, *then* $\mu[U] \not\approx 0$.

(1.3.7) **DEFINITION:**

Let $(\mathbb{V}, \mathrm{Meas}(\mu), \mu)$ *be a hyperfinite measure space. If* $f : \mathbb{V} \longrightarrow [-\infty, \infty]$ *is a function and* $g : \mathbb{V} \longrightarrow {}^*\mathbb{R}$ *is internal and* $f = \tilde{g}$ μ*-almost surely, then we call* g *a* μ-lifting *of* f.

If f has a μ-lifting, g, then $\{v : f(v) \neq \tilde{g}(v)\}$ is μ-measurable with measure zero. Since $\tilde{g}$ is Loeb measurable, f is μ-measurable.

The projection map from the set of internal functions $\mathbb{V} \longrightarrow {}^*\mathbb{R}$ to the set of measurable functions $\mathbb{V} \longrightarrow [-\infty, \infty]$ is not surjective, even ignoring sets of zero μ-measure.

(1.3.8) **EXAMPLE:**

Let n be an infinite natural number, $\mathbb{V} = \left\{\frac{k}{n} : k \in {}^*\mathbb{Z}, |k| \leq n^2\right\}$, and $\mu(v) = \frac{1}{n}$ for each $v \in \mathbb{V}$. Then the indicator function of the set of finite numbers $\mathcal{O} \cap \mathbb{V}$ is measurable, since $\mathcal{O} \cap \mathbb{V} = \bigcup_n \{v \in \mathbb{V} : |v| \leq n\}$ is a Loeb set. This indicator function f has no lifting.

Suppose $g : \mathbb{V} \longrightarrow {}^*\mathbb{R}$ is an internal function with $\tilde{g} = f$ almost everywhere. Then there is an internal set $A \subseteq \mathbb{V}$ of measure $\mu[A] < 1$, such that for each $v \in \mathbb{V} \setminus A$, $g(v) \approx f(v)$. Then,

$$g^{-1}((\tfrac{1}{2},2)) \cup A \supseteq \mathbb{V} \cap \mathcal{O} \quad \text{and} \quad g^{-1}((-\tfrac{1}{2},\tfrac{1}{2})) \cup A \supseteq \mathbb{V} \setminus \mathcal{O}.$$

Since the left hand side is internal and for all $h \in {}^{\sigma}\mathbb{N}$, $\mathbb{V} \cap [-h,h]$ is contained in it, we can deduce that

$$g^{-1}((\tfrac{1}{2},2)) \cup A \supseteq \mathbb{V} \cap [-h,h] \quad \text{for some infinite } h \in {}^*\mathbb{N}.$$

For the same reason, since $g^{-1}((-\frac{1}{2}\ \frac{1}{2})) \cup A$ contains $\mathbb{V} \cap [k,n^2]$ for all infinite $k \in {}^*\mathbb{N}$.

$$g^{-1}((-\tfrac{1}{2},\tfrac{1}{2})) \cup A \supseteq \mathbb{V} \cap [k,n^2] \quad \text{for some finite } k \in {}^*\mathbb{N}.$$

Hence, $A \supseteq \mathbb{V} \cap [k,h]$, which is impossible since $\mu[A] < 1$.

(1.3.9) THE FINITE FUNCTION LIFTING LEMMA:

If $f : \mathbb{V} \longrightarrow [-\infty,\infty]$ *is* μ*-measurable and has* μ*-finite carrier,* $\mu\{f \neq 0\} < +\infty$ *(in particular, whenever the measure* μ *is limited), then there is an internal* $g : \mathbb{V} \longrightarrow {}^*\mathbb{R}$ *such that* $f = \tilde{g}$ a.s. *Moreover, if* $|f| \leq b$ a.s., *then we can choose* g *such that* $|g| \leq b$.

PROOF:

Let c be an unlimited positive number $d_0 = -c$, $d_1 = +c$, and $\{d_i \mid 2 \leq i \in \mathbb{N}\}$ be a dense subset of $\mathbb{R}$. (If $|f| \leq b$, let the latter set be dense in $[-b,b]$.) For each $m \in \mathbb{N}$

define the sets

$$F_0^m = \{v : f(v) = -\infty\}, \quad F_1^m = \{v : f(v) = +\infty\}$$

and

$$F_i^m = \{v : |f(v)| \geq \frac{1}{m} \ \& \ |f(v)-d_i| < \frac{1}{m}\}, \quad \text{for } i \geq 1.$$

By hypothesis these sets are measurable and for each m $\mu[\bigcup_i F_i^m] < \infty$. Hence for each m there is an $h \in \mathbb{N}$ such that $\mu[\bigcup_{i>h} F_i^m] < \frac{1}{m}$. Choose internal sets $A_i^m \subseteq F_i^m$ for $0 \leq i \leq h$ so that

$$\sum_{i=1}^{h} \mu[F_i^m \backslash A_i^m] < \frac{1}{m}.$$

Use intersections of these sets to define a partition of $\bigcup_{i=1}^{h} A_i^m$ $\{B_j^m : 0 \leq j \leq k\}$. Each B_j^m is a subset of some F_i^m, so we may define an internal finite choice function $g_m : \bigcup_{j=1}^{k} B_j^m \longrightarrow \{d_0, \cdots, d_h\}$ such that if $v \in B_j^m$ and $g_m(v) = d_i$, then $B_j^m \subseteq F_i^m$. (For example, take the maximal i such that $B_j^m \subseteq F_i^m$.) The internal function g_m may be extended to $\mathbb{V}$ by taking $g_m(v) = 0$ if $v \notin \bigcup_{j=1}^{k} B_j^m$. Suppose that we interpret the inequality $|\pm c \mp \infty| < \frac{1}{m}$ to be true, so that whenever $v \in \bigcup_{j=1}^{k} B_j$, $|g_m(v)-f(v)| < \frac{1}{m}$. Also, whenever $|f(v)| < \frac{1}{m}$, then $g_m(v) = 0$, so $|g_m(v)-f(v)| < \frac{1}{m}$. Thus we see that

$$\mu[|g_m - f| \geq \frac{1}{m}] \leq \frac{2}{m}.$$

This shows that the sequence of projections $\tilde{g}_m$ converges to f

in measure. Consider another $n \in \mathbb{N}$ with $m < n$ and the associated internal $g_n : \mathbb{V} \longrightarrow {}^*\mathbb{R}$. In order to have $|g_m(v)-g_n(v)| \geq \frac{2}{m}$, either we have $|g_m(v)-f(v)| \geq \frac{1}{m}$ or $|g_n(v)-f(v)| \geq \frac{1}{m}$. Therefore we see that for $m < n$,

$$\mu[|g_m - g_n| \geq \tfrac{2}{m}] \leq \tfrac{4}{m}.$$

Now we use countable comprehension to extend $\{g_m\}$ to an internal sequence and select an infinite n such that for all $m < n$, $\mu[|g_m - g_n| \geq \frac{2}{m}] \leq \frac{4}{m}$. This shows that $\tilde{g}_m$ converges to $\tilde{g}_n$ in measure. We already know that $\tilde{g}_m \longrightarrow f$, hence $\tilde{g}_n = f$ (μ-a.e.).

The reader should check the proof just given to see that we actually have shown the following.

(1.3.10) THE EXTENDED FINITE LIFTING THEOREM:

Suppose that f *does not have finite carrier, but at least for each* m *in* $\mathbb{N}$, $\mu\{|f| > \frac{1}{m}\} < \infty$. *Then there still exists an internal* g *such that* $\tilde{g} = f$ *a.e.*

The extension applies to $f(x) = \frac{1}{x}$ on the line, for example. Recall that the indicator function of the finite part of the line has no lifting.

(1.3.11) DEFINITIONS & REMARKS:

We say that an internal scalar function f *is finite almost surely* if

$$\mu[|\tilde{f}| = \infty] = 0.$$

This is equivalent to

$$\mu[|f| > h] \approx 0,$$

for each infinite h in ${}^*\mathbb{R}^+$. It is also equivalent to

$$\underset{k}{\text{S-lim}}\ \mu[|f| > k] = 0,$$

that is, for each finite positive ϵ, there is a finite h such that $k > h$ implies $\mu[|f| > k] < \epsilon$.

We say that an internal scalar function f is *infinitesmial almost surely if* $\mu[\tilde{f} \neq 0] = 0$. This is equivalent to

$$\mu[|f| > \tfrac{1}{k}] \approx 0$$

for all finite k in ${}^\sigma\mathbb{N}$ and by Robinson's sequential lemma there is an infinite n such that the infinitesimal condition holds for $0 \leq k \leq n$ in ${}^*\mathbb{N}$. Therefore, this is equivalent to the condition that f is *nearly surely infinitesimal*, that is, there is an internal set U with $\mu(U) \approx 0$ and if v is not in U, then $f(v)$ is infinitesimal. In other words, $\tilde{f}$ is *zero almost surely if and only if* f *is infinitesimal nearly surely*. This is not the case with finiteness.

Now we look at the property of an internal function f being *nearly surely finite*, that is, such that there is an internal set U with $\mu[U] \approx 0$ and if v is not in U, then $f(v)$ is finite. If we let $b = {}^*\sup[|f(v)| : v \notin U]$ then b is finite, since the internal set is bounded by every infinite number. The standard part $st(b) = \tilde{b}$ is an essential bound for $\tilde{f}$,

$$\mu[|\tilde{f}| > \tilde{b}] = 0,$$

so that $\tilde{f} \in L^{\infty}(\mu)$, the space of all essentially bounded μ-measurable functions. (Recall that g is essentially bounded by b if $|g(v)| \leq b$ except on a μ-null set.)

If f is finite n.s., then $\tilde{f}$ is in $L^{\infty}(\mu)$ and when μ is limited the finite lifting theorem says every g in $L^{\infty}(\mu)$ has a μ-lifting f that is finite n.s. Also, if $\tilde{f} \in L^{\infty}(\mu)$ then $\tilde{f} = \tilde{g}$ for some n.s. finite g and it follows from remarks above that $f \approx g$ n.s.

Finite a.s. only means that $\mu[|f| = \infty] = 0$, so it is weaker than finite n.s. Certainly finite n.s. is weaker than uniformly finite if there are sets of infinitesimal measure. Unfortunately, f can be finite n.s. and still have an infinite integral, for example, take the space of Example (1.1.11), uniform counting on n elements, and let $f = 0$ except at one point where it equals n^n.

(1.3.12) **EXTENSION TO ALGEBRAS** (con't.):

For $\mu, \mathcal{A}, \mathbb{V}$ *as in* (1.1.8(c)), *if* $f : \mathbb{V} \longrightarrow [-\infty, \infty]$ *is a* Loeb($\mathcal{A}$)*-measurable* [*or* μ*-measurable*] *function, show that there is an internal* $\mathcal{A}$*-measurable function* g *such that* $\tilde{g} = f$ a.s. *Moreover, if* f *is bounded by* b, *we may choose* g *bounded by* b.

A topological space $\mathbb{M}$ is called a Polish space if the topology is given by some complete separable metric, that is if there is a countable dense subset of $\mathbb{M}$ and a Cauchy complete metric ρ on $\mathbb{M}$ that induces the topology.

(1.3.13) METRIC LIFTING:

Suppose $f : \mathbb{V} \longrightarrow \mathbb{M}$ *is a* μ*-measurable function with values in a Polish space* $\mathbb{M}$ *where* μ *is a limited hyperfinite measure on* $\mathbb{V}$*. Show that there is an internal* $g : \mathbb{V} \longrightarrow {}^*\mathbb{M}$ *such that* $\mu[\rho(f,g) \not\approx 0] = 0$.

This exercise can be solved by rephrasing the proof of the scalar lifting theorem in terms of $({}^*\mathbb{M}, \rho)$ rather than $({}^*\mathbb{R}, |\cdot|)$.

(1.4) Hyperfinite Integration

Our next question is the relationship between the sum of an internal function f, $\Sigma[f(v)\delta\mu(v) \mid v \in \mathbb{V}]$, and the integral of its projection, $\int_{\mathbb{V}} \tilde{f}(v)d\mu(v)$.

We always have the inequality $\int_{\mathbb{V}} |\tilde{f}|d\mu \lesssim \Sigma_{\mathbb{V}}|f(v)|\delta\mu(v)$ (see (1.4.9)), but the following examples show that the converse inequality is not always true, even in case the sum is finite. We build our examples on the space of Example (1.3.8) which we may 'visualize' as an infinite discrete line. Let $n \in {}^*\mathbb{N}$ be unlimited, let $\mathbb{V} = \{v \in {}^*\mathbb{R} \mid v = \frac{k}{n}$, for $k \in {}^*\mathbb{Z}$ with $|k| \leq n^2\}$ with constant weight function $\delta\mu(v) = \frac{1}{n}$. (Any unlimited nonatomic space has similar examples.)

(1.4.1) EXAMPLE:

For $\mathbb{V}$ and μ as above, the constant infinitesimal function $f(v) = \frac{1}{n}$ satisfies $\Sigma\, f(v)\delta\mu(v) = 2 + \frac{1}{n^2}$ and $\tilde{f}(v) = 0$, so $\int \tilde{f}(v)d\mu(v) = 0$.

(1.4.2) EXAMPLE:

For $\mathbb{V}$ and μ as above, the function g such that $g(0) = n$, $g(v) = 0$, if $v \neq 0$ ($g(v) = nI_{\{0\}}(v)$, an unlimited multiple of the indicator function) satisfies $\Sigma\, g(v)\delta\mu(v) = 1$, while $\tilde{g}(v) = 0$ μ-a.s., so $\int \tilde{g}(v)d\mu(v) = 0$. Notice that this remains true when restricted to the limited measure subspace ${}^*[0,1] \cap \mathbb{V}$.

It is not enough to have the sum of f near the integral

of $\tilde{f}$. We want the structure of the space of such functions to be like the standard L^1-space, or equivalently, want the function $F(f,A) = \Sigma[f(v)\delta\mu(v) \mid v \in A]$ to be near the function $G(\tilde{f},A) = \int_A \tilde{f} d\mu$. We define the appropriate space of SL^1 of S-integrable functions below in (1.4.4).

(1.4.3) DEFINITIONS OF FL^1, S-AC & S-MC:

Two obvious necessary conditions for f to be in SL^1 are:

(FL^1) $\Sigma[f(v)\delta\mu(v) : v \in \mathbb{V}]$ *is finite (or limited)*

and

(S-AC) $\Sigma[f(v)\delta\mu(v) : v \in A] \approx 0$ *whenever* A *is internal and* $\mu[A] \approx 0$.

The first condition says that the internal L^1-norm is finite. The second condition is "standard absolute continuity." The reason the second condition must hold is the absolute continuity of the projected integral, the integral of $\tilde{f}$ over A is zero. S-absolute continuity fails for Example (1.4.2). For unlimited measures, a third condition corresponding to the standard monotone continuity of integrals is also necessary:

(S-MC) $\Sigma[|f(v)|\delta\mu(v) : v \in A] \approx 0$ *whenever* A *is internal and* f *is infinitesimal on* A.

This must hold since the integral of $\tilde{f}$ on A is the integral of zero. This condition fails for Example (1.4.1). We shall show that the three conditions imply that f is close to $\tilde{f}$ in all senses of integration and (S-MC) is not needed when $\mu[\mathbb{V}]$

is limited.

(1.4.4) DEFINITION:

An internal function $f : \mathbb{V} \longrightarrow {}^*\mathbb{R}$ *is S-integrable if it satisfies the conditions* (FL^1), (S-AC) *and* (S-MC) *above. We denote the space of S-integrable functions by* $SL^1(\mu)$. *If* f *is internal,* $\|f\| = \Sigma[|f(v)|\delta\mu(v) : v \in \mathbb{V}]$ *is called the internal 1-norm of* f *with respect to* μ.

In the case of probability spaces, we have the following characterizations of $SL^1(\mu)$.

(1.4.5) PROPOSITION:

Let $\mu[\mathbb{V}]$ *be limited and* $f : \mathbb{V} \longrightarrow {}^*\mathbb{R}$ *internal. The following are equivalent:*

(a) $f \in SL^1(\mu)$,

(b) f *satisfies* (FL^1) *and* (S-AC),

(c) *For every infinite positive number* k,

$$\Sigma[|f(v)|\delta\mu(v) : |f(v)| > k] \approx 0.$$

PROOF:

(a) ⇒ (b) by definition.

(b) ⇒ (c): Since f satisfies (FL^1), for any infinite $k \in {}^*\mathbb{R}^+$, $\{v : |f(v)| > k\}$ is an internal set of infinitesimal measure. Then (c) follows from (S-AC).

(c) ⇒ (a): Let k be any infinite positive number. Then

$$\begin{aligned}&\Sigma[|f(v)|\delta\mu(v) : v \in \mathbb{V}]\\ &\quad = \Sigma[|f(v)|\delta\mu(v) : |f(v)| > k] + \Sigma[|f(v)|\delta\mu(v) : |f(v)| \le k]\\ &\quad \lesssim k\mu[\mathbb{V}] \le k\mu[\mathbb{V}] + 1.\end{aligned}$$

Since $\mu[\mathbb{V}]$ is finite and the internal inequality is valid for all infinite positive k, (FL^1) holds.

If A is an internal set of infinitesimal measure, for each infinite positive k, reasoning as above,

$$\Sigma[|f(v)|\delta\mu(v) : v \in A] \lesssim k\mu[A].$$

Taking $k = \mu[A]^{-\frac{1}{2}}$, we get (S-AC).

Finally, if A is internal and $f(v) \approx 0$ for all $v \in A$, then for all standard positive ϵ,

$$\Sigma[|f(v)|\delta\mu(v) : v \in A] \le \mu[A]\epsilon.$$

Since $\mu[\mathbb{V}]$ is finite, (S-MC) follows.

(1.4.6) EXERCISE:

Suppose that μ *is a nonatomic limited measure. Show that* (S-AC) *implies* (FL^1), *so* $f \in SL^1(\mu)$ *if and only if* f *satisfies* (S-AC).

If $\delta\mu(v_0) \not\approx 0$, *for some* v_0, *then* (S-AC) *does not imply* (FL^1).

An indicator function of an internal set of finite μ-measure is S-integrable and $\mu[A] \approx \mu(A)$ by our construction of μ. We now build on this fact by taking limits just as if we were constructing the standard integral.

The space $SL^1(\mu)$ of S-integrable functions is an external space with an internal seminorm for $\|f\| = \Sigma[|f(v)|\delta\mu(v) : v \in \mathbb{V}]$. The quotient of $SL^1(\mu)$ obtained by identification of the functions with infinitesimal integral, $f \equiv 0$ iff $\|f\| \approx 0$, is isometric to $L^1(\mu)$.

Suppose $f : \mathbb{V} \longrightarrow {}^*\mathbb{R}$ is internal, but only takes limited values. Since f is internal and every infinite b satisfies $|f(v)| < b$ for all v, f must actually be uniformly finitely bounded. For the rest of the section we shall refer to functions satisfying the hypotheses of (1.4.7) as *μ-finite functions*.

(1.4.7) PROPOSITION:

Suppose that f *is internal, takes only limited values and that* $\mu[\{f \neq 0\}]$ *is limited. Then* f *is S-integrable and* $\Sigma[f(v)\delta\mu(v) : v \in \mathbb{V}] \approx \int_{\mathbb{V}} \tilde{f}d\mu$.

PROOF:

The first part is easy by all the finitesness assumptions. We may assume that f is positive (and finish the general case from $f = f^+ - f^-$). Let M be a finite real number greater than $\mu[\{f \neq 0\}]$, and choose a finite sequence $0 = y_o < y_1 < \cdots < y_m$ in ${}^\sigma\mathbb{R}$, with $y_m > \max(f)$, such that $y_i - y_{i-1} < \varepsilon/3M$ and $\mu(\tilde{f}^{-1}(y_i)) = 0$ for all $i = 1,\cdots,m$ (this can be done because for $y > 0$, $\tilde{f}^{-1}(y) \subseteq \{\tilde{f} \neq 0\} \subseteq \{f \neq 0\}$, and then there can only be a countable number of such y with $\mu(\tilde{f}^{-1}(y)) > 0$).

Then, if we define the four sums

$$\underline{S}(\mu) = \sum_1^m y_{i-1}\mu[f^{-1}(y_{i-1},y_i]], \quad \overline{S}(\mu) = \sum_1^m y_i\mu[f^{-1}(y_{i-1},y_i]]$$

$$\underline{S}(\mu) = \sum_1^m y_{i-1}\mu(\tilde{f}^{-1}(y_{i-1},y_i]), \quad \overline{S}(\mu) = \sum_1^m y_i\mu(\tilde{f}^{-1}(y_{i-1},y_i]),$$

we have

$$\underline{S}(\mu) \le \Sigma\, f(v)\delta\mu(v) \le \overline{S}(\mu), \quad \underline{S}(\mu) \le \int \tilde{f}d\mu \le \overline{S}(\mu)$$

and

$$\overline{S}(\mu) - \underline{S}(\mu) \le \frac{\epsilon}{3M}\,\mu[f^{-1}(0,y_m]] = \frac{\epsilon}{3M}\,\mu[\{f \ne 0\}] < \frac{\epsilon}{3}$$

$$\overline{S}(\mu) - \underline{S}(\mu) \le \frac{\epsilon}{3M}\,\mu(\tilde{f}^{-1}(0,y_m]) = \frac{\epsilon}{3M}\,\mu(\{\tilde{f} \ne 0\}) < \frac{\epsilon}{3}.$$

For any $i = 2,\cdots,m$,

$$\tilde{f}^{-1}(y_{i-1},y_i) \subseteq f^{-1}(y_{i-1},y_i) \subseteq f^{-1}[y_{i-1},y_i] \subseteq \tilde{f}^{-1}[y_{i-1},y_i].$$

Therefore,

$$\begin{aligned}\mu(\tilde{f}^{-1}[y_{i-1},y_i]) &= \mu(\tilde{f}^{-1}(y_{i-1},y_i]) = \mu(\tilde{f}^{-1}(y_{i-1},y_i)) \\ &\le \mu(f^{-1}(y_{i-1},y_i)) \approx \mu[f^{-1}(y_{i-1},y_i)] \\ &\le \mu[f^{-1}(y_{i-1},y_i]] \le \mu[f^{-1}[y_{i-1},y_i]] \\ &\approx \mu(\tilde{f}^{-1}[y_{i-1},y_i]) \le \mu(\tilde{f}^{-1}[y_{i-1},y_i]).\end{aligned}$$

Then, $\underline{S}(\mu) \approx \underline{S}(\mu)$. and hence $|\Sigma\, f(v)\delta\mu(v) - \int \tilde{f}d\mu| < \epsilon$. This completes the proof.

When "viewed" with standard tolerances, the space $SL^1(\mu)$ "looks like" $L^1(\mu)$. The next lemma begins to make this remark precise. We say that a sequence $\{f_m : m \in {}^\sigma\mathbb{N}\}$ of internal functions is an *S-Cauchy* sequence in the 1-norm if for every standard $\epsilon > 0$ there exists a standard m such that for all $j,k \geq m$, $\|f_j - f_k\| < \epsilon$. We say that the sequence f_m has the internal function f as its S-limit, S-lim $f_m = f$, in 1-norm if for every standard $\epsilon > 0$, there is a standard m such that for all standard $j \geq m$, $\|f_j - f\| < \epsilon$.

(1.4.8) **LEMMA:**

(a) *Every* $f \in SL^1(\mu)$ *is the S-limit of a sequence of* μ*-finite functions. Moreover, every* S-*Cauchy sequence in* $SL^1(\mu)$ *has an* S-*limit in* $SL^1(\mu)$.

(b) $L^1(\mu)$ *is the completion of the set of bounded measurable functions,* g, *with* μ*-finite carrier* $\mu\{g \neq 0\} < \infty$.

PROOF:

(a) Let $\{f_m : m \in {}^\sigma\mathbb{N}\}$ be an S-Cauchy sequence of functions in $SL^1(\mu)$, that is, for each p in ${}^\sigma\mathbb{N}$, there exists an $m_p \geq p$ in ${}^\sigma\mathbb{N}$ such that for all $j,k \geq m_p$, $\Sigma[|f_j(v) - f_k(v)|\delta\mu(v) : v \in \mathbb{V}] \leq \frac{1}{p}$. Extend the sequence f_m to an internal sequence, $\{f_m : m \in {}^*\mathbb{N}\}$. For each finite p in ${}^\sigma\mathbb{N}$, the internal set

$$\{n \in {}^*\mathbb{N} : (\forall\ j,k \in {}^*\mathbb{N})(m_p \leq j,k \leq n \Rightarrow \Sigma[|f_j(v) - f_k(v)|\delta\mu(v)] < \tfrac{1}{p})\}$$

contains an infinite n_p. The countable family of intervals of natural numbers from m_p to n_p therefore has the finite intersection property. Saturation lets us pick an infinite

$n \leq n_p$ for all $p \in {}^\sigma\mathbb{N}$ (see (0.4.2)). Then if $f = f_n$, S-$\lim \|f - f_m\| = 0$. Next we show that $f \in SL^1(\mu)$.

The property (FL^1) is clear. To prove the absolute continuity (S-AC), take any $p \in {}^\sigma\mathbb{N}$. Then for some finite m,

$$\left| \sum_A |f(v)|\delta\mu(v) - \sum_A |f_m(v)|\delta\mu(v) \right| \leq \sum_A \big||f(v)| - |f_m(v)|\big|\delta\mu(v)$$
$$\leq \sum_{\mathbb{V}} |f(v) - f_m(v)|\delta\mu(v) \leq \frac{1}{2p},$$

so that if A is an internal set with $\mu[A] \approx 0$,

$$\sum_A |f(v)|\delta\mu(v) \leq \frac{1}{2p} + \sum_A |f_m(v)|\delta\mu(v) < \frac{1}{p}.$$

Therefore, $\Sigma[|f(v)|\delta\mu(v) : v \in A] \approx 0$.

Now assume $f(v) \approx 0$ for all $v \in A$, where A is a certain internal subset of $\mathbb{V}$, and let p, m be as in the last paragraph. Set $g_m(v) = f_m(v)$ if $|f_m(v)| \leq |f(v)|$, and $g_m(v) = f(v)$ if $|f_m(v)| > |f(v)|$. The following facts are are obvious: for each $v \in \mathbb{V}$, $|f(v) - g_m(v)| \leq |f(v) - f_m(v)|$; for $v \in A$, $g_m(v) \approx 0$ and g_m is S-monotone continuous. Hence, by inequalities as above,

$$\sum_A |f(v)|\delta\mu(v) \leq \frac{1}{2p} + \sum_A |g_m(v)|\delta\mu(v) < \frac{1}{p}.$$

Thus, $\Sigma[|f(v)|\delta\mu(v) : v \in A] \approx 0$.

To finish the proof of (a) we are to show that any $f \in SL^1(\mu)$ is the limit of a μ-finite sequence. For each $n \in {}^*\mathbb{N}$, define

$$f_n(v) = \begin{cases} 0 & \text{if } |f(v)| < \frac{1}{n} \\ n \cdot \operatorname{sgn} f(v) & \text{if } |f(v)| > n \\ f(v) & \text{if } \frac{1}{n} \leq |f(v)| \leq n. \end{cases}$$

It is easy to see that for finite n, f_n is μ-finite. Moreover, for each infinite n,

$$n\mu[\{v \in \mathbb{V} : |f(v)| > n\}] \leq \sum_{v \in \mathbb{V}} |f(v)|\delta\mu(v) \in \mathcal{O},$$

hence

$$\Sigma[|f(v)-f_n(v)|\delta\mu(v) : v \in \mathbb{V}] \leq \Sigma[|f(v)|\delta\mu(v) : f(v) \neq f_n(v)]$$
$$= \Sigma[|f(v)|\delta\mu(v) : |f(v)| > n] + \Sigma[|f(v)|\delta\mu(v) : |f(v)| < \tfrac{1}{n}] \approx 0.$$

Therefore, $\|f-f_n\| \longrightarrow 0$ as $n \longrightarrow \infty$.

(b) The proof is basically the same as the last part of (a): pick $f \in L^1(\mu)$ and for each $n \in {}^\sigma\mathbb{N}$, define f_n as above. The Finite Lifting Theorem (1.3.9) and the Dominated Convergence Theorem guarantee that $\|f-f_n\| \longrightarrow 0$.

The next lemma shows that the projection is a contraction.

(1.4.9) LEMMA:

If $f : \mathbb{V} \longrightarrow {}^*\mathbb{R}$ *is an internal function,* $\int_{\mathbb{V}} |\tilde{f}| d\mu \lesssim \sum_{\mathbb{V}} |f(v)|\delta\mu(v)$ *(in* ${}^*[0,+\infty]$ *with* $\infty \approx \beta$ *if* β *is positive infinite).*

PROOF:

If $\Sigma[|f(v)|\delta\mu(v) : v \in \mathbb{V}]$ is finite (otherwise the relation is trivial), for each $n \in {}^\sigma\mathbb{N}$ define f_n as in the proof of (1.4.8)(a). Then each f_n is finite, has finite carrier, $|f_n| \le |f|$ and $(|\tilde{f}_n|) \uparrow |\tilde{f}|$ pointwise. By the Monotone Convergence Theorem and the Proposition (1.4.7),

$$\int_{\mathbb{V}} |\tilde{f}| d\mu = \lim \int_{\mathbb{V}} |\tilde{f}_n| d\mu,$$

and

$$\int_{\mathbb{V}} |\tilde{f}_n| d\mu \approx \sum_{\mathbb{V}} |f_n(v)| \delta\mu(v) \le \sum_{\mathbb{V}} |f(v)| \delta\mu(v).$$

(1.4.10) DEFINITION:

Given $g \in L^1(\mu)$, *a function* $f \in SL^1(\mu)$ *such that* $\tilde{f} = g$ μ-a.s., *is called an S-integrable* μ*-lifting of* g.

S-integrable liftings are more than just close almost surely, the error only sums to an infinitesimal. Every integrable function has an S-integrable lifting.

(1.4.11) THE INTEGRABLE LIFTING LEMMA:

The projection of internal S-integrable functions maps $SL^1(\mu)$ *onto* $L^1(\mu)$. *Moreover, if* $f, g \in SL^1(\mu)$, *then the norm in* SL^1 *satisfies* $\|f-g\| \approx \|\tilde{f}-\tilde{g}\|$ *and* $\int_{\mathbb{V}} \tilde{f}(v) d\mu(v) \approx \Sigma_{\mathbb{V}} f(v) \delta\mu(v)$.

PROOF:

We know that the mapping $f \longrightarrow \tilde{f}$ is continuous by (1.4.9). By the Dominated Convergence Theorem, each g in $L^1(\mu)$ is the L^1-limit of its truncation $g_m(v) = g(v) I_{\{\frac{1}{m} \le |g| \le m\}}(v)$, as indicated in (1.4.8)(b). Each g_m has a lifting f_m with $|f_m| \le m$, by (1.3.9), and by (1.4.7) $f_m \longrightarrow \tilde{f}_m$ is isometric.

Since $\int |g_m - \tilde{f}_m| d\mu = 0$, f_m is an S-Cauchy sequence, with S-limit f, say, by (1.4.8)(a). By continuity, $\tilde{f} = g$ μ-a.s.

We know $\int \tilde{f}_m d\mu \approx \Sigma[f_m \delta\mu]$ and $\int \tilde{f}_m d\mu \longrightarrow \int \tilde{f} d\mu$ while S-lim $\Sigma[|f-f_m|\delta\mu] = 0$, hence $\int_{\mathbb{V}} \tilde{f} d\mu \approx \Sigma[f(v)\delta\mu(v) : v \in \mathbb{V}]$.

INFINITESIMAL HULLS:

The Lebesgue space of the hyperfinite measure, $L^1(\mu)$, can be identified with part of the norm infinitesimal hull of the internal space ${}^*L^1(\mu)$ (Stroyan & Luxemburg [1976, chapter 10]). The norm infinitesimal hull of ${}^*L^1(\mu)$ consists of finite mod infinitesimal internal ${}^*L^1$-functions; the continuity conditions locate a classical L^1 space inside the norm hull. We shall denote the finite elements by

$$FL^1(\mu) = \{f : \Sigma|f|\delta\mu \text{ is a finite scalar}\}.$$

We denote the infinitesimal elements by

$$IL^1(\mu) = \{f : \Sigma|f|\delta\mu \approx 0\}.$$

The hull is the quotient space $FL^1(\mu)/IL^1(\mu)$.

Henson and Moore [1974] show that Kakutani's characterization of abstract L-spaces implies there is a measure $\hat{\mu}$ (not hyperfinite μ) such that

$$FL^1(\mu)/IL^1(\mu) = L^1(\hat{\mu}).$$

Hence we have the following embeddings and projections:

$$\begin{array}{ccccc} IL^1(\mu) & \subset & FL^1(\mu) & \xrightarrow{\ \wedge\ } & L^1(\hat{\mu}) \\ \| & & \cup & & \downarrow \\ IL^1(\mu) & \subset & SL^1(\mu) & \xrightarrow{\ \sim\ } & L^1(\mu). \end{array}$$

(1.4.12) **PROPOSITION:**

(a) *If* $f, g \in SL^1(\mu)$, *then* $\hat{f} = \hat{g}$ *iff* $\tilde{f} = \tilde{g}$ a.e.

(b) *If* $f \in SL^1(\mu)$, $g \in FL^1(\mu)$ *and* $\hat{f} = \hat{g}$, *then* $\tilde{f} = \tilde{g}$ a.e. *and* $g \in SL^1(\mu)$.

PROOF:

Part (a) is just another way of saying that the projection is an S-isometry, and part (b) follows from (a) and the obvious facts that $IL^1(\mu) \subseteq SL^1(\mu)$ is a vector space. Here is a direct proof. By definition of the map $\wedge$, $\hat{f} = \hat{g}$ means $\Sigma[\,|f(v)-g(v)|\delta\mu(v) : v \in \mathbb{V}] \approx 0$; then by Lemma (1.4.10), $\int_{\mathbb{V}} |\tilde{f}-\tilde{g}|d\mu = 0$; thus, $\tilde{f} = \tilde{g}$ a.e.

If $A \subseteq \mathbb{V}$ is an internal set of infinitesimal measure, $\Sigma[\,|f(v)|\delta\mu(v) : v \in A] \approx 0$, so that

$$\begin{aligned} \Sigma_A |g(v)|\delta\mu(v) &\approx \Sigma_A(|g(v)|-|f(v)|)\delta\mu(v) \lesssim \Sigma_A |g(v)-f(v)|\delta\mu(v) \\ &\lesssim \Sigma_{\mathbb{V}} |g(v)-f(v)|\delta\mu(v) \approx 0. \end{aligned}$$

Finally, let A be an internal subset of $\mathbb{V}$ where g only takes infinitesimal values. Then, $f(v) \approx 0$ for all v in $A \backslash N$, where $N \subseteq A$, $\mu(N) = 0$. Choose a sequence of internal sets $(A_p \mid p \in {}^{\sigma}\mathbb{N})$ with $N \subseteq A_p \subseteq A$, $\mu[A_p] < 1/p$. By the S-absolute continuity of f, for each $\epsilon \in {}^{\sigma}\mathbb{R}^+$ there is a $p \in {}^{\sigma}\mathbb{N}$ such that

$$\Sigma[|f(v)|\delta\mu(v) : v \in A_p] < \epsilon.$$

Since f has S-MC, on $A \setminus A_p$ its sum is infinitesimal, hence $\Sigma[|f(v)|\delta\mu(v) : v \in A] \lesssim \epsilon$, so that it is infinitesimal. Now, the same inequalities used in proving (S-AC) give us $\Sigma[|g(v)|\mu(v) : v \in A] \approx 0$.

We know after (1.4.9), (1.4.10) that the projection $FL^1(\mu) \xrightarrow{\sim} L^1(\mu)$ is a contraction and is isometric on $SL^1(\mu)/IL^1(\mu)$. As a matter of fact, the isometry property completely characterizes the space of S-integrable functions $SL^1(\mu)$.

(1.4.13) THEOREM:

If $f : \mathbb{V} \longrightarrow {}^*\mathbb{R}$ *is internal, the following are equivalent:*

(a) $f \in SL^1(\mu)$.

(b) $\tilde{f} \in L^1(\mu)$ *and* $\int_{\mathbb{V}} |\tilde{f}|d\mu \approx \Sigma[|f(v)|\delta\mu(v) : v \in \mathbb{V}]$.

(c) $\tilde{f} \in L^1(\mu)$ *and* $\int_{\mathbb{V}} |\tilde{f}|d\mu \gtrsim \Sigma[|f(v)|\delta\mu(v) : v \in \mathbb{V}]$.

PROOF:

(a) implies (b) is part of the claim of Theorem (1.4.11). On the other hand, Lemma (1.4.9) proves that (b) and (c) are equivalent.

Assume $\tilde{f} \in L^1(\mu)$ is the projection of an internal function $f \notin SL^1(\mu)$, and $\|f\| \approx \|\tilde{f}\|$. Then obviously $f \in FL^1(\mu)$, hence either (S-AC) or (S-MC) are false for f. But this is impossible:

Let $A \subseteq \mathbb{V}$ be an internal set. According to Lemma (1.4.9),

$$\int_{\mathbb{V}} |\tilde{f}|\,d\mu = \int_{\mathbb{V}\setminus A} |\tilde{f}|\,d\mu + \int_A |\tilde{f}|\,d\mu \leq \text{st}\ \Sigma_{\mathbb{V}\setminus A} |f(v)|\,\delta\mu(v) + \int_A |\tilde{f}|\,d\mu.$$

Suppose that $\Sigma[\,|f(v)|\,\delta\mu(v) : v \in A] \not\approx 0$, and $\mu[A] \approx 0$ or $f(v) \approx 0$ for v in A. In either case,

$$\int_A |\tilde{f}|\,d\mu = 0 < \text{st}\ \Sigma_A |f(v)|\,\delta\mu(v),$$

hence

$$\int_{\mathbb{V}} |\tilde{f}|\,d\mu < \text{st}\ \Sigma_{\mathbb{V}\setminus A} |f(v)|\,\delta\mu(v) + \text{st}\ \Sigma_A |f(v)|\,\delta\mu(v)$$
$$= \text{st}\ \Sigma_{\mathbb{V}} |f(v)|\,\delta\mu(v),$$

contrary to our hypothesis $\|\tilde{f}\| \approx \|f\|$.

(1.4.14) **PROPOSITION:**

If f *is S-integrable,* g *is internal and* $|g(v)| \leq |f(v)|$ *for all* v, *then* g *is S-integrable.*

PROOF:

Exercise. Also, show that it is not sufficient to assume $|g(v)| \leq |f(v)|$ a.s. or n.s. (Make g big on a small set.)

(1.4.15) **ASIDE:**

Using the hull completeness theorem (Stroyan & Luxemburg, [1976], (10.1.20)), it is immediate that $SL^1(\mu)/IL(\mu)$ is complete, thus shortening the proof of (1.4.8)(a). On the other hand, by (1.4.12)(b), $SL^1(\mu)/IL^1(\mu)$ is a true normed subspace of $L^1(\hat{\mu})$. What is more: using the Integrable Lifting Theorem, Lemma (1.4.9) and (1.4.12)(a), one easily shows that there is a projection P of norm one from $L^1(\hat{\mu})$ onto $SL^1(\mu)/IL^1(\mu)$, so that the latter is a complemented subspace of the former. Finally (1.4.13) tells us something about the geometrical location of $SL^1(\mu)/IL^1(\mu)$ within $L^1(\hat{\mu})$: $\|P\hat{f}\| < \|\hat{f}\|$ whenever f is not S-integrable.

(1.4.16) **EXTENSION TO ALGEBRAS** (con't):

For $\mu, \mathcal{A}$, *etc. as in* (1.1.8)(c) *and* (1.3.12), *show that a* μ*-integrable function has an internal* $\mathcal{A}$*-measurable* S*-integrable lifting.*

The following lemma is Hoover and Perkins' [1980] formulation of an old uniform integrability criterion (due to de la Valee-Poussin?) similar to Burkholder, Davis and Gundy's [1972, Lemma 5.1]. We shall need it in chapter 7.

(1.4.17) LEMMA:

Let μ *be an internal limited positive measure. Let* $f : \mathbb{V} \longrightarrow {}^*\mathbb{R}$ *be an internal function. Then* f *is* S*-integrable if and only if there is a convex, increasing, internal function* $\Phi : {}^*[0,\infty) \longrightarrow {}^*[0,\infty)$ *with* $\Phi(0) = 0$ *such that*

(a) $\sup[\frac{x}{\Phi(x)} : x \geq n] \approx 0$ *for infinite* n,

(b) $\Phi(2u) \leq 4\Phi(u)$ *for all* $u \geq 0$,

(c) $\Sigma[\Phi(|f(v)|)\delta\mu(v) : v \in \mathbb{V}]$ *is finite.*

PROOF:

Sufficiency of this condition is quite easy. If such a Φ exists and n is infinite then by (a) there is a $\sigma(n) \approx 0$ so that

$$|f(v)| \le \sigma(n)\Phi(|f(v)|) \quad \text{whenever} \quad |f(v)| \ge n.$$

Summing, we obtain from (c),

$$\Sigma[|f(v)|\delta\mu(v) : |f(v)| \ge n]$$
$$\le \sigma(n)\Sigma[\Phi(|f(v)|)\delta\mu(v) : |f(v)| \ge n] \approx 0.$$

This shows that (1.4.5)(c) holds and f is S-integrable. (Notice that we did not require (b) or convexity, though they are important in chapter 7.)

Conversely, suppose that f is S-integrable. We must construct Φ. We will use two conditions, (c.1) and (c.2), on a number b in order to construct an internal sequence $\{a_j : j \in {}^*\mathbb{N}\}$. The conditions are:

(c.1)(b,j): $$b - a_j > 2a_j$$

&

(c.2)(b,j): $$\Sigma[|f(v)|\delta\mu(v) : |f(v)| \ge b] \le \frac{\Sigma[|f(v)|\delta\mu(v) : v \in \mathbb{V}]}{2^{j+1}}.$$

Let $a_0 = 0$ and choose $a_1 = 1 + \inf[b \in {}^*\mathbb{R} : (c.2)(b,0)]$. The internal inf is finite by (1.4.5)(c) and (c.2)$(a_1,0)$ holds because a_1 is greater than the inf. Next, select $a_2 = 1 + \inf[b \in {}^*\mathbb{R}$: (c.1)(b,1) & (c.2)(b,1)]. Again, a_2 is finite by S-integrability of f and (c.1)$(a_2,1)$ and (c.2)$(a_2,1)$ both hold. This is an internal process which we may use to select an internal sequence $\{a_j : j \in {}^*\mathbb{N}\}$ such that (c.1)(a_{j+1},j) and (c.2)(a_{j+1},j) hold for all j in ${}^*\mathbb{N}$.

(The inf is always nonempty.) When j is finite a_j is finite and when j is infinite a_j is infinite.

Define $\varphi(t) = j$ for $t \in [a_{j-1}, a_j)$ and define Φ by the internal formula

$$\Phi(x) = \int_0^x \varphi(t)dt.$$

Condition (c.1) implies that for all t, $\varphi(2t) \leq 2\varphi(t)$, since if $t \in [a_{j-1}, a_j)$ then $2t \leq a_{j+1}$, so $\varphi(2t) \leq (\frac{j+1}{j})\varphi(t) \leq 2\varphi(t)$. This fact about φ gives us conclusion (b) of the lemma because

$$\Phi(2x) = \int_0^{2x} \varphi(t)dt = 2\int_0^x \varphi(2s)ds \leq 4\int_0^x \varphi(s)ds = 4\Phi(x).$$

Conclusion (a) and the preliminary conditions are easy. Since φ is positive and increasing, Φ is increasing and convex. Clearly, $\Phi(0) = 0$. Since $\Phi(x)/x$ is an average value of φ over $[0,x]$ and more than half of that average is the last j with $a_j \leq x$, we see that $\Phi(x)/x$ is increasing and infinite whenever x is infinite. This forces

$$\sigma(n) = \sup[\frac{x}{\Phi(x)} : x \geq n] \approx 0 \quad \text{for infinite } n.$$

Finally, $\Phi(x) \leq kx$ when $x \leq a_k$, so, by (c.2)(a_j, j),

$$\Sigma[\Phi(|f(v)|)\delta\mu(v):v \in \mathbb{V}] = \begin{cases} \Sigma[\Phi(|f(v)|)\delta\mu(v):a_0 \leq |f(v)| < a_1] \\ \qquad + \\ \sum_{j=1}^{\infty} \Sigma[\Phi(|f(v)|)\delta\mu(v):a_j \leq |f(v)| < a_{j+1}] \end{cases}$$

$$\leq \begin{cases} \Sigma[|f(v)|\delta\mu(v):a_0 \leq |f(v)| < a_1] \\ \qquad + \\ \sum_{j=1}^{\infty} \Sigma[(j+1)|f(v)|\delta\mu(v):a_j \leq |f(v)|] \end{cases}$$

$$\leq \left[1 + \sum_{k=2}^{\infty} \frac{k}{2^k}\right] \Sigma[|f(v)|\delta\mu(v):v \in \mathbb{V}].$$

This shows that the sum of $\Phi(|f|)\delta\mu$ is finite and completes the proof of our lemma.

(1.5) Absolutely Continuous Measures & Conditional Expectation

Until now we have dealt only with positive real measures, but now we want to consider general (signed) finite real measures (or even complex measures). Let $\mathbb{V}$ be a *finite set and let $\mathcal{F}$ be a sigma algebra containing the internal subsets of $\mathbb{V}$, so $\mathcal{F} \supseteq \mathrm{Loeb}(\mathbb{V})$, the Loeb algebra. Recall that a finite measure is a function $\nu : \mathcal{F} \longrightarrow \mathbb{R}$ such that ν is countably additive, that is, if $F = \cup F_k$ is a disjoint countable union, then $\nu[F] = \sum_{k=1}^{\infty} \nu[F_k]$, the finitely absolutely convergent real series.

(1.5.1) EXAMPLE (Keisler):

As the reader has no doubt guessed, in section (2.3) we will show that every Borel measure on $\mathbb{R}^d$ has a hyperfinite representation in terms of a weight function $\delta\mu$ on an S-dense *finite set. However, not every probability measure on Loeb($\mathbb{V}$) is hyperfinite, that is, equal to the extension of an internal measure $\mu : {}^*\mathcal{B}(\mathbb{V}) \longrightarrow {}^*[0,\infty)$.

Let $\mathbb{V}$ be *finite and let $\mathcal{U}$ be a free ultrafilter in ${}^*\mathcal{B}(\mathbb{V})$, the algebra of internal subsets of $\mathbb{V}$. The indicator function of $\mathcal{U}$ $[u(U) = 1$ if $U \in \mathcal{U}$, $u(W) = 0$ if $W \notin \mathcal{U}]$ is a finitely additive measure on the algebra of internal sets, $u(\emptyset) = 0$ and

$$u\left(\bigcup_{k=1}^{m} A_k\right) = \sum_{k=1}^{m} u(A_k),$$

for finitely many disjoint internal sets A_k, $1 \leq k \leq m \in {}^\sigma\mathbb{N}$.

By (0.4.2) we know that no disjoint strictly countable union of internal sets is internal, so that the hypotheses of Caratheodory's extension theorem (Wheeden & Zygmund [1977, section 11.5], for example) are fulfilled. Thus u has a countably additive extension $\bar{u}$ to Loeb $(\mathbb{V})$.

Since $\mathcal{U}$ is a free ultrafilter, no finite sets are in $\mathcal{U}$ and $u(\{v\}) = 0$ for every $v \in \mathbb{V}$. If the extension $\bar{u} = \mu$, the hyperfinite extension of some internal $\mu : {}^*\mathcal{B}(\mathbb{V}) \longrightarrow {}^*[0,\infty)$, then $\delta\mu(v) \approx 0$ for every $v \in \mathbb{V}$. Since $u(V) = 1$, $\Sigma[\delta\mu(v) : v \in \mathbb{V}] \approx 1$. Enumerate $\mathbb{V}$ by an internal map $v(k)$, $1 \leq k \leq n$ and define

$$m = \max[k: \Sigma[\delta\mu(v(h)) : 1 \leq h \leq k] \leq 1/2].$$

The number m is defined internally, so it exists. The extension $\bar{u}(\{v : v = v(k), 1 \leq k \leq m\})$ takes only the values zero and one while $\mu[\{v : v = v(k), 1 \leq k \leq m\}] = 1/2$. This proves $\bar{u} \neq \mu$ as claimed.

Recall that two finite measures σ, λ over the same sigma algebra are said to be *mutually singular*, $\sigma \perp \lambda$, if there is a measurable N with $\lambda(N) = 0$ such that $\sigma(M) = 0$ whenever M is measurable and $M \subseteq \mathbb{V} \backslash N$. We shall see that for every hyperfinite measure μ, the measures $\bar{u}$ and μ are mutually singular over Loeb$(\mathbb{V})$.

Let μ be any (positive) limited hyperfinite measure on $\mathbb{V}$ (induced from an internal weight function $\delta\mu$). Recall that a finite measure $\upsilon : \text{Loeb}(\mathbb{V}) \longrightarrow \mathbb{R}$ is called *absolutely continuous with respect to* μ, denoted $\upsilon \ll \mu$, if $\mu[N] = 0$ implies $\upsilon[N] = 0$. Since $\mu[\mathbb{V}] < \infty$, $\upsilon \ll \mu$ is equivalent to the condition that for every standard $\epsilon > 0$, there is a standard $\vartheta > 0$, such that for $M \in \text{Loeb}(\mathbb{V})$, $\mu[M] < \vartheta$ implies $|\upsilon[M]| < \epsilon$.

If $\upsilon \ll \mu$ on $\text{Loeb}(\mathbb{V})$, then υ extends to the μ-measurable sets, because

$$\text{Meas}(\mu) = \{M \subseteq \mathbb{V} : \exists\ U,W \in \text{Loeb}(\mathbb{V}), U \subseteq M \subseteq W \ \&\ \mu[W\setminus U] = 0\},$$

the μ-completion of $\text{Loeb}(\mathbb{V})$. Since $\mu[W\setminus U] = 0$ implies $\upsilon[W\setminus U] = 0$, we may extend $\upsilon[M] = \upsilon[W] = \upsilon[U]$, using finiteness.

(1.5.2) PROPOSITION:

If υ *is a finite measure on* $\text{Loeb}(\mathbb{V})$ *and* $\upsilon \ll \mu$, *for some limited hyperfinite measure* μ *on* $\mathbb{V}$, *then there is an internal* f *in* $SL^1(\mu)$ *such that*

$$\upsilon[U] \approx \Sigma[f(v)\delta\mu(v) : v \in U]$$

for every internal $U \subseteq \mathbb{V}$. *Moreover,* f *is almost unique; if* g *has the property above, then* $\|f-g\| \approx 0$ *in* $SL^1(\mu)$ *and any* h *in* $SL^1(\mu)$ *with* $\|f-h\| \approx 0$ *also represents* υ *on internal sets.*

PROOF:

We shall base our proof on a well-known classical result,

the Lebesgue Decomposition Theorem, stated for the reader's convenience as (1.5.3) below. The proposition (1.5.2) follows easily from the Integrable Lifting Lemma (1.4.11) and the Radon-Nikodym part of (1.5.3). A complete proof of the proposition can be fashioned on a partition argument like the one that Wheeden and Zygmund [1977, section 10.3] use to prove Lebesgue Decomposition. We feel that the slight changes do not justify an independent proof, so we appeal to the classical fact.

Proposition (1.5.2) shows that ν is the limited hyperfinite measure with "signed weight function" $f(v)\delta\mu(v)$. If we already knew that $\nu = \boldsymbol{\alpha}$ for a positive limited hyperfinite measure given by a weight function $\delta\alpha(v)$ and if we have the weaker hypothesis that $\delta\mu(v) = 0$ implies $\delta\alpha(v) = 0$, then we may define

$$f(v) = \begin{cases} \dfrac{\delta\alpha(v)}{\delta\mu(v)}, & \delta\mu(v) \neq 0 \\ 0, & \delta\mu(v) = 0. \end{cases}$$

In order to have $f \in SL^1(\mu)$, we need to know the external condition that whenever $\mu[A] \approx 0$, then $\alpha[A] \approx 0$, because $\Sigma[f(v)\delta\mu(v) : v \in A] = \alpha[A]$ (see (1.4.5.b)). In general, if $\delta\alpha$ is not positive, we can work with the separate positive internal measures α^+ and α^- with weight functions $\max[0,\delta\alpha(v)]$ and $\max[0,-\delta\alpha(v)]$. We need to know that $\alpha^+[\mathbb{V}]$ and $\alpha^-[\mathbb{V}]$ are both limited in order for f to satisfy FL^1. We also need the infinitesimal absolute continuity. If μ is nonatomic, then we only need $\mu[A] \approx 0$ implies $\alpha[A] \approx 0$, by (1.4.6).

(1.5.3) **LEBESGUE DECOMPOSITION THEOREM:**

Let $(X,\mathcal{F},\lambda)$ *be a finite positive measure space and let* ν *be a finite measure on* $\mathcal{F}$. *Then there is a unique decomposition*

$$\nu[F] = \alpha[F] + \sigma[F], \quad \textit{for} \quad F \in \mathcal{F}$$

where α *and* σ *are measures,* $\alpha << \lambda$ *and* $\sigma \perp \lambda$. *Moreover, there is an a.s. unique* $f \in L^1(\lambda)$ *and a null* N, $\lambda[N] = 0$, *such that*

$$\alpha[F] = \int_F f d\lambda \qquad \textit{and} \qquad \sigma[F] = \nu[F \cap N],$$

for $F \in \mathcal{F}$. *The function* f *is called the Radon-Nikodym derivative of* α *with respect to* λ.

We conclude the discussion of example (1.5.1) by showing that $\overline{u} \perp \mu$ for every hyperfinite measure. Let $\mu = \lambda$ be a nonzero bounded hyperfinite measure, $\mathcal{F} = \text{Loeb}(\mathbb{V})$ and $\nu = \overline{u}$ from (1.5.1). Let $N \in \text{Loeb}(\mathbb{V})$ be as in the decomposition theorem. Either $\overline{u}[N] = 0$, in which case $\sigma = 0$, or $\overline{u}[N] = 1$. Let g be a μ-S-integrable lifting of f where $\alpha = \int f d\mu$. If $\sigma = 0$, then $\overline{u} = \alpha$ and α is the hyperfinite measure induced by the weights $g(v)\delta\mu(v)$, contrary to the reasoning in (1.5.1). Thus $\overline{u}[N] = \overline{u}[\mathbb{V}] = 1$ and $\mu[N] = 0$, so $\overline{u} \perp \mu$.

(1.5.4) **NOTATION:**

When $\mathbb{V}$ is *finite and $\delta\mu = \delta P : \mathbb{V} \longrightarrow {}^*[0,\infty)$ satisfies $\Sigma[\delta P(v) : v \in \mathbb{V}] \approx 1$ then $\mu = P$ is a hyperfinite probability measure. The letter P looks more like "probability." There is also a custom of calling integrals "expected values." If $f : \mathbb{V} \longrightarrow {}^*\mathbb{R}$ is internal,

$$E[f] = \Sigma[f(v)\delta P(v) : v \in \mathbb{V}]$$

is the internal expected value of f. Let $g : \mathbb{V} \longrightarrow \mathbb{R}$ be P-integrable, then

$$E[g] = \int_{\mathbb{V}} g dP.$$

One important use of Radon-Nikodym derivatives [see (1.5.3)] is in showing that conventional conditional expectations exist.

(1.5.5) **DEFINITION:**

Let $(\mathbb{V}, \text{Meas}(P), P)$ *be a hyperfinite probability space, let* $\mathcal{F} \subseteq \text{Meas}(P)$ *be a sigma algebra and let* $g : V \longrightarrow \mathbb{R}$ *be P-integrable. The (external) conditional expectation of* g *given* $\mathcal{F}$,

$$E[g|\mathcal{F}]$$

is the a.s. unique $\mathcal{F}$*-measurable function* $h : \mathbb{V} \longrightarrow \mathbb{R}$

such that
$$\left.\begin{array}{l} E[h(v)I_F(v)] = E[g(v)I_F(v)] \\ \text{or} \displaystyle\int_F h dP = \int_F g dP \end{array}\right\} \quad \textit{for } F \textit{ in } \mathcal{F}.$$

($I_F(v) = 0$ *if* $v \notin F$, $= 1$ *if* $v \in F$, *is the indicator function.*)

If we let $\nu[F] = \int_F g(v)d\mu(v)$ for F in $\mathcal{F}$ and restrict μ to $\mathcal{F}$, then h from the definition is the Radon-Nikodym derivative and $\sigma = 0$ because $\nu << \mu$.

(1.5.6) **EXAMPLE:**

Let $\mathbb{V} = \Omega$ and $\mu = P$ as in (0.2.7-10), (0.3.7) and (0.3.12). A time t in $\mathbb{T}$ determines an equivalence relation on Ω as follows. Let $\omega^t = \omega|[0,t] = (\omega_0, \omega_{\delta t}, \cdots, \omega_t)$, the restriction. Two sample points $\omega, \nu \in \Omega$ are equivalent if $\omega^t = \nu^t$, that is, if they agree up to the time t. We may denote the class by $[\omega^t] = \{\nu \in \Omega : \omega^t = \nu^t\}$. We shall want to know internal conditional expectations like

$$E[f|[\omega^t]].$$

(1.5.7) **DEFINITION:**

Let $\mathbb{V}$ *be* **finite, let* $\delta P : V \longrightarrow {}^*[0,1]$ *satisfy* $\Sigma[\delta P(v) : v \in V] \approx 1$ *and let* $f : \mathbb{V} \longrightarrow {}^*\mathbb{R}$ *be internal. Let* ρ *be an internal equivalence relation on* $\mathbb{V}$ *with the equivalence class of* v *denoted* $\rho(v)$. *The (internal) conditional expectation of* f *given* ρ *is the function* $v \longrightarrow E[f|\rho(v)]$, *where*

$$E[f|\rho(v)] = \Sigma[f(u)\delta P(u) : u\rho v]/P[\rho(v)].$$

An internal equivalence relation ρ on V induces a sigma subalgebra of the Loeb algebra, $\mathrm{Loeb}(\rho) \subseteq \mathrm{Loeb}(\mathbb{V})$, defined to be the smallest sigma algebra containing the internal sets U with the property that if $u \rho v$ and $v \in U$, then $u \in U$.

Any equivalence relation ρ (external or not) induces a sigma subalgebra of a sigma algebra $\mathcal{F}$. For any set $W \subseteq \mathbb{V}$, let $W(\rho) = \{u : u \rho w \text{ for some } w \in W\}$. The sigma algebra is

$$\mathcal{F}(\rho) = \{F \in \mathcal{F} : F = F(\rho)\},$$

the set of ρ-closed $\mathcal{F}$-sets.

(1.5.8) **EXERCISE:**

Prove that $\mathcal{F}(\rho)$ *is a sigma algebra. Let* ρ *be internal and prove that* $\mathrm{Loeb}(\rho) = \mathrm{Loeb}(\mathbb{V})(\rho)$.

An easy proof of this exercise can be given by using the separation theorem (2.2.3).

(1.5.9) **REMARK:**

Let P be a *finite probability and *let* ρ *be an internal equivalence relation.* The sigma algebra

$$\mathrm{Meas}(\mathbf{P})(\rho) = \{M \in \mathrm{Meas}(\mathbf{P}) : M = M(\rho)\}$$

can also be thought of as a sort of completion. We know that if M is in $\mathrm{Meas}(\mathbf{P})$ and $M = M(\rho)$, then for every finite natural number k, there exist internal sets U, W contained in $\mathbb{V}$ such that $U \subseteq M \subseteq W$ and $P[W \setminus U] < \frac{1}{k}$. The internal set

$U(\rho) = \{v : \exists\, u \in U \;\&\; u\rho v\}$ contains U and since $M = M(\rho)$ $U \subseteq U(\rho) \subseteq M$. Also, if $N = V\backslash M = M^c$, $N(\rho) = N$ and $W^c \subseteq N$, then $U \subseteq U(\rho) \subseteq M \subseteq [W^c(\rho)]^c \subseteq W$. Therefore, a set S is in $Meas(P)(\rho)$ if and only if $S = S(\rho)$ and for every finite k, there exist internal ρ-closed sets $U = U(\rho)$, $W = W(\rho)$ such that

$$U \subseteq S \subseteq W \quad \text{and} \quad P[W\backslash U] < \frac{1}{k}.$$

(1.5.10) **PROPOSITION:**

Let $(\mathbb{V}, Meas(P), P)$ *be a hyperfinite probability space and let* ρ *be an internal equivalence relation. Let* $Meas(P,\rho) = Meas(P)(\rho)$ *from* (1.5.9). *If* f *is an S-integrable lifting of* g, *then* $E[f\,|\,\rho(v)]$ *is an S-integrable lifting of* $E[g\,|\,Meas(P,\rho)]$.

PROOF:

We know that if $F(v) = E[f\,|\,\rho(v)]$, then $u\rho v$ implies $F(u) = F(v)$. Therefore the sets $\{\tilde{F}(v) < r\}$ are ρ-closed and Loeb, hence in $Meas(P,\rho)$. It is easy to see that $F(v)$ is S-integrable, e.g., use $\Sigma[F(u)\delta P(u) : F(u) > k]$, k infinite.

We know that $\int_U g\,dP \approx \Sigma[f(v)\delta P(v) : v \in U]$ whenever U is internal. If U is also ρ-closed, $U(\rho) = U$, then

$$\Sigma[f(v)\delta P(v) : v \in U] = \Sigma[\Sigma[f(v)\delta P(v) : v \in \rho(u)] : u \in U'],$$

where U' is an internal selection of one representative from each ρ equivalence class in U. The second sum equals

$$\Sigma[E[f\,|\,\rho(u)]\delta P(\rho(u)) : u \in U']$$

by definition of the internal conditional expectation. The

latter sum equals

$$\Sigma[E[f|\rho(u)]\delta P(u) \ : \ u \in U].$$

Hence $F(v)$ has the property that

$$\int_U g dP = \int_U \tilde{F} dP$$

for all internal ρ-closed sets U. The remark (1.5.9) completes the proof.

(1.5.11) EXERCISE:

Let g *be* P-*integrable and let* f *be an* S-*integrable lifting of* g. *What is the relationship between* $E[g|\text{Loeb}(\mathbb{V})]$ *and* f?

(1.6) Weak Compactness and SL^1

We believe that the following result can be a useful lemma in probability.

(1.6.1) THE DUNFORD-PETTIS CRITERION:

Let $(\mathbb{V}, Meas(P), P)$ *be a hyperfinite probability space. If* H *is an internal set of S-integrable functions,* $H \subseteq SL^1(P)$, *then the set of its projections,* $\tilde{H} \subseteq L^1(P)$, *is* $\sigma(L^1(P), L^\infty(P))$*-weakly relatively compact, that is, has compact closure in the weak topology.*

The reason for this is the following.

(1.6.2) LEMMA:

If H *is an internal set of* S-integrable *functions* $H \subseteq SL^1(\mu)$, *then*

(a) *The set of projections* $\tilde{H}$ *is bounded in* L^1*norm;*

(b) *For every standard positive* ϵ, *there is a standard positive* ϑ, *such that if* $A \subseteq \mathbb{V}$ *is internal, then*

$$\mu[A] < \vartheta \text{ implies } \Sigma[|h(v)|\delta\mu(v) : v \in A] < \epsilon \quad \text{for all } h \in H.$$

PROOF:

(a) The set $\{\Sigma[|h(v)|\delta\mu(v) : v \in \mathbb{V}] : h \in H\}$ is internal and only contains finite numbers, so its supremum exists and has to be finite.

(b) Follows from the S-absolute continuity condition (S-AC): given $\epsilon \in {}^\sigma\mathbb{R}^+$, the set

$$\{\vartheta \in {}^*\mathbb{R}^+ : (\forall A \in {}^*\mathcal{B}(V))[\mu[A] < \vartheta \Rightarrow (\forall h \in H)\ \Sigma[|h(v)|\delta\mu(v) : v \in A] < \epsilon]\}$$

is internal and contains all positive infinitesimals, so it contains a standard positive ϑ.

The Eberlein-Smulian Theorem (Dunford & Schwartz [1958, p. 430]) renders the following consequence about weak sequential compactness:

(1.6.3) **COROLLARY:**

Let $(\mathbb{V}, \mathrm{Meas}(P), P)$ *be a hyperfinite probability space, and* H *an internal set of* S-*integrable functions. Then for every sequernce* h_n *from* H, *there is an* S-*integrable* g *and a subsequence* h_{n_k} *so that for every finite internal* f,

$$\underset{k}{\text{S-lim}}\ \Sigma[h_{n_k}(v)f(v)\delta P(v) : v \in \mathbb{V}] = \Sigma[g(v)f(v)\delta P(v) : v \in \mathbb{V}]$$

The projected set need not always be weakly closed.

(1.6.4) **PROPOSITION:**

If H *is an internal subset of* $SL^1(P)$, *then* $\tilde{H}$ *is closed in* $L^1(P)$ *for the norm topology. If, in addition,* H *is* S-*convex, then* $\tilde{H}$ *is also weakly compact.*

PROOF:

Since in any Banach space the weak and the norm closure agree on convex sets, the second part follows from the first and from the last theorem and its corollary.

We leave the proof of norm closure as an exercise.

The following is an example of an internal set of S-integrable functions on a probability space whose projection is not weakly closed:

(1.6.5) **EXAMPLE:**

Let $\mathbb{V} = \{v_1,\cdots,v_s\}$, s an infinite factorial, be a probability space with uniformly distributed weights. For each $n \in {}^*\mathbb{N}$, define an internal function $h_n : \mathbb{V} \longrightarrow {}^*\mathbb{R}$ as follows:

$$h_n(v_i) = +1 \quad \text{if} \quad i \equiv 0,\cdots,n-1 \pmod{2n}$$
$$h_n(v_i) = -1 \quad \text{if} \quad i \equiv n,\cdots,2n-1 \pmod{2n}.$$

Not only are all the h_n internal, but the set $H = \{h_n | n \in {}^*\mathbb{N}\}$ is internal too. Moreover, since $|h_n| = 1$ for all n (incidentally, h_n equals the constant function 1 from $n = s$ on), the sum of $|h_n|$ and the integral of $|\tilde{h}_n|$ $(= |h_n|)$ are both infinitely close to 1; hence, $H \subseteq SL^1(P)$.

Observe that all the subsets of H of the form

$$H(p,r) = \{h_n \in H | p < n < r\}$$

with p finite, r infinite, are also internal; and their projections $\tilde{H}(p,r)$ contain all but a finite number of terms of the sequence $(\tilde{h}_m | m \in {}^\sigma\mathbb{N})$.

Assume that $\tilde{H}$ is weakly sequentially compact. Then a subsequence of $(\tilde{h}_m | m \in {}^\sigma\mathbb{N})$ converges weakly to some $\tilde{h}_n$. We prove next that this subsequence cannot have a subsequence with a weak limit within each $\tilde{H}(p,r)$, so that some of these projections of internal sets are not weakly sequentially compact. Thus it is neither weakly compact nor weakly closed.

If n is finite, the set $\tilde{H}(2n,s)$ is not sequentially compact, since no function in it is almost equal to $\tilde{h}_n$ (and the weak topology is Hausdorff). Let us prove this: if $h_q \in H(2n,s)$, let $k = [q/n] \geq 2$; then the set

$$A_0 = \{v_i \in V | i = 1,\cdots,q \;\&\; h_q(v_i) \neq h_n(v_i)\}$$

$$\supseteq [n,2n) \cup [3n,4n) \cup \cdots \cup \begin{cases} [(k-1)n,kn) \\ [(k-2)n,(k-1)n) \end{cases}$$

depending on whether k is even or odd. Then $\mu[A_0] \geq \frac{1}{3}\,\mu[\{v_i | i = 1,\cdots,q\}]$; and a similar argument can be carried over to every interval $\{v_i | i = jq+1,\cdots,(j+1)q\}$. For $j = 0,\cdots,s/q-1$,

$A_j = \{v_i \in V | i = jq+1,\cdots,(j+1)q \;\&\; h_q(v_i) \neq h_n(v_i)\}$ thus $\mu[A_j] \geq \frac{1}{3}\frac{q}{s}$. Therefore, $\mu[\{v \in V | h_q(v) \neq h_n(v)\}] = \mu[A_1] + \cdots + \mu[A_{s/q}] \geq \frac{1}{3}\frac{q}{s}\frac{s}{q} = \frac{1}{3}$.

If n is infinite, the set $\tilde{H}(1,[n/2])$ is not sequentially compact, since no function in it is almost equal to $\tilde{h}_n$ (to prove this, take $h_q \in H(1,[n/2])$ and proceed as before, inverting the roles of q and n).

(1.6.6) QUESTION:

What is the closure of the sequence $\{h_n\}$?

CHAPTER 2: MEASURES AND THE STANDARD PART MAP

This chapter makes several connections between hyperfinite measures and classical constructions in Euclidean spaces. Many of the results have generalizations to topological spaces as noted below. The idea of the chapter is to measure all points with an appropriate hyperfinite measure that lie near points being measured by the classical measure. This means that we use the inverse of the standard part map. We begin with Lebesgue measure because a complete measure is easier to treat than the naked Borel algebra which follows.

Let

$$\mathcal{O}^d = \{r \in {}^*\mathbb{R}^d \mid r = (r_1,\cdots,r_d) \textit{ with each } r_j \textit{ limited}\}$$

denote the Cartesian product of the limited scalars. Recall the definition of the standard part map from section (0.3).

(2.1) Lebesgue Measure

Let $\mathbb{T}^d$ be a *finite subset of $^*\mathbb{R}^d$. (It need not be a d-fold Cartesian product.) We say that $\mathbb{T}^d$ is S-dense if every standard point of $\mathbb{R}^d$ is close to some point of $\mathbb{T}^d$,

$$\{st(t) \mid t \in \mathbb{T}^d \cap \mathcal{O}^d\} \supseteq \mathbb{R}^d.$$

We say that an internal weight function $\delta\alpha : \mathbb{T}^d \to {}^*[0,\infty)$ *approximates volumes of standard rectangles* in $\mathbb{R}^d$ if for every bounded standard rectangle, for example,

$$I = \{r \in \mathbb{R}^d \mid a_j \leq r_j < b_j\},$$

the d-dimensional volume of I is approximated by α,

$$\prod_{j=1}^{d} (b_j - a_j) = \text{d-vol}(L)$$
$$\approx \Sigma[\delta\alpha(t) \mid t \in \mathbb{T}^d \cap {}^*I] = \alpha[\mathbb{T}^d \cap {}^*I].$$

Here is a good exercise for beginners at infinitesimal analysis: Show that for every S-dense *finite set $\mathbb{T}^d \subseteq {}^*\mathbb{R}^d$, there is a weight function that approximates standard rectangles. (See Appendix 1.)

An nice example of a weight function and S-dense set is to let $h \in {}^*\mathbb{N}$ be an infinite integer, $n = h!$ and take

$$\mathbb{T}^d = \{t \in {}^*\mathbb{R}^d \mid t = \left(\frac{k_1}{n}, \cdots, \frac{k_d}{n}\right), \ k_j \in {}^*\mathbb{Z} \ \& \ |k_j| \leq n^2\}$$

and

$$\delta\alpha(t) \equiv \left(\frac{1}{n}\right)^d.$$

Notice that $\mathbb{T}^d$ contains all standard rational points, because $\frac{p}{q} = \frac{p \cdot \Pi[k : 1 \leq k \leq h \ \& \ k \neq q]}{n}$ is an internal expression. It follows that for rectangles $I = \{r : a_j \leq r_j < b_j\}$ with rational edges, $a_j, b_j \in \mathbb{Q}$, we have $\text{d-vol}(I) = \alpha[{}^*I \cap \mathbb{T}^d]$. The infinitesimal volume approximation for all standard bounded rectangles follows by finite rational approximation.

Notice that the internal measure α does not represent d-dimenstional Lebesgue measure λ by the formula $\lambda(S) = \alpha({}^*S \cap \mathbb{T}^d)$ for all sets S. For example, $\mathbb{Q}^d$ has

Lebesgue measure zero, yet ${}^*\mathbb{Q}^d \supset \mathbb{T}^d$ in the $\mathbb{T}^d$-example just given. Also, we cannot generally say that $\lambda(S) = \alpha[{}^\sigma S \cap \mathbb{T}^d]$ for the hyperfinite measure, because $\mathbb{T}^d$ need not contain any standard points.

Throughout this chapter we use the following notational conventions. If $r \in {}^*\mathbb{R}^d$, $r = (r_1,\cdots,r_d)$, then $st(r) = (st(r_1),\cdots,st(r_d))$, provided each component is limited, $r_j \in \mathcal{O}$, or $r \in \mathcal{O}^d$. If $S \subseteq {}^*\mathbb{R}^d$, then $st(S)$ denotes the set of standard parts of limited vectors from S,

$$st(S) = \{st(s) \mid s \in S \cap \mathcal{O}^d\}.$$

We usually discuss st restricted to $\mathbb{T}^d$ because we are comparing hyperfinite measures and classical ones. In this context we write $st^{-1}(E)$ for the inverse of the restriction,

$$st^{-1}(E) = \{t \in \mathbb{T}^d \cap \mathcal{O}^d \mid st(t) \in E\}.$$

In particular, our formula for Lebesgue measure is $\lambda(L) = \alpha[st^{-1}(L)]$ with this convention in effect.

The next result plays an important role in our proofs.

(2.1.1) **PROPOSITION:**

The standard part of an internal subset of ${}^*\mathbb{R}^d$ *is closed.*

PROOF:

Let $r \in closure(st[B])$ for an internal $B \subseteq {}^*\mathbb{R}^d$. Then for each finite natural number m, $st[B] \cap \{s \in \mathbb{R}^d \mid |r-s| \le \frac{1}{m}\}$

$\neq \emptyset$ hence $B \cap \{s \in {}^*\mathbb{R}^d : |r-s| < \frac{2}{m}\} = B_m \neq \emptyset$. The countable decreasing chain B_m has nonempty intersection by saturation and if $b \in \cap B_m$, then $st(b) = r$. (The generalization of this to arbitrary topological spaces can be found in Stroyan & Luxemburg [1976], section 8.3.)

Note: Perhaps it would be better to say that st^{-1} is measure preserving in our next result, but we have used the customary terminology.

(2.1.2) THEOREM:

Suppose $\delta\alpha : \mathbb{T}^d \longrightarrow {}^*[0,\infty)$ *is a* **finite weight function that approximates volumes of standard rectangles in* $\mathbb{R}^d$ *as above. Then* st *is a measure-preserving map from the Lebesgue measure space* $(\mathbb{R}^d, \text{Leb}, \lambda)$ *into the hyperfinite space* $(\mathbb{T}^d, \text{Meas}(\alpha), \alpha)$, *specifically,* $L \subseteq \mathbb{R}^d$ *is Lebesgue measurable if and only if* $st^{-1}(L)$ *is* α*-measurable, and then* $\lambda(L) = \alpha[st^{-1}(L)]$.

PROOF:

First, the limited points of $\mathbb{T}^d$ form an α-measurable set because they are a countable union of internal sets,

$$\mathcal{O}^d \cap \mathbb{T}^d = \bigcup_m \{t \in \mathbb{T}^d : |t| \leq m\} \in \text{Loeb}(\mathbb{T}^d).$$

Next, we establish three general claims which we use in the proof of each implication of the theorem.

<u>Claim 1)</u> *If* I *is a bounded open rectangle, then*

$$st^{-1}(I) \in \text{Meas}(\alpha) \text{ and } \lambda(I) = \alpha[st^{-1}(I)].$$

Since I is open, whenever $r \in {}^{\sigma}I$ and $t \approx r$, since ${}^{*}I$ contains a finite neighborhood of r, $t \in {}^{*}I$. Thus, $st^{-1}(I) \subseteq {}^{*}I \cap \mathbb{T}^{d}$ and $\overline{\alpha}[st^{-1}(I)] \leq st(\alpha[{}^{*}I \cap \mathbb{T}^{d}]) = \lambda(I)$.

We may write I as a countable increasing union of compact rectangles, $I = \cup I_m$. Since I_m is closed, ${}^{*}I \cap \mathbb{T}^{d} \subseteq st^{-1}(I_m)$ and $\underline{\alpha}[st^{-1}(I_m)] \geq st(\alpha[{}^{*}I_m \cap \mathbb{T}^{d}]) = \lambda(I_m)$. Moreover, we have the inclusions

$$\bigcup_m {}^{*}I_m \cap \mathbb{T}^{d} \subseteq \bigcup_m st^{-1}(I_m) = st^{-1}(I) \subseteq {}^{*}I \cap \mathbb{T}^{d}.$$

Now we use rectangle approximations to see that

$$\overline{\alpha}[st^{-1}(I)] \leq \lambda(I) \leq \lim \underline{\alpha}[st^{-1}(I_m)] \leq \underline{\alpha}[st^{-1}(I)].$$

Hence, $st^{-1}(I)$ is α-measurable with α-measure $\lambda(I)$ as claimed. (It is easy to see that $st^{-1}(I) \in \text{Loeb}(\mathbb{T}^{d})$, see (2.2.8).)

Claim 2) *The family of sets* $\Sigma_1 = \{L \subseteq \mathbb{R}^{d} : st^{-1}(L) \in \text{Meas}(\alpha)\}$ *is a sigma algebra containing the Borel algebra* $\text{Borel}(\mathbb{R}^{d})$.

This claim is easy to verify, because inverse functions "commute" with set algebra operations, so Σ_1 is a sigma algebra. Claim 1) implies that Σ_1 contains all open rectangles, thus the whole Borel algebra.

Next, suppose that $L \subseteq \mathbb{R}^{d}$ is a Lebesgue set of measure zero, $\lambda(L) = 0$. Then for every standard $\epsilon > 0$, there is a countable cover of L by open rectangles of total measure less than ϵ. By Claim 1) it follows that $\alpha[st^{-1}(L)] = 0$, so

$st^{-1}(L)$ is an α-null set and $L \in \Sigma_1$ in particular.

The collection $\Sigma_2 = \Sigma_1 \cap \text{Leb}$ is a sigma algebra, but since Σ_1 contains $\text{Borel}(\mathbb{R}^d)$ and all Lebesgue null sets, $\Sigma_2 = \text{Leb}$, that is, if $L \in \text{Leb}$, then $st^{-1}(L) \in \text{Meas}(\alpha)$.

The family of sets where two defined measures agree is a sigma algebra. Thus, $\Sigma_3 = \{L \in \text{Leb} : \alpha[st^{-1}(L)] = \lambda(L)\}$ is a sigma subalgebra of Leb which contains all Lebesgue null sets and contains all open rectangles. Hence again $\Sigma_3 = \text{Leb}$ and this proves the next claim.

<u>Claim 3</u>)

$$\text{Leb} = \{L \subseteq \mathbb{R}^d : L \in \text{Leb}, st^{-1}(L) \in \text{Meas}(\alpha) \ \& \ \alpha[st^{-1}(L)] = \lambda(L)\}.$$

The third claim makes one implication of the theorem trivial because if L is a Lebesgue set, then L belongs to the right-hand-side of claim 3, hence satisfies the conclusion.

Now we prove the reverse implication of the theorem. It suffices to prove that whenever $L \subseteq \mathbb{R}^d$ is a bounded set with $st^{-1}(L) \in \text{Meas}(\alpha)$, then $\alpha[st^{-1}(L)] = \lambda(L)$ and $L \in \text{Leb}$ in particular. The reason this suffices is because we may treat the general case as a countable union $L = \bigcup_m (L \cap I_m)$ where the I_m are bounded rectangles. We shall establish Lebesgue measurability of bounded α-measurable sets by classical inner and outer approximation. This is the step where Proposition (2.1.1) is used.

Let L be a set contained in a bounded rectangle I with $st^{-1}(L) \in \text{Meas}(\alpha)$. Let ϵ be a standard positive tolerance. We know that there is an internal $B \subseteq st^{-1}(L)$ with $\alpha[st^{-1}(L)]$

$\leq \alpha[B]+\epsilon$. The set $st(B)$ is compact (by 2.1.1), so by claim 3, $\alpha[B] \leq \alpha[st^{-1}(st\ B)] = \lambda[st(B)]$. Therefore, the Lebesgue inner measure of L is at least $\alpha[st^{-1}(L)]$, because $st(B) \subseteq L$ and $\lambda[st(B)] \geq \alpha[st^{-1}(L)]-\epsilon$ with ϵ arbitrary.

Now we apply a similar argument to the set $I\backslash L$ and an arbitrary $\epsilon > 0$, obtaining a compact set $st(A) \subseteq I\backslash L$ with $\lambda[st(A)] \geq \alpha[st^{-1}(I\backslash L)]-\epsilon$. The set $I\backslash st(A)$ covers L and satisfies $\lambda[I\backslash st(A)] \leq \alpha[st^{-1}(L)]+\epsilon$. Thus, the Lebesgue outer measure of L is at most $\alpha[st^{-1}(L)]$, since ϵ is arbitrary. But this means that the inner and outer Lebesgue measures of L coincide, making L Lebesgue measurable with measure $\alpha[st^{-1}(L)]$. This completes the proof.

We shall be most interested in the next definition in the case where g is internal and the set M equals the set of limited vectors, $M = \mathcal{O}^d \cap \mathbb{T}^d$. If g is α-almost S-continuous on $\mathcal{O}^d \cap \mathbb{T}^d$, then the real-valued function $\hat{g} : st(U) \to \mathbb{R}$ is well-defined by $\hat{g}(st(t)) = st\ g(t)$.

(2.1.3) **DEFINITION:**

a) *Let* $\mathbb{T}^d$ *and* α *be as above. We say that a function* $g : \mathbb{T}^d \longrightarrow {}^*\mathbb{R}$ *is* α*-almost* S*-continuous on an* α*-measurable set* $M \subseteq \mathbb{T}^d$ *provided there is an* α*-measurable* $U \subseteq M$ *with* $\alpha[M\backslash U] = 0$ *such that whenever* $s,t \in U$ *satisfy* $s \approx t$, *then* $g(s) \approx g(t)$ *and both* $g(s)$ *and* $g(t)$ *are limited.*

b) *Let* $f : \mathbb{R}^d \longrightarrow \mathbb{R}$ *be a function. Suppose that* $g : \mathbb{T}^d \to {}^*\mathbb{R}$ *is internal. We say that* g *is an* α*-almost* S*-continuous lifting of* f *provided that*

$$\text{st } g(t) = f(\text{st}(t)) \quad a.e. \text{ on } \mathcal{O}^d \cap \mathbb{T}^d.$$

Suppose there is a set $U \subseteq \mathcal{O}^d \cap \mathbb{T}^d$ such that $\alpha[(\mathcal{O}^d \cap \mathbb{T}^d)\backslash U] = 0$ and whenever $t \in U$, $\text{st}[g(t)] = f(\text{st}(t))$. Then g is S-continuous and takes limited values on U because if $s, t \in U$ and $s \approx t$, then $\text{st}[g(t)] = f(\text{st}(t)) = f(\text{st}(s)) = \text{st}[g(s)]$. In other words, g is α-almost S-continuous on the set of limited vectors.

Fussy readers will say, "This isn't a special case of the liftings in chapter 1." They're right, this lifting is two legged, so the following diagram 'almost commutes':

$$\begin{array}{ccc} \mathcal{O}^d \cap \mathbb{T}^d & \xrightarrow{\ g\ } & {}^*\mathbb{R} \cap \mathcal{O} \\ \text{st} \downarrow & & \downarrow \text{st} \\ \mathbb{R}^d & \xrightarrow[\ f\ ]{} & \mathbb{R} \end{array}$$

Note the standard part on both sides instead of equality on the left. There are many kinds of 'liftings' in this book—all you can be sure they have in common is that they are internal objects with special properties that correspond to measurable objects under various kinds of projections built from standard parts. We strive for simple properties of *finite objects.

(2.1.4) ANDERSON'S LUSIN THEOREM:

Let $\mathbb{T}^d$ *and* α *be as above.*

a) $f : \mathbb{R}^d \to \mathbb{R}$ *is Lebesgue measurable if and only if* f *has an (internal)* α*-almost S-continuous lifting* g.

b) $f : \mathbb{R}^d \longrightarrow \mathbb{R}$ *is Lebesgue integrable if and only if* f *has an almost S-continuous lifting* g *in* $SL^1(\alpha)$ *such that*

$$\int_{\mathbb{T}^d \setminus \mathcal{O}^d} \tilde{g}(t)d\alpha(t) = 0.$$

In this case, for every Lebesgue measurable $L \subseteq \mathbb{R}^d$,

$$\int_{st^{-1}(L)} \tilde{g}(t)d\alpha(t) = \int_L f(x)d\lambda(x).$$

PROOF:

(a) If such a g exists then for $\mathbb{T}^d \cap \mathcal{O}^d \supseteq U$ with null difference, $U \cap st^{-1}\{r : f(r) < a\} = \{t \in U : \tilde{g}(t) < a\}$ and f is Lebesgue measurable by theorem (2.1.2).

Conversely, since $\{t \in \mathbb{T}^d \cap \mathcal{O}^d : f(st(t)) < a\} = st^{-1}\{f(r) < a\}$ we see that $f(st(\cdot))$ is α-measurable. Thus, by the Finite Function Lifting Theorem, for each finite natural number k, there is an internal function g_k such that $\tilde{g}_k(t) = f(st(t))$ a.e. on $\{|t| \leq k\}$. We may choose the sequence g_k to be progressive extensions, that is, $h < k$ implies $g_h(t) = g_k(t)$ on $|t| \leq h$, by using the internal definition principle. Extend g_k to an internal sequence and select an infinite n such that $m < n$ implies $g_m(t) = g_n(t)$ when $|t| \leq m$. Let $g = g_n$. The function $\tilde{g}(t)$ agrees with $f(st(t))$ on $\mathcal{O}^d \cap \mathbb{T}^d$ a.e. Now if $s \approx t$ and $\tilde{g}(s) = f(st(s)) = \tilde{g}(t)$, then $g(s) \approx g(t)$, so this proves (a).

Note: the set where g is S-continuous and lifts f cannot be internal unless f is actually uniformly continuous, since the infinitesimal hull of an internal function in an internal set is uniformly continuous. (See Appendix 1 for basic S-continuity.)

To prove (b) first consider the case where $f \geq 0$ is Lebesgue integrable then apply this case to f^+ and f^-. Let $I_{[-m,m]}(x)$ denote the indicator function of the interval $[-m,m]$. Using part (a), we may choose a sequence g^m of almost S-continuous liftings of $f(x)I_{[-m,m]}(x)$ such that $g^{m+1}(t) = g^m(t)$ for $|t| \leq m$ and $g^m(t) = 0$ for $|t| > m$. The truncations $g_m(t) = \min[m, g^m(t)]$ are bounded (hence S-integrable) almost S-continuous liftings of $f_m(x) = \min[m, (f(x)I_{[-m,m]}(x))]$ and they form an S-Cauchy $SL^1(\alpha)$-sequence because (2.1.2) says $\int \tilde{g}_m(t)d\alpha(t) = \int f_m(x)d\lambda(x)$ and dominated convergence says f_m is a convergent $L^1(d\lambda(x))$-sequence.

Extend $g^m(t)$ to an internal sequence satisfying $g^n(t) = g^m(t)$ for $n > m \geq |t|$ and maintain the internal truncation formula $g_n(t) = \min[n, g^n(t)]$ for infinite n. S-completeness of $SL^1(\alpha)$ says there is an infinite n such that $g_n(t) = g(t)$ is the $SL^1(\alpha)$ limit of $g_m(t)$. Certainly $g \in SL^1(\alpha)$ and g is an almost S-continuous lifting of f. It remains to show that $\int_{\mathbb{T}^d \setminus \mathcal{O}^d} \tilde{g}(t)d\alpha(t) = 0$, but we know that $\tilde{g}_m \longrightarrow \tilde{g}$ in $L^1(\alpha)$ and each $\tilde{g}_m$ satisfies this integral formula so the integral formula follows for $\tilde{g}$. The rest is a consequence of (2.1.2).

Conversely, suppose such a g exists. By part (a) and (2.1.2) $\int |f(x)|d\lambda(x) = \int_{\mathcal{O}^d} |\tilde{g}(t)|d\alpha(t) < \infty$.

(2.2) Borel and Loeb Sets

In this section we use some basic theory of abstract analytic sets or Souslin sets to give a relationship between Borel sets and Loeb sets. Some of this machinery will also be important later in the study of hyperfinite stochastic processes.

Let Seq denote the set of all finite sequences of natural numbers. Let $\mathcal{F}$ be a family of subsets of a set X. We say that a mapping $F : \text{Seq} \to \mathcal{F}$ is a *Souslin scheme*. In other words, a Souslin scheme attaches an $\mathcal{F}$-set, F_s, to each finite sequence, s. We may think of the sets F_s as attached to the nodes of a tree which branches infinitely many times as each sequence $(s_1, \cdots, s_m)$ is increased to $(s_1, \cdots, s_m, s_{m+1})$.

If $F : \text{Seq} \to \mathcal{F}$ is a Souslin scheme, then the *kernel* of F is the set

$$S = \bigcup_{\sigma}[\bigcap_{m}(F_{\sigma|m} : m \in \mathbb{N}) \mid \sigma \in \mathbb{N}^{\mathbb{N}}],$$

where the union ranges over all infinite sequences of natural numbers, $\sigma \in \mathbb{N}^{\mathbb{N}}$, and $\sigma|m$ denotes the finite sequence $(\sigma_1, \cdots, \sigma_m) = \sigma|m$. In terms of the tree interpretation of F, the set S is the union "along the top" of the intersections "up each branch."

(2.2.1) DEFINITION:

If $\mathcal{F}$ *is any family of subsets of a set* X, *then a set* S *is said to be derived from* $\mathcal{F}$ *by the Souslin operation if*

$$S = \bigcup_{\sigma} \bigcap_{m} F_{\sigma|m}$$

is the kernel of a Souslin scheme from $\mathcal{F}$. *The collection of these sets is denoted* Sous($\mathcal{F}$).

Here are some basic observations about the Souslin operation. Countable unions and intersections are special cases of the Souslin operation. Moreover, Sous(Sous($\mathcal{F}$)) = Sous($\mathcal{F}$). Hence Sous($\mathcal{F}$) is closed under countable unions and intersections.

There is no loss in generality when studying Sous($\mathcal{F}$) to assume that $\mathcal{F}$ is closed under finite unions and intersections, that is, if $\overline{\mathcal{F}}$ is the closure of $\mathcal{F}$ under finite unions and intersections, then Sous($\overline{\mathcal{F}}$) = Sous($\mathcal{F}$). Suppose that $\mathcal{F}$ is closed under finite intersections and F is a Souslin scheme. The mapping, $s \to G_s$, given by

$$G_s = \bigcap_{k=1}^{m} F_{s|k},$$

where $s = (s_1, \cdots, s_m)$, is a *decreasing* Souslin scheme, that is, if $s = t|m$, then $G_s \supseteq G_t$. We have the same kernel,

$$S = \bigcup_{\sigma} \bigcap_{m} G_{\sigma|m} = \bigcup_{\sigma} \bigcap_{m} F_{\sigma|m},$$

so every set $S \in$ Sous($\mathcal{F}$) may be derived from $\mathcal{F}$ by a decreasing Souslin scheme.

(2.2.2) DEFINITION:

A family $\mathcal{F}$ *of subsets of a set* X *is called a paving of* X *if* $\mathcal{F}$ *is nonempty and closed under finite unions and intersections. A paving* $\mathcal{F}$ *is said to be semicompact if every countable subset of* $\mathcal{F}$ *which has the finite intersection property has nonempty intersection.*

There are two basic pavings in this book: the family of compact subsets and the family of internal sets. If $\text{Kpt}(\mathbb{R}^d)$ denotes the family of all compact subsets of $\mathbb{R}^d$, then $\mathcal{F} = \text{Kpt}(\mathbb{R}^d)$ is a semicompact paving. The sets derived from the compact sets by the Souslin operation are called the *analytic* sets,

$$\text{Sous}(\text{Kpt}(\mathbb{R}^d)) = \text{Anal}(\mathbb{R}^d).$$

Analytic sets may also be characterized as the continuous images of Borel sets or continuous images of the irrationals, see Dellacherie & Meyer [1978] or Kuratowski [1966].

Let $\mathbb{V}$ be an internal set. The family of all internal subsets ${}^*\mathbb{B}(\mathbb{V}) = \mathcal{F}$ is a semicompact paving by the saturation property of section (0.4). We refer to the sets derived from the internal sets by the Souslin operation as Henson sets,

$$\text{Sous}({}^*\mathbb{B}(\mathbb{V})) = \text{Hens}(\mathbb{V}).$$

Each of these two pavings has the property that complements of $\mathcal{F}$-sets are $\text{Sous}(\mathcal{F})$-sets (open sets are countable unions of compacts). Therefore, they contain the sigma algebras generated by $\mathcal{F}$,

$$\text{Anal}(\mathbb{R}^d) \supseteq \text{Borel}(\mathbb{R}^d)$$

$$\text{Hens}(\mathbb{V}) \supseteq \text{Loeb}(\mathbb{V}).$$

If $\mathcal{F}$ is a family of sets, let $\cap\cup(\mathcal{F})$ denote the closure of $\mathcal{F}$ under countable union and countable intersection. The next result is an abstract form of a classical theorem.

(2.2.3) LUSIN'S SEPARATION THEOREM:

Suppose $\mathcal{F}$ *is a semicompact paving of* X. *If* $A,B \in \text{Sous}(\mathcal{F})$ *are disjoint, then there exist disjoint* $C,D \in \cap\cup(\mathcal{F})$ *such that* $A \subseteq C$ *and* $D \supseteq B$.

PROOF:

See Dellacherie and Meyer [1978, III.14].

Two immediate applications are as follows. If $A \subseteq \mathbb{R}^d$ and $\mathbb{R}^d \setminus A$ are both analytic, then $A \in \text{Borel}(\mathbb{R}^d)$. This is the classical Lusin result. If $H \subseteq \mathbb{V}$ and $\mathbb{V}\setminus H$ are both Henson sets, then $H \in \text{Loeb}(\mathbb{V})$. We shall see other applications in later chapters.

Now we begin the specific study of Loeb sets on an S-dense internal set $\mathbb{T}^d \subseteq {}^*\mathbb{R}^d$, with $\text{st}(\mathbb{T}^d) = \mathbb{R}^d$.

(2.2.4) LEMMA:

If $A_1 \supseteq A_2 \supseteq \cdots$ *is a decreasing sequence of internal subsets of* $\mathbb{T}^d$, *then* $\text{st}(\bigcap_m A_m) = \bigcap_m \text{st}(A_m)$.

PROOF:

It is sufficient to show that $\bigcap_m \text{st}(A_m) \subseteq \text{st}(\bigcap_m A_m)$, since the other inclusion is trivial. Let $r \in \bigcap_m st(A_m)$ be an arbitrary point. For each standard m, there is a $t_m \in A_m$ such that $st(t_m) = r$. Use the comprehension principle (0.4.3) to extend $\{(t_m, A_m) : m \in \mathbb{N}\}$ to be an internal sequence $\{(t_n, A_n) : n \in {}^*\mathbb{N}\}$. The set

$$\{n \in {}^*\mathbb{N} : (\forall m \in {}^*\mathbb{N})[m \leq n \Rightarrow (t_m \in A_m \ \& \ |t_m - r| < \tfrac{1}{m})]\}$$

is internal and contains all finite indices $n \in {}^\sigma\mathbb{N}$, hence it contains an infinite n. This satisfies $t_n \in \bigcap_m A_m$ and $st(t_n) = r$.

(2.2.5) **PROPOSITION:**

a) $A \in Anal(\mathbb{R}^d) \Rightarrow st^{-1}(A) \in Hens(\mathbb{T}^d)$,

b) $H \in Hens(\mathbb{T}^d) \Rightarrow st(H) \in Anal(\mathbb{R}^d)$.

PROOF:

If K is a compact set, then $st^{-1}(K)$ is a countable intersection of the internal sets

$$I_m = \{t \in \mathbb{T}^d : dist(t, {}^*K) < \tfrac{1}{m}\}.$$

Therefore, if A is the kernel of a compact Souslin scheme, K_s, then

$$st^{-1}(A) = \bigcup_\sigma \bigcap_m st^{-1}(K_{\sigma|m})$$

is a Henson set because $Sous(Hens(\mathbb{T}^d)) = Hens(\mathbb{T}^d)$. This proves part a.

As noted above, every Henson set H may be derived from a decreasing Souslin scheme, $I_{\sigma|m} \supseteq I_{\sigma|(m+1)}$. Then Lemma (2.2.4) shows that

$$st(H) = \bigcap_{\sigma} \bigcup_{m} st(I_{\sigma|m}).$$

Proposition (2.1.1) proves that the sets $st(I_{\sigma|m})$ are closed. Since $Sous(Anal(\mathbb{R}^d)) = Anal(\mathbb{R}^d)$, $st(H)$ is analytic.

(2.2.6) **THEOREM:**

Let $\mathbb{T}^d$ *be an S-dense internal subset of* ${}^*\mathbb{R}^d$. *A set* $B \subseteq \mathbb{R}^d$ *is a Borel subset of* $\mathbb{R}^d$ *if and only if its inverse standard part* $st^{-1}(B)$ *is a Loeb subset of* $\mathbb{T}^d$.

PROOF:

Let $B \subseteq \mathbb{R}^d$ be a Borel set, so that both B and $\mathbb{R}^d \backslash B$ are analytic. By Proposition (2.2.5), the sets $st^{-1}(B)$, $st^{-1}(\mathbb{R}^d \backslash B) = [\mathbb{T}^d \backslash st^{-1}(B)] \cap \mathcal{O}^d$ and $\mathbb{T}^d \backslash st^{-1}(B)$ are Henson sets. The Separation Theorem (2.2.3) shows that $st^{-1}(B)$ is Loeb.

Conversely, if $B \subseteq \mathbb{R}^d$ and $st^{-1}(B) \in Loeb(\mathbb{T}^d)$, then both $st^{-1}(B)$ and $\mathbb{T}^d \backslash st^{-1}(B)$ are Henson sets and so, by Proposition (2.2.5), $B = st(st^{-1}(B))$ and $\mathbb{R}^d \backslash B = st[\mathbb{T}^d \backslash st^{-1}(B)]$ are analytic. Again, the Separation Theorem (2.2.3) shows that B is Borel.

The reader should note that we have not given a correspondence between Loeb sets and Borel sets. Sets which are inverse standard parts of standard sets are closed under the infinitesimal relation $t \approx s$, so they are not arbitrary Loeb sets.

In the following discussion it is convenient to introduce some modern set-theoretical notation. If $\mathbb{V}$ is an internal set, let

$$\Pi_0^0(\mathbb{V}) = \Sigma_0^0(\mathbb{V}) = {}^*\mathbb{B}(\mathbb{V}).$$

If X is a topological space, let

$$\Pi_1^0(X) = \{C \mid C \textit{ is closed in } X\}$$

$$\Sigma_1^0(X) = \{U \mid U \textit{ is open and a countable union of closed sets}\}.$$

In both cases, continue inductively with countable operations:

$$\Sigma_\alpha^0 = \{\cup A_m \mid A_m \in \bigcup_{\beta<\alpha} \Pi_\beta^0\},$$

$$\Pi_\alpha^0 = \{\cap A_m \mid A_m \in \bigcup_{\beta<\alpha} \Sigma_\beta^0\}.$$

These sets generate the Loeb and Borel sigma algebras,

$$\text{Loeb}(\mathbb{T}^d) = \bigcup_{\alpha<\omega_1} \Pi_\alpha^0(\mathbb{T}^d) = \bigcup_{\alpha<\omega_1} \Sigma_\alpha^0(\mathbb{T}^d)$$

$$\text{Borel}(\mathbb{R}^d) = \bigcup_{\alpha<\omega_1} \Pi_\alpha^0(\mathbb{R}^d) = \bigcup_{\alpha<\omega_1} \Sigma_\alpha^0(\mathbb{R}^d)$$

where ω_1 is the first uncountable ordinal. Two older names for some of these families are $\Pi_2^0(\mathbb{R}^d) = G_\delta$ and $\Sigma_2^0(\mathbb{R}^d) = F_\sigma$.

We have already seen that standard parts of Henson sets are analytic. Theorem (2.2.8) shows that every analytic set is actually the standard part of an internal $\sigma\delta$-set, just two levels up the sigma algebra hierarchy. We need two technical results to prove that theorem.

(2.2.7) **LEMMA:**

If A *is an internal subset of* $\mathbb{T}^d$, *and* S_2 *is a nonempty closed subset of* $S_1 = st(A)$ *with* $A \cap st^{-1}(x)$ *infinite for each* $x \in S_1$, *then there exist disjoint internal* A_1, A_2 *contained in* A *such that* $st(A_j) = S_j$, *and for each* $x \in S_j$, $A_j \cap st^{-1}(x)$ *is infinite* $(j=1,2)$.

PROOF:

Let S_1', S_2' be denumerable dense subsets of S_1, S_2, respectively. Choose (external) disjoint sets D_1, D_2 contained in A such that $st(D_j) = S_j'$ and for each $x \in S_j'$ there are exactly a countable number of elements in $D_j \cap st^{-1}(x)$ $(j=1,2)$. Notice that D_1, D_2 are countable.

Enumerate these sets $D_j = \{d_m^j : m \in \mathbb{N}\}$ and extend the sequences d_m^j to internal sequences with values in A $(j=1,2)$. For each finite m we have the internal condition

$$(\forall k < m)[dist(D_k^2, {}^*S_2) < \frac{1}{m} \ \& \ d_k^2 \notin \{d_1^1, \cdots, d_m^1\}],$$

hence this holds for some infinite n. Let $A_1 = \{d_k^1 : k \leq n\}$ and $A_2 = \{d_k^1 : k \leq n\}$.

This lemma has the following extension.

(2.2.8) **LEMMA:**

If B *is an internal subset of* $\mathbb{T}^d$ *and* $\{C_m \mid m \in \mathbb{N}\}$ *is a sequence of closed subsets of* $st(B) = C$, *where* $B \cap st^{-1}(x)$ *is infinite for every* $x \in C$, *then there is a disjoint sequence* $\{B_m \mid m \in \mathbb{N}\}$ *of internal subsets of* B *such that for all* m, $st(B_m) = C_m$ *and* $B_m \cap st^{-1}(x)$ *is infinite for every* $x \in C_m$.

PROOF:

We apply the previous lemma inductively. Let $S_1 = C$ and $S_2 = C_1$. Choose A_1 and A_2 as in Lemma (2.2.7) and let $B_1 = A_2$. Then A_1 is disjoint from B_1 and we may apply the lemma again to $S_1 = st(A_1) = C$ and $C_2 = S_2$, splitting A_1 into two disjoint sets A_2 and B_2 with $st(A_2) = C$ and $st(B_2) = C_2$ and each infinite-to-one. Also, B_2 is disjoint from B_1.

For the induction assume that disjoint sets $B_1, \cdots, B_m$ have been chosen along with a disjoint A_m such that $st(A_m) = C$, $st(B_j) = C_j$ and each is infinite-to-one. Apply Lemma (2.2.7) to the set $A = A_m$ and $S_2 = C_{m+1}$, splitting it into two disjoint internal sets B_{m+1} and A_{m+1} such that $st(B_{m+1}) = C_{m+1}$, $st(A_{m+1}) = C$, and each is infinite-to-one. Since A_m is disjoint from $B_1, \cdots, B_m$, B_{m+1} is disjoint from them.

Now we are ready to characterize analytic sets.

(2.2.9) **THEOREM:**

Let $\mathbb{T}^d$ *be an S-dense internal subset of* ${}^*\mathbb{R}^d$. *Every analytic* $A \in \mathrm{Anal}(\mathbb{R}^d)$ *is the standard part of some* $B \in \Pi_2^0(\mathbb{T}^d)$, $A = \mathrm{st}(B)$.

PROOF:

Let A be given as the kernel of a decreasing Souslin scheme of closed sets,

$$A = \bigcup_{\sigma} \bigcap_{m} F_{\sigma|m}.$$

We will apply Lemma (2.2.8) to define a decreasing Souslin scheme of internal sets B_s, $s \in \mathrm{Seq}$, satisfying

a) $\{B_m \mid m \in \mathbb{N}\}$ is a disjoint for sequences of length 1.

b) $\mathrm{st}(B_s) = F_s$, for all $s \in \mathrm{Seq}$.

c) $B_s \supseteq B_{sk}$, for all $s \in \mathrm{Seq}$, $k \in \mathbb{N}$.

d) $B_{sh} \cap B_{sk} = \emptyset$, for all $s \in \mathrm{Seq}$, $h,k \in \mathbb{N}$ with $h \neq k$.

To start, apply Lemma (2.2.8) with $C = \mathbb{R}^d$ and $C_m = F_m$. This gives us the internal sets B_s for sequences of length 1. Once a B_s has been chosen, apply the lemma again with $B = B_s$ and $C_m = F_{sm}$.

Since our internal scheme is decreasing, we have

$$A = \bigcup_{\sigma} \bigcap_{m} \mathrm{st}(B_{\sigma|m}) = \mathrm{st}(\bigcup_{\sigma} \bigcap_{m} B_{\sigma|m}),$$

by (2.2.4). Thus we need to show that

$$\bigcup_{\sigma} \bigcap_{m} B_{\sigma|m} = \bigcap_{m} \bigcup_{n} A_m^n$$

for internal sets A_m^n. Conditions (a), (c) and (d) on B_s imply that whenever s and t are sequences of the same length, then B_s and B_t are disjoint. This implies that

$$\bigcup_{\sigma} \bigcap_{m} B_{\sigma|m} = \bigcap_{n} \bigcup_{s \in \mathbb{N}^n} B_s$$

and completes the proof.

In light of this result one might ask how far through the family of sets Σ_α^0 and Π_α^0 one must go in order to generate the Loeb algebra. In the case of the real numbers it is known that one must take the union all the way to ω_1 in order to generate the Borel algebra. Kunen has shown the following result.

(2.2.10) THEOREM:

Let $\mathbb{T}$ *be an S-dense internal subset of* ${}^*I = {}^*[0,1]$. *For* $A \subseteq I$ *and* $\alpha \geq 1$

a) $A \in \Pi_\alpha^0(I)$ *if and only if* $st^{-1}(A) \in \Pi_\alpha^0(\mathbb{T})$,

and

b) $A \in \Sigma_\alpha^0(I)$ *if and only if* $st^{-1}(A) \in \Sigma_\alpha^0(\mathbb{T})$.

The following is a sketch of the proof kindly sent to us by K. Kunen. K. Kunen and A. Miller plan to publish the topological fact (2.2.11) in a paper tentatively titled, "Borel and projective sets from the point of view of compact sets."

The difficult part of the theorem reduces to showing that if $st^{-1}(A) \in \Sigma_\alpha^0(\mathbb{T})$, then $A \in \Sigma_\alpha^0(I)$. First, we may consider

$$st^{-1}(A) = \varphi(K_0, K_1, K_2, \cdots)$$

to be a Σ_α^0-combination of a sequence of internal sets $\{K_m\}$. Next, we topologize the problem by letting X be the Stone space of the Boolean algebra of internal sets, ${}^*\mathbb{B}(\mathbb{T})$. This gives us a commutative diagram consisting of the natural inclusion $i : \mathbb{T} \longrightarrow X$ and a lifting f of st:

$$\begin{array}{ccc} X & \xrightarrow{f} & I \\ i \uparrow & & \| \\ \mathbb{T} & \xrightarrow[st]{} & I \end{array}$$

We take

$$C = \varphi(N_0, N_1, N_2, \cdots) \in \Sigma_\alpha^0(X)$$

where N_m is the clopen set in X determined by K_m. Saturation implies that $f^{-1}(A) = C$, since if $u \in f^{-1}(A) \setminus C$, there is a $t \in \mathbb{T}$ such that $st(t) = f(u)$ and $t \in K_m$ if and only if $u \in N_m$. Now $t \notin \varphi(K_0, K_1, \cdots) = st^{-1}(A)$, so $f(u) = st(t) \in A$. The contradiction shows that $f^{-1}(A) = C$.

The following theorem of Miller and Kunen completes the proof.

(2.2.11) THEOREM:

Let X and Y be compact Hausdorff spaces, let $f : X \longrightarrow Y$ be continuous and onto and let $A \subseteq Y$. Then $A \in \Sigma_\alpha^0(Y)$ if and only if $f^{-1}(A) \in \Sigma_\alpha^0(X)$.

(2.2.12) **REMARK:**

Let Σ be a sigma algebra on a set X. We say that a set $U \subseteq X$ is *universally* Σ *measurable* if whenever μ is a finite measure on Σ, then U is in the μ-completion of Σ. In other words, there is an $S \in \Sigma$ such that the symmetric difference has zero outer measure, $\bar{\mu}[S \triangledown U] = 0$. An argument based on the Souslin operation can be used to show that sets in $\mathrm{Sous}(\Sigma)$ are universally Σ measurable. This also follows from Choquet's theorem, see Dellacherie & Meyer [1978, III.33(a)]. In particular, analytic sets are universally Borel measurable and Henson sets are universally Loeb measurable.

(2.3) Borel Measures

Suppose that $\mathbb{T}^d$ is an S-dense *finite subset of $^*\mathbb{R}^d$ and $\delta\alpha : \mathbb{T}^d \longrightarrow {}^*[0,\infty)$ is a limited internal weight function. Since Theorem (2.2.6) says that for every Borel set $B \subseteq \mathbb{R}^d$, $st^{-1}(B)$ is a Loeb set, the hyperfinite extension α defines a Borel measure μ_α by

$$\mu_\alpha[B] = \alpha[st^{-1}(B)].$$

The mapping from internal measures α to Borel measure μ_α is not unique. For example, if $\delta\alpha$ is the indicator function of one point $s \approx 0$, while $\delta\beta$ is the indicator of a different point $t \approx 0$, then both μ_α and μ_β are unit mass at zero because $st^{-1}\{0\}$ contains both s and t. The purpose of this section is to show how to start with a Borel measure μ and find an internal *finite measure α such that $\mu = \alpha \circ st^{-1}$.

(2.3.1) LEMMA:

Let μ be a finite positive Borel measure on $\mathbb{R}^d$. Let $\mathbb{T}^d$ be a *finite S-dense subset of $^*\mathbb{R}^d$. Then there is an infinite natural number $h \in {}^*\mathbb{N}$ and an internal weight function $\delta\alpha_\mu : \mathbb{T}^d \longrightarrow {}^*[0,\infty)$ such that whenever $k_1, \cdots, k_d \in {}^*\mathbb{Z}$ make $\left(\frac{k_1}{h!}, \cdots, \frac{k_d}{h!}\right)$ limited, then

$$\alpha_\mu\left[\mathbb{T}^d \cap \prod_{i=1}^{d} \left[\frac{k_i - 1}{h!}, \frac{k_i}{h!}\right)\right] = {}^*\mu\left[\prod_{i=1}^{d} \left[\frac{k_i - 1}{h!}, \frac{k_i}{h!}\right)\right].$$

PROOF:

Consider the internal property of a natural number m:

$$\Phi(m) = (\exists S \in {}^*\mathcal{B}(\mathbb{T}^d))[(\forall k_1, \cdots, k_d \in {}^*\mathbb{Z})\Big[|k_i| < (m!)^2$$

$$\Rightarrow S \cap \prod_{i=1}^{d} \left[\frac{k_i - 1}{m!}, \frac{k_i}{m!}\right] \text{ is a singleton}\Big].$$

This simply says that there is a finite sample from one of each of a finite number of boxes. Since Φ is internal it must also hold at an infinite h. Denote the corresponding set by S and define $\delta\alpha_\mu$ by

$$\delta\alpha_\mu(t) = \begin{cases} {}^*\mu\left[\prod_{i=1}^{d}\left[\frac{k_i-1}{h!}, \frac{k_i}{h!}\right]\right], & \text{if} \quad t \in S \cap \prod_{i=1}^{d}{}^{2}\left[\frac{k_i-1}{h!}, \frac{k_i}{h!}\right] \\ 0 & , \quad \text{otherwise.} \end{cases}$$

This internally defined function satisfies the lemma because

$$\Sigma\left[\delta\alpha_\mu(t) \ : \ t \in \prod_{i=1}^{d}\left[\frac{k_i-1}{h!}, \frac{k_i}{h!}\right]\right] = \delta\alpha_\mu(s)$$

where s is the single point in the box.

The reason that we chose $h!$ a factorial in the lemma is because now all standard rationals $p \in {}^{\sigma}\mathbb{Q}$ may be expressed as $\frac{k}{h!}$ and this gives α_μ the right measure on all standard rational-edge rectangles.

(2.3.2) LEMMA:

For every open bounded rectangle R *with rational edges,* $st^{-1}(R)$ *is Loeb and* $\alpha_\mu(st^{-1}(R)) = \mu(R)$.

PROOF:

The first part is included in Theorem (2.2.6). If $p,q \in {}^{\sigma}\mathbb{Q}^d$, let

$$I[p,q) = \{x \in {}^{\sigma}\mathbb{R}^d \mid p_i \leq x_i < q_i, \ i = 1,\cdots,d\}$$

(we will also use the notations $I[p,q], I(p,q)$, with the obvious meaning). $I[p,q)$ can be decomposed in a *finite disjoint union of rectangles $\prod_{i=1}^{d} \left[\frac{k_i - 1}{h!}, \frac{k_i}{h!}\right]$ (recall that all rational numbers are of the form $\frac{k}{h!}$); hence

$$\mu(I[p,q)) = \alpha_\mu[\mathbb{T}^d \cap {}^*I[p,q)].$$

On the other hand, it is clear that

$$st^{-1}(I[p,q]) \supseteq st^{-1}(I(p,q)).$$

therefore

$$\alpha_\mu(st^{-1}(I[p,q])) \geq \mu(I[p,q)) \geq \alpha_\mu(st^{-1}(I(p,q))).$$

Now, let R be an open bounded rectangle with rational edges, $R = I(p,q)$. Then,

$$R = \cup I[p + \tfrac{1}{n}, q - \tfrac{1}{n}) = \cup I[p + \tfrac{1}{2n}, q - \tfrac{1}{2n}) = \cup I[p + \tfrac{1}{n}, q - \tfrac{1}{n}],$$

hence

$$\mu(R) = \lim \mu(I[p + \frac{1}{n}, q - \frac{1}{n})) \leq \lim \boldsymbol{\alpha}_{\mu}(st^{-1}(I[p + \frac{1}{n}, q - \frac{1}{n}]))$$

$$\leq \lim \boldsymbol{\alpha}_{\mu}(st^{-1}(I(p + \frac{1}{2n}, q - \frac{1}{2n})))$$

$$\leq \lim \mu(I[p + \frac{1}{2n}, q - \frac{1}{2n})) = \mu(R),$$

and

$$\boldsymbol{\alpha}_{\mu}(st^{-1}(R)) = \lim \boldsymbol{\alpha}_{\mu}(st^{-1}(I[p + \frac{1}{n}, q + \frac{1}{n}])) = \mu(R).$$

(2.3.3) **THEOREM:**

If $\mathbb{T}^d$ *is an S-dense* **finite subset of* $^*\mathbb{R}^d$, μ *is a Borel measure on* $\mathbb{R}^d$ *and* α_{μ} *is as above, then* $\mu[B] = \boldsymbol{\alpha}_{\mu}[st^{-1}(B)]$ *for every Borel set* B.

PROOF:

The family

$$\Sigma = \{B \in \mathrm{Borel}(\mathbb{R}^d) \mid \mu[B] = \boldsymbol{\alpha}[st^{-1}(B)]\}$$

is a sigma algebra containing all rational edge rectangles. Hence, $\Sigma = \mathrm{Borel}(\mathbb{R}^d)$.

(2.3.4) **THEOREM:**

Let $\mathbb{T}^d$ *be an S-dense* **finite subset of* $^*\mathbb{R}^d$ *with a* **finite measure* α. *Let* μ *be a Borel measure such that* $\mu[B] = \boldsymbol{\alpha}[st^{-1}(B)]$ *for* $B \in \mathrm{Borel}(\mathbb{R}^d)$.

a) A *set* $M \subseteq \mathbb{R}^d$ *is* μ*-measurable if and only if* $st^{-1}(M)$ *is* $\boldsymbol{\alpha}$*-measurable with* $\boldsymbol{\alpha}[st^{-1}(M)] = \mu[M]$.

b) A *function* $f : \mathbb{R}^d \longrightarrow \mathbb{R}$ *is* μ*-measurable if and only if* $f(st(t))$ *is* $\boldsymbol{\alpha}$*-measurable.*

PROOF:

If M is μ-measurable, then there are Borel sets B_1, B_2 such that the symmetric difference satisfies $M \triangledown B_1 \subseteq B_2$ and $\mu[B_2] = 0$. Then we have $\alpha[st^{-1}(B_2)] = 0$, so $st^{-1}[M]$ is α-measurable with measure $\mu[B_2]$.

The remainder of the proof can be finished in the same way as the end of Theorem (2.1.2). If $st^{-1}(M)$ is α-measurable, then for each standard $\epsilon > 0$ there is an internal $I \subseteq st^{-1}(M)$ with $\alpha[I] > \alpha[st^{-1}(M)]-\epsilon$. The set $st(I)$ is closed and contained in M, so $\alpha[I] \le \alpha[st^{-1}(st(I))] = \mu[st(I)] \le \alpha[st^{-1}(M)]$, forcing the inner μ-measure of M to be $\alpha[st^{-1}(M)]$. A similar argument shows that the μ-inner measure of $\mathbb{R}^d \setminus M$ equals $\alpha[st^{-1}(\mathbb{R}^d \setminus M)]$ and finiteness of μ forces M to be μ-measurable because the μ-outer measure of M is $\alpha[st^{-1}(M)]$.

The reader should notice that it is easier to show that a completed Borel algebra lifts via st^{-1} to α-measurable sets by an argument like Theorem (2.1.2) than it is to give Henson's result (2.2.6). We could have done section (2.3) without (2.2) if we used only complete measures.

The function version of this theorem is left as an exercise. The following is an integrated version.

(2.3.5) A CHANGE OF VARIABLES THEOREM:

Let $\mu = \alpha \circ st^{-1}$ *as above. If* $f : \mathbb{R}^d \longrightarrow \mathbb{R}$ *is* μ*-integrable, then* $\int_B f(r)d\mu(r) = \int_{st^{-1}(B)} f(st(t))d\alpha(t)$.

PROOF:

If f is an indicator function of a μ-measurable set,

then the result follows from (2.3.4)(a). Otherwise break f into positive and negative parts f^+, f^-. Let g_n be a sequence of simple functions such that g_n increases to f^+ (resp. f^-). We know that

$$\int g_n(r)d\mu(r) = \int g_n(st(t))d\alpha(t)$$

and

$$\int g_n(r)d\mu(r) \longrightarrow \int f^+(r)d\mu(r).$$

Since $\alpha[\{t : g_n(st(t)) \longrightarrow f(st(t))\}] = \alpha[st^{-1}\{r : g_n(r) \longrightarrow f(r)\}]$, we may apply the Monotone Convergence Theorem to complete the proof.

When $f(r)$ is a Borel or μ-measurable function on $\mathbb{R}^d$, the function $f(st(t))$ is a Loeb or α-measurable function on $\mathbb{T}^d$, but it need not be internal. If we seek an internal replacement g for f, we want to make the diagram

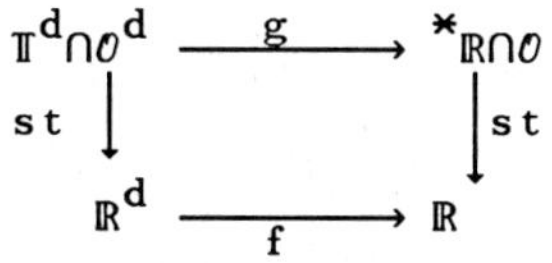

$$\begin{array}{ccc} \mathbb{T}^d \cap \mathcal{O}^d & \xrightarrow{\ g\ } & {}^*\mathbb{R} \cap \mathcal{O} \\ st \downarrow & & \downarrow st \\ \mathbb{R}^d & \xrightarrow[\ f\ ]{} & \mathbb{R} \end{array}$$

α-almost commute. See (2.1.3) for the definition of an α-almost S-continuous lifting.

(2.3.6) THEOREM:

Let $\mathbb{T}^d$ *be an S-dense* **finite subset of* $^*\mathbb{R}^d$ *with a* **finite measure* α. *Let* μ *be a completed Borel measure such that* $\mu = \alpha \circ st^{-1}$.

a) *A function* $f : \mathbb{R}^d \longrightarrow \mathbb{R}$ *is* μ*-measurable if and only if there is an internal function* $g : \mathbb{T}^d \longrightarrow {}^*\mathbb{R}$ *with* $\alpha[\{t \in \mathbb{T}^d \mid st(g(t)) \neq f(st(t))\}] = 0$.

b) *A function* $f : \mathbb{R}^d \longrightarrow \mathbb{R}$ *is* μ*-integrable if and only if there is an* α*-S-integrable* $g : \mathbb{T}^d \longrightarrow {}^*\mathbb{R}$ *with* $\alpha[st(g(t)) \neq f(st(t))] = 0$. *In this case*

$$\int_M f(r)d\mu(r) = \int_{st^{-1}(M)} \tilde{g}(t)d\alpha(t)$$

for every μ*-measurable set* M.

PROOF:

Left as an exercise with the hint to apply a chapter 1 lifting to the function $f(st(t))$.

(2.3.7) THEOREM:

Let $\mu = \alpha \circ st^{-1}$ *as above. If* $f : \mathbb{T}^d \longrightarrow {}^*\mathbb{R}$ *is an* α*-S-integrable internal function, then there is a Borel measurable* $g : \mathbb{R}^d \longrightarrow \mathbb{R}$ *such that*

$$\int_B g d\mu = \int_{st^{-1}(B)} \tilde{f} d\alpha \approx \Sigma[f(t)\alpha(t) : t \in A],$$

for all Borel sets B *and internal* A *such that* $\alpha[A \triangledown st^{-1}(B)] = 0$.

PROOF:

Let $\upsilon(B) = \int_{st^{-1}(B)} \tilde{f}(t)d\alpha(t)$ define a standard Borel measure. Since $\upsilon << \mu$ we know that there is a μ-a.s. unique Borel measurable Radon-Nikodym derivative $g = \frac{d\upsilon}{d\mu}$ such that $\int g d\mu = \upsilon(B)$.

(2.3.8) **EXAMPLE:**

Rapidly oscillating internal functions cannot be liftings of Borel functions. For example, let $\delta t = \frac{1}{n}$ for some infinite $n \in {}^*\mathbb{N}$ and let $\mathbb{T}$ be the uniform hyperfinite space $\{k \cdot \delta t \mid 0 < k \leq n,\ k \in {}^*\mathbb{N}\}$ with weights $\delta\alpha(t) = \delta t$ for all $t \in \mathbb{T}$. The function $f : \mathbb{T} \longrightarrow \{0,1\}$ given by

$$f(t) = \begin{cases} 0, & \text{if } \frac{t}{\delta t} \text{ is odd} \\ 1, & \text{if } \frac{t}{\delta t} \text{ is even} \end{cases}$$

has the property that for any interval of finite length, $[a,b)$,

$$\Sigma[f(t)\delta t \ : \ t \in \mathbb{T} \cap {}^*[a,b)] \approx \tfrac{1}{2}(b-a).$$

Hence, $\int_B \frac{1}{2}\, dr = \int_{st^{-1}(B)} f(t)d\alpha(t)$, for all Borel sets B. In some sense, the constant $\frac{1}{2}$ is as close as a Borel function can be to f.

(2.4) Weak Standard Parts of Measures

This section is related to probabilistic "convergence in distribution" below.

(2.4.1) NOTATION:

Let $BC(\mathbb{R}^d)$ *denote the Banach space of bounded continuous real-valued functions defined on* $\mathbb{R}^d$. *The uniform norm* $\|\varphi-\psi\|_u = \sup[|\varphi(x)-\psi(x)| : x \in \mathbb{R}^d]$ *induces the uniform infinitesimal relation on* ${}^*[BC(\mathbb{R}^d)]$ *given by* $\varphi \overset{u}{\approx} \psi$ *if and only if* $\varphi(x) \approx \psi(x)$ *for all* $x \in {}^*\mathbb{R}^d$. *A function* $A : {}^*BC(\mathbb{R}^d) \longrightarrow {}^*\mathbb{R}$ *is (uniform-norm) S-continuous if* $\varphi \overset{u}{\approx} \psi$ *implies* $A(\varphi) \approx A(\psi)$.

Now let $\mathbb{T}^d \subseteq {}^*\mathbb{R}^d$ be a *finite set and let $\delta\alpha : \mathbb{T}^d \longrightarrow {}^*\mathbb{R}$ be an internal signed weight function. We may view the measure α as acting on ${}^*BC(\mathbb{R}^d)$ by letting

$$\alpha\langle\varphi\rangle = \Sigma[\delta\alpha(t)\varphi(t) : t \in \mathbb{T}^d].$$

(2.4.2) PROPOSITION:

For α *and* $\mathbb{T}^d$ *as above, suppose* $\Sigma[|\delta\alpha(t)| : t \in \mathbb{T}^d]$ *is finite. Then* $\varphi \overset{u}{\approx} \psi$ *implies* $\alpha\langle\varphi\rangle \approx \alpha\langle\psi\rangle$, *that is,* $\alpha\langle\ \rangle$ *is S-continuous on* ${}^*BC(\mathbb{R}^d)$.

PROOF:

Since $\varphi(x) \approx \psi(x)$ for all x in ${}^*\mathbb{R}^d$, there is an infinitesimal $\iota > |\varphi(x)-\psi(x)|$ for all x in ${}^*\mathbb{R}^d$. Then

$$|\alpha\langle\varphi\rangle-\alpha\langle\psi\rangle| \lesssim \Sigma[|\delta\alpha(t)||\varphi(t)-\psi(t)| : t \in \mathbb{T}^d]$$

$$\lesssim \iota\Sigma[|\delta\alpha)t)|] \approx 0.$$

The S-continuity of α means that we can associate α with an element of the continuous dual of $BC(\mathbb{R}^d)$. We can also associate α with a countably additive measure α on $\mathbb{T}^d$. We saw in the last section how every countably additive Borel measure can be represented by such α's and this section explores one aspect of the opposite direction.

(2.4.3) **EXAMPLE:**

Let t_0 be an unlimited or infinite element of $\mathbb{T}^d$ and let $\delta\alpha : \mathbb{T}^d \longrightarrow [0,1]$ be the indicator function of $\{t_0\}$. Consider the externally defined standard functional $\beta : BC(\mathbb{R}^d) \longrightarrow \mathbb{R}$ given by $\beta(\varphi) = st(\alpha\langle{}^*\varphi\rangle) = st({}^*\varphi(t_0))$ for standard φ. The norm of β is clearly 1. *There is no countably additive measure on* $\mathbb{R}^d$ *that represents* β, since if $\beta(\varphi) = \int\varphi d\mu$ and we pick the monotone decreasing sequence

$$\varphi_n(x) = \begin{cases} 0 & , \quad |x| \lesssim n \\ |x|-n & , \quad n \lesssim |x| \lesssim n+1, \\ 1 & , \quad n+1 \lesssim |x| \end{cases}$$

then $\beta(\lim \varphi_n) = \lim \beta(\varphi_n)$ by the monotone convergence theorem for μ. This is not possible because ${}^*\varphi_n(t_0) = 1$ for finite n, while $\lim \varphi_n \equiv 0$. The "problem" with the α in this example is that it is carried on the unlimited points.

Let $M(\mathbb{R}^d)$ denote the space of finitely or countably additive measures on $\mathbb{R}^d$. Every continuous linear functional on $BC(\mathbb{R}^d)$ can be represented by a *finitely additive* measure and Loeb [1972] shows how to represent all of those on *certain* *finite sets. We will content ourselves with studying positive internal α's on our set $\mathbb{T}^d$ (we can always apply these results to $\alpha^+ - \alpha^-$) since our modest aim is construction of some interesting processes, not representation of all possible ones.

(2.4.4) DEFINITION:

Let $\mathbb{T}^d$ *be a* **finite subset of* $^*\mathbb{R}^d$ *and let* $\delta\alpha : \mathbb{T}^d \longrightarrow {}^*[0,\infty)$ *be a positive internal weight function. We say* α *has near-standard carrier provided that for each internal set* A *containing only unlimited points of* $\mathbb{T}^d$, $\alpha[A] \approx 0$. *We say* α *is an* S-*tite measure if* $\alpha[\mathbb{T}^d]$ *is finite and* α *has near-standard carrier.*

It follows that α has near-standard carrier if and only if $\underline{\alpha}(\mathbb{T}^d \setminus \mathcal{O}^d) = 0$. This is a more compact way of saying it, but we can do even better: $\mathcal{O}^d$ is Loeb, so $\mathbb{T}^d \setminus \mathcal{O}^d$ is Loeb as well. Then by (1.2.25), α has near-standard carrier if and only if $\alpha(\mathbb{T}^d \setminus \mathcal{O}^d) = 0$.

The generalization of this notion to arbitrary Tychonoff spaces can be found in Anderson and Rashid [1978] and Loeb [1979a] as well as in the special extension treated in chapter 5 below for the path spaces C[0,1] and D[0,1]. In non-locally compact spaces a distinction must be made between "limited" and "near-standard" or "unlimited" and "remote," and it is worse than that because the "Baire" vs. "Borel" distinction also enters. (Loeb [1979a] has a nice universal measurability result in Borel sets.) These technicalities distract the uninitiated from the central infinitesimal analysis (and aren't substantially harder for the initiated). Moreover, this form is useful to us in the description of the law of a process.

The next result says that S-tite measures are weakly near countably additive standard measures. See Stroyan & Luxemburg [1976, chapts. 8 & 10] for the infinitesimal functional analysis jargon; we will show that the infinitesimal relation holds.

(2.4.5) **PROPOSITION:**

If α *is an* S-*tite measure on a* **finite set* $\mathbb{T}^d \subseteq {}^*\mathbb{R}^d$, *then the countably additive measure* $\mu_\alpha = \alpha \circ st^{-1}$ *is the (weak-star)* $\sigma(M(\mathbb{R}^d), BC(\mathbb{R}^d))$-*standard part of* α, *that is, for each standard* $\varphi \in BC(\mathbb{R}^d)$

$$\Sigma\, \varphi(t)\delta\alpha(t) \approx \int \varphi(x)d\mu_\alpha(x).$$

PROOF:

First of all, what does near-standard carrier have to do with it? Clearly, $\alpha \circ st^{-1}$ is *always* a Borel measure—but notice that it is zero for example (2.4.3), whereas $\sigma_{BC(\mathbb{R}^d)}$ does not "see it" as zero.

We always have $\Sigma\varphi(t)\alpha(t) \approx \int \tilde{\varphi}(t)d\alpha(t)$ by general results of Chapter 1, since φ and α are finite.

Since $\alpha(\mathbb{T}^d)$ is limited and for every infinite n, $\alpha[|t| > n] \approx 0$, for standard positive ϵ there is a finite m such that $n > m$ implies $\alpha[|t| \le n] > \alpha(\mathbb{T}^d)-\epsilon$. Thus,

$\alpha[\mathcal{O}^d \cap \mathbb{T}^d] = \alpha[st^{-1}(\mathbb{R}^d)] \approx \alpha[\mathbb{T}^d]$, so $\int \tilde{\varphi}(t)d\alpha(t) = \int_{\mathcal{O}^d} \tilde{\varphi}(t)d\alpha(t)$.

A standard bounded continuous function is finite and S-continuous at each (near-standard or) limited t; this means $\tilde{\varphi}(t) = \varphi(st(t))$, or $\tilde{\varphi}$ is constant on the infinitesimal neighborhood of t. Combining this with the last remark, we see that $\tilde{\varphi}(\cdot) = \varphi(st(\cdot))$, [a.e. α]. Now we know $\int \tilde{\varphi}(t)d\alpha(t) = \int \varphi(st(t))d\alpha(t)$. The Change of Variables Theorem (2.3.5) shows that $\int \varphi(st(t))d\alpha(t) = \int \varphi(x)d\mu_\alpha(x)$, so we see that $\Sigma\varphi(t)\alpha(t) \approx \int \varphi(x)d\mu_\alpha(x)$. (Notice that st^{-1}(Borel) is Loeb.)

(2.4.6) PROPOSITION:

Let $\mathbb{T}^d \subseteq {}^*\mathbb{R}^d$ be *finite and let $\{\alpha_k : k \in {}^\sigma\mathbb{N}\}$ be a sequence of positive internal functions on $\mathbb{T}$. The following are equivalent:

a) α_k converges weakly to a standard countably additive finite Borel measure μ, that is, for each $\varphi \in BC(\mathbb{R}^d)$,

$$\text{S-lim } \Sigma\varphi(t)\alpha_k(t) = \int \varphi(x)d\mu(x).$$

b) For each internal extension $\{\alpha_k : k \in {}^*\mathbb{N}\}$ to a sequence of functions on $\mathbb{T}^d$, there exists an infinite n such that for all infinite $m \leq n$, α_n is S-tite and $\alpha_m(st^{-1}(\ \ )) = \alpha\ (st^{-1}(\ \ )) = \mu$.

PROOF:

(b) implies (a) by the last result and the internal definition principle, since all sufficiently small infinite elements are within epsilon.

Notice that we do not assume that the α_k are S-tite for finite k; the linear functionals $\alpha_k\langle\varphi\rangle = \Sigma\varphi(t)\alpha_k(t)$ may only be finitely additive when k is finite.

The converse part is left as an exercise with these suggestions: μ is regular, so it is approximately carried on a compact set. The compact set's indicator function can be approximated by continuous functions. This leads to S-titeness. The continuous functions can distinguish different Borel measures.

CHAPTER 3: PRODUCTS OF HYPERFINITE MEASURES

Let U and V be *finite sets and let $\delta\mu : U \longrightarrow {}^*[0,\infty)$ and $\delta\upsilon : V \longrightarrow {}^*[0,\infty)$ be internal weight functions with $\mu[U] = \Sigma\delta\mu(u)$ and $\upsilon[V] = \Sigma\delta\upsilon(v)$ limited in $\mathcal{O}$. We may apply the procedure of chapter 1 to μ and υ and then form the complete product measure associated with the hyperfinite measures $\boldsymbol{\mu}$ and $\boldsymbol{\upsilon}$, $\boldsymbol{\mu} \times \boldsymbol{\upsilon}$ on $U \times V = \{(u,v) : u \in U, v \in V\}$. Alternately, we may first construct the internal product, letting

$$\delta\pi((u,v)) = \delta\mu(u)\delta\upsilon(v),$$

then form the *finite measure $\pi = \Sigma\delta\pi$ and the hyperfinite extension, $\boldsymbol{\pi}$. This is a measure on $U \times V$ and we know $\pi[A\times B] = \mu[A]\upsilon[B]$ for internal $A \subseteq U$, $B \subseteq V$. This is just a simple property of *summation. The measure $\boldsymbol{\pi}$ is an extension of $\boldsymbol{\mu} \times \boldsymbol{\upsilon}$, but it is a proper extension (in general) and yet still has a Fubini-type theorem. This chapter simply explains the last statement. We only work with bounded hyperfinite measures.

(3.1) Anderson's Extension

We begin with a brief description of the complete product of two bounded hyperfinite measure spaces $(U,\mathrm{Meas}(\boldsymbol{\mu}),\boldsymbol{\mu})$ and $(V,\mathrm{Meas}(\boldsymbol{\upsilon}),\boldsymbol{\upsilon})$. The reader can find the details of this construction in Royden [1968], for example.

(3.1.1) **DEFINITION:**

The complete product $\mu \times \nu$ *on* $U \times V$ *of two bounded hyperfinite measures is the unique countably additive extension (Caratheodory extension) of the function* $\mu(A)\nu(B) = \mu \times \nu(A \times B)$ *defined for measurable rectangles* $A \times B$, *for* A *in* $\text{Meas}(\mu)$ *and* B *in* $\text{Meas}(\nu)$. *We shall denote the complete sigma algebra containing the measurable rectangles by* $\mathcal{C}(\mu \times \nu)$.

To apply Caratheodory extension one must show that $\mu \times \nu$ is countably additive in the case where a countable union of disjoint rectangles is a rectangle. Anderson [1976] observed:

(3.1.2) **PROPOSITION:**

The complete product measurable sets, $\mathcal{C}(\mu \times \nu) \subseteq \text{Meas}(\boldsymbol{\pi})$, *where* $\pi = \mu \times \nu$ *is the internal product. Moreover, the restriction,*

$$\boldsymbol{\pi} \mid \mathcal{C}(\mu \times \nu) = \mu \times \nu,$$

the hyperfinite extension $\boldsymbol{\pi}$ *of the internal product,* $\pi = \mu \times \nu$, *is an extension of the complete product of the separate hyperfinite extensions* μ *and* ν.

PROOF:

Take A in $\text{Meas}(\mu)$ and B in $\text{Meas}(\nu)$. We know by (1.2.13) that there are internal sets $C \subseteq U$ and $D \subseteq V$, such that $\mu[A \triangledown C] = \nu[B \triangledown D] = 0$. For internal sets,

$$\pi[C \times D] = \mu[C] \cdot \nu[D],$$

so that

$$\pi[C \times D] = \mu[A]\nu[B] = \mu[C]\nu[D].$$

Moreover, we know π is a complete measure and

$$\mu \times \nu[(A \times B) \triangledown (C \times D)] = 0,$$

because $(A \times B) \triangledown (C \times D) \subseteq [(A \triangledown C) \times (B \cup D)] \cup [(A \cup C) \times (B \triangledown D)]$ and both measures of a rectangle with one zero measure component are zero. (For each $\epsilon > 0$, there are internal sets E, F such that $E \supseteq N$, $F \supseteq M$ and $\mu[E] < \epsilon$ while $\nu[F] \leq 1$, so $\pi[E \times F] < \epsilon \cdots$).

(3.1.3) DEFINITION:

Let $C \subseteq U \times V$. *The sections of* C *are:*

$$C^u = \{v \in V : (u,v) \in C\}, \quad \textit{for each} \quad u \in U$$

and

$$C_v = \{u \in U : (u,v) \in C\}, \quad \textit{for each} \quad v \in V.$$

Let $f : U \times V \longrightarrow [-\infty,\infty]$ *be a function. The sections of* f *are:*

$$f^u : V \longrightarrow [-\infty,\infty] \quad \textit{given by} \quad f^u(v) = f(u,v,),$$
$$\textit{for each} \quad u \in U$$

and

$$f_v : U \longrightarrow [-\infty,\infty] \quad \textit{given by} \quad f_v(u) = f(u,v),$$
$$\textit{for each} \quad v \in V.$$

Our next result is a prelude to a Fubini theorem.

(3.1.4) **PROPOSITION:**

Let Loeb(U), Loeb(V), Loeb(U × V) *denote the Loeb algebras of the respective* *finite *sets as above.*

a) *If* $C \in$ Loeb(U × V), *then each section satisfies* $C^u \in$ Loeb(V) *and* $C_v \in$ Loeb(U).

b) *If* $f : U \times V \longrightarrow [-\infty,\infty]$ *is* Loeb(U × V)-*measurable, then each section* f^u *is* Loeb(V)-*measurable and each* f_v *is* Loeb(U)-*measurable.*

Let μ, ν *and* π *be as above.*

c) *If* $C \in$ Meas(π), *then for almost all* u *in* U, $C^u \in$ Meas(ν) *and for almost all* v *in* V, $C_v \in$ Meas(μ).

d) *If* $f : U \times V \longrightarrow [-\infty,\infty]$ *is* π-*measurable, then*

for almost all u *in* U, f^u *is* ν-*measurable*

and

for almost all v *in* V, f_v *is* μ-*measurable.*

PROOF:

Part a) is proved by observing that the collection of all sets $S \subseteq U \times V$ such that for all u,v both $S^u \in$ Loeb(V) and $S_v \in$ Loeb(U) hold, is a sigma algebra containing the internal sets.

Part b) follows easily from part a).

Part c) can be shown by using (1.2.13). Let D be internal and satisfy $\pi[C \triangledown D] = 0$. We know that the sections of an internal set are internal, so it would be sufficient to prove that $\nu[C^u \triangledown D^u] = 0$ for almost all u in U. We will prove this for an arbitrary null set.

Let N satisfy $\pi[N] = 0$ and let W_m be a decreasing sequence of internal sets $W_m \supseteq N$ with $\pi[W_m] < 1/m^2$, for finite m. By summation we see that

$$\mu[\{u : \nu[W_m^u] > 1/m\}] < 1/m,$$

because

$$\Sigma[\delta\mu(u)\delta\nu(v) : (u,v) \in W_m \ \&\ \nu[W_m^u] > \tfrac{1}{m}] \geq \tfrac{1}{m} \cdot \mu[\{u : \nu[W_m^u] > \tfrac{1}{m}\}].$$

Since $N \subseteq W_m$, $\{u : \bar{\nu}[N^u] > 1/m\} \subseteq \{u : \nu[W_m^u] > 1/m\}$, so that the outer measure $\bar{\mu}\{u : \bar{\nu}[N^u] > 1/m\} \leq 1/m$ for every finite m. This means that the outer measure $\bar{\mu}\{u : \bar{\nu}[N^u] > 0\} = 0$. Hence almost all sections of a set of measure zero themselves have measure zero, so this concludes the proof of part c).

Part d) follows easily from part c), $\{(u,v) : f(u,v) < r\}^u = \{v : f(u,v) < r\} = \{v : f^u(v) < r\}$.

(3.2) Hoover's Strict Inclusion

One of the most interesting hyperfinite probabilities is the uniform type $\delta\mu(u) = 1/{}^{\#}[U]$, for all u in U, an infinite *finite set. The following example of D. N. Hoover shows that the *internal* subsets of $U \times V$ are not all measurable with respect to $\mu \times \nu$, or in $\mathscr{C}(\mu \times \nu)$, when U and V are uniform probabilities. This description of Hoover's example comes from notes of Keisler and uses the following basic facts of probability. (Independence arguments give a more intuitive, but less direct proof.)

(3.2.1) **MARKOV'S INEQUALITY:**

Let (Ω, P) *be a* **finite probability and let* $g : \Omega \longrightarrow {}^*\mathbb{R}$ *be internal. For* $b > 0$, $P[g \geq b] \leq E[g^+]/b$.

PROOF:

$$\begin{aligned}\Sigma[g^+(\omega)P(\omega) : \omega \in \Omega] &\geq \Sigma[g(\omega)P(\omega) : g(\omega) \geq b] \\ &\geq b\Sigma[P(\omega) : g(\omega) \geq b] \\ &\geq bP[g(\omega) \geq b].\end{aligned}$$

(3.2.2) **CHEBYSHEV'S INEQUALITY:**

Let (Ω, P) *be a* **finite probability and let* $f : \Omega \longrightarrow {}^*\mathbb{R}$ *be internal. For any* m *in* $^*\mathbb{R}$ *and* $a > 0$,

$$P[|f - m| \geq a] \leq \frac{E[(f-m)^2]}{a^2}$$

in particular, if $m = E[f]$ *and if we denote the variance of* f *by* $V[f] = E[(f-E[f])^2]$, *then*

$$P[|f-E[f]| \geq a] \leq \frac{V[f]}{a^2}.$$

PROOF:

Let $g(\omega) = (f(\omega)-m)^2$ and let $b = a^2$. Since $g^+ = g$ and $\{\omega : g(\omega) \geq b\} = \{\omega : |f(\omega)-m| \geq a\}$, Markov's inequality gives the result.

Let $\mathbb{T} = \{t \in {}^*\mathbb{R} : 0 < t \leq 1,\ t = k\delta t,\ k \in {}^*\mathbb{N}\}$, where $\delta t = 1/n$ for an infinite n in ${}^*\mathbb{N}$. Let $\Omega = \{-1,1\}^{\mathbb{T}}$ be the set of internal functions $\omega : \mathbb{T} \longrightarrow \{-1,1\}$. We take $U = \mathbb{T}$ and $\delta\mu(t) = \delta t$ for all t, while $V = \Omega$ and $\delta\upsilon(\omega) = \delta P(\omega) = 1/2^n$, so both U and V have uniform probabilities. Now $\delta\pi = \delta\mu\cdot\delta\upsilon = \delta t\cdot\delta P$.

(3.2.3) HOOVER'S HALF:

We call the set

$$H = \{(t,\omega) : \omega(t) = 1\}$$

Hoover's half of $\mathbb{T} \times \Omega$. *It is not measurable with respect to* $\mu \times \upsilon$, *but it is internal, hence* $H \in \text{Loeb}(\mathbb{T} \times \Omega)$ *and* H *is* π*-measurable.*

We suggest that the reader draw a 'picture' of $\mathbb{T} \times \Omega$ as sequences of -1's and 1's lined up over the $\mathbb{T}$-axis. For example, let $\omega_h(k\delta t) = (-1)^{[\![\frac{h+k-1}{k}]\!]}$, where $[\![x]\!]$ denotes the greatest integer less than or equal to x, $1 \leq k \leq n$,

$1 \leq h \leq 2^n$. The signs alternate in the first column, change every second time in the second column, every third row in the third, and so on. Show that $\pi[H] = \frac{1}{2}$.

(3.2.4) **LEMMA:**

If S *is an internal subset of* $\mathbb{T}$ *and* Λ *is an internal subset of* Ω, *then* $\pi[H \cap (S \times \Lambda)] \approx \frac{1}{2}\, \delta t[S]P[\Lambda]$.

PROOF:

Note $\delta t[S] = \mu[S] = {}^{\#}[S]\delta t$. We may assume $\delta t[S] \not\approx 0$. We define

$$g(\omega) = \frac{1}{{}^{\#}[S]} (\Sigma[\omega(s)\sqrt{\delta t} : s \in S])^2$$

so that in case $S \times \{\omega\} \subseteq H$ (that is, $\omega(s) = 1$ for all s in S) then $g(\omega) = \delta t[S]$. (Note the relationship between $B^2(t)$ from (0.2.8), (0.3.12) and g. We develop separate simple estimates here.) The expected value of $g(\omega)$ is:

$$E[g] = \frac{\delta t}{{}^{\#}[S]} \sum_{\omega} \delta P(\omega)(\Sigma[\omega^2(s) : s \in S] + \Sigma[\omega(s)\omega(t) : s \neq t \text{ in } S])$$

$$= \delta t + \frac{\delta t \delta P(\omega)}{{}^{\#}[S]} \sum_{s \neq t} \sum_{\omega} \omega(s)\omega(t) = \delta t.$$

The second sum is zero because it runs through all of Ω. Applying Markov's inequality to g, we see that

$$P[g \geq \epsilon] \leq \delta t/\epsilon,$$

so we let $\epsilon = \sqrt{\delta t}$ and obtain:

$$P[g < \sqrt{\delta t}] \geq 1 - \sqrt{\delta t}.$$

Thus for $\Omega' = \{\omega : g(\omega) < \sqrt{\delta t}\}$ and $\Lambda' = \Lambda \cap \Omega'$ we have $P[\Lambda'] = P[\Lambda \cap \Omega'] \approx P[\Lambda]$. Also, for λ in Λ',

$$g(\lambda) = ({}^{\#}[(S \times \{\lambda\}) \cap H] - {}^{\#}[(S \times \{\lambda\})\setminus H])^2 \frac{\delta t}{{}^{\#}[S]} < \sqrt{\delta t}$$

so that

$$\left|\frac{{}^{\#}[(S \times \{\lambda\}) \cap H]}{{}^{\#}[S]} - \frac{{}^{\#}[S \times \{\lambda\})\setminus H]}{{}^{\#}[S]}\right| < \left[\frac{\sqrt{\delta t}}{\delta t[S]}\right]^{\frac{1}{2}}.$$

Therefore, for each λ in Λ'.

$${}^{\#}[(S \times \{\lambda\}) \cap H]\delta t \approx {}^{\#}[(S \times \{\lambda\}\setminus H]\delta t$$

and

$$\pi[(S \times \Lambda) \cap H] \approx \pi[(S \times \Lambda)\setminus H]$$

while

$$\pi[(S \times \Lambda)] = \pi[(S \times \Lambda) \cap H] + \pi[(S \times \Lambda)\setminus H].$$

This proves the assertion.

(3.2.5) PROPOSITION:

Hoover's half, $H \subseteq \mathbb{T} \times \Omega$, *is not* $\mu \times \nu = \delta t \times P$-*measurable,* $H \notin \mathscr{C}(\mu \times \nu)$.

PROOF:

Since π is an extension of $\mu \times \nu$, if H is measurable, $(\delta t \times P)[H] = \frac{1}{2}$.

The last lemma extends easily to measurable rectangles by (1.2.13); for measurable S' and Λ' take internal S and Λ

with $\delta t[S \triangledown S'] = P[\Lambda \triangledown \Lambda'] = 0$, so $\pi[(S' \times \Lambda') \cap H] \approx \pi[(S \times \Lambda) \cap H] \approx \frac{1}{2}(\delta t \times P)[S \times \Lambda] \approx \frac{1}{2}(\delta t \times P)[S' \times \Lambda']$. The measure $\beta[S' \times \Lambda'] = \pi[(S' \times \Lambda') \cap H]$ on measurable rectangles also extends to $\mathscr{C}(\delta t \times P)$, so if H is measurable, $\beta[H] = \frac{1}{2}(\delta t \times P)[H] = \pi[H]$, a contradiction.

(3.2.6) EXERCISE:

Give another proof that Hoover's half H is nonmeasurable in the complete product along the following lines. If $S \subseteq \mathbb{T}$ and $\Lambda \subseteq \Omega$ are internal sets and $S \times \Lambda \subseteq H$, then $\delta t \times P[S \times \Lambda] \approx 0$ because $\Lambda \subseteq \bigcap_{t \in S} \{\lambda \in \Omega : \lambda(t) = 1\}$ and $P[\Lambda] \leq \prod_{t \in S} P[\lambda(t) = 1]$, by independence, see section (4.3). Either S is finite or $P[\Lambda] \leq 1/2^{\#[S]}$, so in either case, $\delta t \times P[S \times \Lambda] \approx 0$. Use the internal computation to show that the $\delta t \times P$-inner measure of H is zero and its outer measure is one.

(3.3) Keisler's Fubini Theorem

Even though π is a strict extension of $\mu \times \nu$ in general, we have already seen that sections of π-measurable functions are respectively ν- or μ-measurable, (3.1.4). We can also do iterated integration.

(3.3.1) KEISLER'S FUBINI THEOREM:

Let $(U, \delta\mu)$, $(V, \delta\nu)$ and $\delta\pi = \delta\mu \cdot \delta\nu$ be as above where μ and ν are limited hyperfinite measures. If $f : U \times V \longrightarrow [-\infty, \infty]$ is π-integrable, then

a) f^u *is ν-integrable μ-a.s.,*

b) $F(u) = \int f(u,v) d\nu(v)$ *is μ-integrable,*

c) $\int F(u) d\mu(u) = \int f(u,v) d\pi(u,v)$.

PROOF:

Recall that the last part of the proof of (3.1.4) showed that if $N \subseteq U \times V$ has $\pi[N] = 0$, then $\nu[N^u] = 0$ μ-a.s. If f is bounded, then f has a bounded π-lifting g by (1.3.9), $g(u,v) \approx f(u,v)$ for $(u,v) \notin N$, $\pi[N] = 0$. The fact about null sections means that g^u is a ν-lifting of f^u a.s. $\mu(u)$. For these u's, $G(u) = \Sigma[g(u,v)\delta\nu(v) : v \in V] \approx \int f^u d\nu$, so $G(u)$ is a μ-lifting of $F(u)$ and

$$\begin{aligned} \int f d\pi &\approx \Sigma[g(u,v)\delta\mu(u)\delta\nu(v) : u \in U, v \in V] \\ &= \Sigma[G(u)\delta\mu(u) : u \in U] \\ &\approx \int F d\mu. \end{aligned}$$

We obtain the general case from the bounded case by treating f^+ and f^- separately, using linearity and the fact that there are bounded functions $f_k \uparrow f^+$, namely, $f_k = \min[f^+, k]$.

We assume that $f \geq 0$ and $f_k = \min[k, f]$. Since f_k is bounded, for each k there is a μ-null set N_k such that f_k^u is ν-integrable for $u \notin N_k$, so $N = \cup N_k$ has μ-measure zero and f_k^u is ν-integrable for all k if $u \notin N$. By monotone convergence, for $u \notin N$,

$$F_k(u) = \int f_k^u d\nu \uparrow \int f(u,v) d\nu(v) = F(u).$$

Each F_k is μ-integrable and satisfies the theorem, so

$$\int f_k d\pi = \int F_k(u) d\mu(u) \uparrow \int F(u) d\mu(u)$$

by monotone convergence again. Since f is π-integrable, the left side tends to $\int f d\pi$ as well, hence $F(u)$ is μ-integrable and

$$\int f(u,v) d\pi(u,v) = \int [\int f(u,v) d\nu(v)] d\mu(u).$$

The general case follows by linearity since we know it now for f^+ and f^-.

(3.3.2) **EXERCISE:**

For μ, ν, π as above, let $A \subseteq U \times V$ be π-measurable and $0 \leq r \leq 1$. Prove that

$$\{v : \mu[A_v] > r\}$$

is ν-measurable. (Clearly one can reverse the roles of u and v in Fubini's theorem.)

When we study stochastic processes below we will have a product $[0,1] \times \Omega$, where $[0,1]$ carries either the Borel algebra or the Lebesgue algebra and Ω is hyperfinite. In fact, Ω will consist of internal sequences $\omega = (\omega_0, \omega_{\delta t}, \cdots, \omega_1)$ and at times we will want to break Ω into pairs $((\omega_0, \cdots, \omega_{t-\delta t}); (\omega_t, \cdots, \omega_1))$. This results in a three-way product,

$$[0,1] \times \Omega_1 \times \Omega_2,$$

with one classical factor and two hyperfinite factors. We can obtain results about three factors from four factors,

$$A \times B \times U \times V,$$

where U and V are hyperfinite and A and B are arbitrary, by taking B equal to a one-point space. This technical convenience makes the results easier to state.

(3.3.3) A MIXED TYPE FUBINI THEOREM:

Let $(U, \mathrm{Meas}(\mu), \mu)$ *and* $(V, \mathrm{Meas}(\nu), \nu)$ *be bounded hyperfinite measure spaces and let* $U \times V$ *and* $\pi = \mu \times \nu$ *be the internal products. Let* $(A, \mathcal{A}, \alpha)$ *and* $(B, \mathcal{B}, \beta)$ *be arbitrary bounded measure spaces and consider the product*

$$A \times B \times U \times V.$$

a) *If* $H \subseteq A \times B \times U \times V$ *is* $(\mathcal{A} \times \mathcal{B} \times \text{Meas}(\pi))$-*measurable, then, except for a* ν-*null set, the sections,*

$$H(b,v) = \{(a,u) \in A \times U : (a,b,u,v) \in H\}$$

are $(\mathcal{A} \times \text{Meas}(\mu))$-*measurable. We may draw the same conclusion about the sections of a* $(\mathcal{A} \times \mathcal{B} \times \text{Meas}(\pi))$-*measurable function;* $f_{(b,v)}(a,u)$ *is* $(\mathcal{A} \times \text{Meas}(\mu))$-*measurable a.s.* ν.

b) *If* $H \subseteq A \times B \times U \times V$ *(or a function* f*) is* $(\alpha \times \beta \times \pi)$-*measurable, then, except for a* $(\beta \times \nu)$-*null set, the sections* $H(b,v)$ *(resp.* $f_{(b,v)}(a,u)$*) are* $(\alpha \times \mu)$-*measurable, a. s.* $\beta \times \nu$*).*

c) *If* $f : A \times B \times U \times V \longrightarrow \mathbb{R}$ *is* $(\alpha \times \beta \times \pi)$-*integrable, then the sections* $f_{(b,v)}(a,u)$ *are* $(\alpha \times \mu)$-*integrable, a.s.* $\beta \times \nu(b,v)$*, the integral* $F(b,v) = \int f_{(b,v)}(a,u)d(\alpha \times \mu)$ *is* $(\beta \times \nu)$-*integrable, and* $\int F(b,v)d(\beta \times \nu) = \int f(a,b,u,v)d(\alpha \times \beta \times \pi)$*. Moreover, if* f *is* $(\mathcal{A} \times \mathcal{B} \times \text{Meas}(\pi))$-*measurable, then* F *is* $(\mathcal{B} \times \text{Meas}(\nu))$-*measurable.*

First we refresh the reader's memory about the distinction between the measurable sets of a product of measures as in (3.1.2), which is always complete, and the sigma algebra $\mathcal{A} \times \mathcal{B}$, where $\mathcal{A}$ and $\mathcal{B}$ are sigma algebras and completeness with respect to measures isn't mentioned. The collection of pairwise disjoint finite unions of measurable rectangles $A \times B$, with A

in $\mathcal{A}$ and B in $\mathcal{B}$ forms an algebra of sets. *The smallest sigma algebra containing these is denoted* $\mathcal{A} \times \mathcal{B}$. A useful set-theoretical fact about this situation is the next result, which the reader can find in Hewitt and Stromberg [1965, 21.6].

(3.3.4) THE MONOTONE CLASS LEMMA:

Let V *be a set and* $\mathcal{A}$ *be an algebra of subsets of* V. *The sigma algebra generated by* $\mathcal{A}$ *is the smallest family* $\mathcal{F}$ *of subsets of* V *that contains* $\mathcal{A}$ *and is monotone complete, that is, satisfies*

a) *if* $F_m \in \mathcal{F}$, $F_m \subseteq F_{m+1}$, *for* $m = 1,2,\cdots$, *then* $\cup F_m \in \mathcal{F}$

and

b) *if* $F_m \in \mathcal{F}$, $F_m \supseteq F_{m+1}$, *for* $m = 1,2,\cdots$, *then* $\cap F_m \in \mathcal{F}$.

The collection of disjoint unions of measurable rectangles is an algebra, so this result says roughly that a monotone property of rectangles is true in $\mathcal{A} \times \mathcal{B}$. The Monotone Class Lemma is the most convenient tool used in proving the incomplete classical Fubini theorem.

PROOF OF (3.3.3):

The class of sets

$$\{H \in \mathcal{A} \times \mathcal{B} \times \mathrm{Meas}(\pi) : (\exists V_0 \in \mathrm{Meas}(\nu))[\nu[V_0] = 0$$
$$\& \ (v \notin V_0 \Rightarrow H(b,v) \in \mathcal{A} \times \mathrm{Meas}(\mu))]\}$$

is a sigma algebra because countable unions of null sets are null and complements, unions and intersections commute with taking sections. If $H = C \times W$ for $C \in \mathcal{A} \times \mathcal{B}$ and $W \in \text{Meas}(\pi)$, then $H(b,v) = C_b \times W_v \in \mathcal{A} \times \text{Meas}(\mu)$ a.s. $\nu(v)$ by (3.1.4) and a simple sigma algebra argument on the first factor. This proves (a).

Next we prove a special case of (c) for indicator functions as a lemma to establish both (b) and the general case of (c). We wish to show that *if* $H \in \mathcal{A} \times \mathcal{B} \times \text{Meas}(\pi)$ *and we define* $\varphi(b,v) = (\alpha \times \mu)[H(b,v)]$, *letting* $\varphi = 0$ *when* $H(b,v)$ *is not* $(\alpha \times \mu)$*-measurable, then* φ *is* $(\mathcal{B} \times \text{Meas}(\nu))$*-measurable and* $\int \varphi d(\beta \times \nu) = (\alpha \times \beta \times \pi)[H]$. The proof of this is based on the Monotone Class Lemma (3.3.4). We claim that the collection of sets in $\mathcal{A} \times \mathcal{B} \times \text{Meas}(\pi)$ that satisfy this property is a monotone class containing disjoint unions of rectangles. The function $\varphi(b,v) = \alpha[C_b]\mu[W_v]$ when $H = C \times W$, and is $(\mathcal{B} \times \text{Meas}(\nu))$-measurable by the classical and hyperfinite Fubini theorems combined. Moreover, those theorems imply the second part of the property for each factor as well. Since disjoint finite unions of measurable rectangles produce disjoint sums, we have the property holding on an algebra. A monotone limit of sets with the property also has the property by the Dominated Convergence Theorem. Thus, (3.3.4) shows that all sets in $\mathcal{A} \times \mathcal{B} \times \text{Meas}(\pi)$ have the property.

Part (b) follows from our special case, because if K is $(\alpha \times \beta \times \pi)$-measurable, then there is an $H \in (\mathcal{A} \times \mathcal{B} \times \text{Meas}(\pi))$ such that $(\alpha \times \beta \times \pi)[K \nabla H] = 0$, or, in other words, a null H' containing the difference. By our property above, the integral of the sections of $K \triangledown H$ is zero and so almost every section of K differs from the section of H by a null set.

Part (c) follows from the special case because any positive integrable f is a monotone limit of linear combinations of indicator functions, $f_n \uparrow f$. The sections of all f_n are measurable a.s., so the sections of f are. If $F_n(b,v) = \int f_n(b,v,\cdot)d,\ (\alpha \times \mu)$, then $F_n \uparrow F$ by the Monotone Convergence Theorem, so F is measurable. Last, $\iint f_n = \int F_n \longrightarrow \int F = \iint f$, so the second property is satisfied. Finally, we may decompose any integrable f into positive and negative parts to obtain the general case of (c).

CHAPTER 4: DISTRIBUTIONS

In this chapter we let (Ω, P) be a *finite probability space and adhere to the expected value notation (1.5.4) rather than the general measure notation. One "definition" of probability is: 'the study of invariants of distribution'. Another "definition" of probability is: 'measure theory'. On hyperfinite spaces equal distributions imply measure-theoretic equivalence by reshuffling the points of the space; moreover, this can be extended to processes. This chapter only gives the basics of distributions.

(4.1) One Dimensional Distributions

In accordance with customs in probability we shall call an internal function $X : \Omega \longrightarrow {}^*\mathbb{R}$ an *internal random variable* and a P-measurable function $Y : \Omega \longrightarrow [-\infty, \infty]$ a *measurable random variable*. The general results of Chapter 1 apply to lifting and projecting random variables.

(4.1.1) DEFINITION:

Let $X : \Omega \longrightarrow {}^*\mathbb{R}$ *be internal. The cumulative distribution function of* X *is the internal function* $F : {}^*\mathbb{R} \longrightarrow {}^*[0,1]$ *given by*

$$F(x) = F_X(x) = P[\{\omega : X(\omega) \leq x\}]$$

and the characteristic function of X *is the internal function* $f : {}^*\mathbb{R} \longrightarrow {}^*\mathbb{C}$ *given by*

$$f(u) = f_X(u) = E[e^{iuX}],$$

the Fourier transform of $F(x)$.

Notice that since Ω is *finite the internal Stieltjes integral with respect to F is a *finite sum, for example, $\int e^{iux} dF(x) = E[e^{iux}]$, in fact they are the same sum. The standard part of the distribution F may be discrete, singular, absolutely continuous or some of each, depending on the infinitesimal jumps. We already saw in Chapter 2 that every Borel measure has a *finite representation.

To standardize distribution functions, we need to view a limit from above with standard tolerances. If $g(y)$ is a function defined for all $y >> x$ ($y > x$ and $y \not\approx x$), we say

$$\underset{y \downarrow x}{\text{S-lim}}\, g(y) = b$$

if $b \in \mathbb{R}$ and for every $\epsilon >> 0$, there is a $\vartheta >> 0$ such that for all y satisfying $x << y << x + \vartheta$, we have $|g(y) - b| < \epsilon$. If g is internal, the reader can show that this is equivalent to the existence of a $z \approx x$ such that $g(z) \approx b$ and whenever $w \geq z$ and $w \approx z$ then $g(w) \approx g(z)$. (Also see Appendix 1.)

(4.1.2) **PROPOSITION:**

For X *and* $F(x)$ *as above, define*

$$\tilde{F}(x) = \tilde{F}(x) = \underset{y \downarrow x}{\text{S-lim}}\, F(y), \quad \textit{for } x \textit{ in } \mathbb{R}.$$

The function $\tilde{F}$ *is increasing, right continuous, takes values between zero and one and satisfies*

$$\tilde{F}(x) = P[\tilde{X} \leq x]$$

(for the projection of X *already defined).*

PROOF:

$\tilde{X}$ is P-measurable and $P[\tilde{X} \leq x] = \lim_{m\to\infty} \text{st } P[X < x+1/m]$. Increasing x makes the sets larger, hence the probabilities larger in both cases. Right continuity of $\tilde{F}$ is monotone convergence of measures.

We have not shown that $\tilde{F}$ is a distribution function in the strict sense because we have not shown that $P[X < -n] \approx 0$ or $P[X > n] \approx 1$ for every infinite n. It is easy to see that this is equivalent to $P[|\tilde{X}| = \infty] = 0$ or "X is finite almost surely." It is also easy to see that it is equivalent to: $\lim_{m\to\infty} \tilde{F}(-m) = 0$ and $\lim_{m\to\infty} \tilde{F}(m) = 1$. The next result lists these and more equivalent conditions—they mean that F is near the standard distribution $\tilde{F}$ (compare Greenwood and Hersh [1975]).

(4.1.3) **PROPOSITION:**

The following are equivalent (notation as above):

(a) X *is finite almost surely,* $P[|\tilde{X}| < \infty] = 1$.

(b) *For positive infinite* b *in* $^*\mathbb{R}$, $F(-b) \approx 0$ *and* $F(b) \approx 1$.

(b') $\lim_{x\to-\infty} \tilde{F}(x) = 0$ *and* $\lim_{x\to+\infty} \tilde{F}(x) = 1$.

(c) *The internal* **finite measure* $dF(x)$ *on* $^*\mathbb{R}$ *has near-standard carrier and*

$$dF \circ st^{-1} = d\tilde{F}$$

as a Borel measure.

(d) *For each standard bounded continuous* $\varphi \in BC(\mathbb{R})$,

$$E[\varphi(X)] \approx E[\varphi(\tilde{X})].$$

(e) *The function* $f(u)$ *is S-continuous at zero.*

(f) *The function* $f(u)$ *is S-continuous for all* u *and* $st(f(u)) = E[\exp(iu\tilde{X})]$ *if* u *is standard.*

PROOF:

(a) iff (b) iff (b') is the point of the remarks preceding the statement of the result.

(b) implies (c) is clear since the dF measure of positive or negative infinite sets is less than the measure of an infinite interval containing them, say $(-\infty, M)$ with $M = {}^*\sup[x : x \in N]$ in the negative case, and that measure is $F(M) \approx 0$. For the second part of (c), notice that $dF\{st^{-1}(-\infty, b]\} = \text{S-}\lim_{m\to\infty} F(b+1/m) = \tilde{F}(b)$ and use the classical representation of a positive Borel measure by a Lebesgue-Stieltjes measure.

(c) implies (d) by (2.4.5) and the general fact that $E[\varphi(\tilde{X})] = \int_{-\infty}^{\infty} \varphi(x)d\tilde{F}(x)$ (as a classical Stieltjes integral) which follows from a simple partitioning argument.

(d) implies (b): Let

$$\varphi_n(x) = \begin{cases} 0 & , \quad |x| < n \\ |x|-n & , \quad n \le |x| \le n+1 \\ 1 & , \quad |x| > n+1 \end{cases}$$

and apply the Monotone Convergence Theorem to $E[\varphi_n(\tilde{X})]$. We see that for finitely positive ϵ there exists finite m such that $E[\varphi_m(X)] < \epsilon$ for all $n > m$. Hence, for every infinite integer $P|X| > n] \approx 0$.

(b) implies (e): If $u \approx 0$ and $P|X| > 1/\sqrt{|u|} \approx 0$, then when $|X| \le 1/\sqrt{|u|}$, $\exp(iuX) \approx 1$. Since $|\exp(iuX)| = 1$ even when $|X| > 1/\sqrt{|u|}$, $E[\exp(iuX)] \approx 1$ so $f(u) \approx f(0)$.

(e) implies (b) follows from the classical result (4.1.4) which follows next.

(4.1.4) INEQUALITY:

$$F([-u,u]^c) \le \alpha u\int_0^{1/u} (1-\text{Re } f(v))dv$$

where

$$\alpha = 1/\inf\left[\left[1 - \frac{\sin t}{t}\right] : |t| \ge 1\right].$$

PROOF OF (4.1.4):

$$u\int_0^{1/u} (1-\text{Re } f(v))dv \ge u\int_0^{1/u}\int(1-\cos(vx))dF(x)dv$$

$$= \int u\int^{1/u} (1-\cos(vx))dvdF(x)$$

$$= \int\left[1 - \frac{\sin(x/u)}{(x/u)}\right]dF(x)$$

$$\geq \int_{|\frac{x}{u}|\geq 1} \inf\left[\left[1 - \frac{\sin(x/u)}{(x/u)}\right]\right] dF(x)$$

$$= \frac{1}{\alpha} F([-u,u]^c).$$

(b) implies (f): Let δ be positive and infinitesimal. The set $\{\omega : |X| > 1/\sqrt{\delta}\}$ has infinitesimal probability, so

$$|f(u+\delta)-f(u)| \leq E[|\exp(iuX)\ (\exp(i\delta X)-1)|]$$

$$\leq E[|\exp(i\delta X)-1|] \approx 0.$$

This concludes the proof of (4.1.3).

The next result is a simple "universality theorem" or "representation theorem" for a single random variable. Keisler [1980] has given an interesting generalization of this idea for stochastic processes.

(4.1.5) PROPOSITION:

Let Ω *be a uniform* **finite probability space,* $\delta P(\omega) = 1/{}^{\#}[\Omega]$ *and let* $G(x)$ *be a standard distribution function. There is an internal* $X : \Omega \longrightarrow {}^*\mathbb{R}$ *such that* $G = \tilde{F}_X$.

PROOF:

For a finite natural number m define $p_{-m^2} = G(-m)$; $p_h = G\left[\frac{h}{m}\right] - G\left[\frac{h-1}{m}\right]$, for $-m^2 < h \leq m^2$; and $p_{m^2+1} = 1-G(m)$. Use these probabilities to form a partition of the unit interval

$u_{-m^2} = 0, \quad u_h = \sum_{i=-m^2}^{h} p_i$. Let $\{u_k\}$ be a sample of these points without repetitions (in case some $p_i = 0$).

Since Ω is *finite there is an internal bijection $u : \Omega \longrightarrow {}^*[0,1]$ onto the points $\frac{k}{n}$, $1 \leq k \leq n$, for some infinite natural number, n. For example, if $\Omega = \{\omega_1, \omega_2, \cdots, \omega_n\}$, $u(\omega_k) = \frac{k}{n}$. Define $X_m(\omega) = \frac{h}{m}$, whenever $u_{h-1} < u(\omega) \leq u_h$. Since repetitions have been removed from the u_h's, these intervals are finite and

$$P\left[\frac{h-1}{m} < X_m \leq \frac{h}{m}\right] \approx p_h, \quad \text{for} \quad -m^2 < h \leq m^2.$$

Define the sequence

$$\sigma_m = \max\left\{P\left[\frac{j}{m} < X_m \leq \frac{h}{m}\right] - G\left(\frac{h}{m}\right) + G\left(\frac{j}{m}\right) \;:\; -m^2 < j < h \leq m^2\right\}$$

so σ_m is internal and infinitesimal for finite m. By Robinson's Sequential Lemma, there is an infinite m so that σ_m is still infinitesimal. Let $X = X_m$ for this infinite m and let $F = F_X$, so

$$F\left(\frac{h}{m}\right) - F\left(\frac{j}{m}\right) \approx G\left(\frac{h}{m}\right) - G\left(\frac{j}{m}\right)$$

for $-m^2 < j < h \leq m^2$ with m infinite. Since G is a standard distribution function $G(-m) \approx 0$ and $G(m) \approx 1$, so $\tilde{F}$ equals G for all x. Recall that $\tilde{F}(x) = \underset{y \downarrow x}{\text{S-lim}}\, F(y)$, see (4.1.2).

This concludes the proof, but we want to add a few words of

caution about the S-limit that we have just used. The limit above is pointwise and G is standard and right continuous by convention, hence if x is standard and $x' > x$ with $x' \approx x$, then $G(x') \approx G(x)$. If we take $x' = [x]_m = {}^*\min\left[\frac{h}{m} : \frac{h}{m} > x\right]$, we know $F([x]_m) - F(-m) \approx F([x]_m) \approx G(x) - G(-\infty) = G(x)$, in other words, $G(x) = \mathrm{st}\, F([x]_m)$, but only for standard x. Comparing two internal distributions with $\tilde{G} = \tilde{F}$ is more difficult as the reader can see by translating jumps an infinitesimal amount. It is easy to show that if $\operatorname{S-lim}_{y \downarrow x} F(y)$ exists, then there exists $x' \approx x$ such that whenever $x'' \approx x$ and $x'' \geq x'$, then $F(x'') \approx F(x') \approx$ S-limit. These difficulties come up in a more serious way for paths in Chapter 5 and for Stieltjes integrals in Chapter 7 where we show (roughly) that the $[x]_m$ trick always works provided we let m be a small enough infinite number (so you miss the whole gap between translated jumps).

(4.1.6) **PROPOSITION:**

Let (Ω, P) *be a uniform* **finite probability space. Suppose that internal random variables* X *and* Y *are finite a.s. and have distribution functions* F *and* G*, respectively, satisfying* $\tilde{F} = \tilde{G}$*. Then there is an internal bijection* $\sigma : \Omega \longrightarrow \Omega$ *such that* $\tilde{X}(\omega) = \tilde{Y}(\sigma(\omega))$ *a.s.*

PROOF:

An internal bijection σ preserves P because all points have the same weight.

Since $\tilde{F} = \tilde{G}$, for each finite natural number m there is a sequence x_h such that $x_h \approx h/m$ and $F(x_h) \approx G(x_h)$, for $-m^2 < h \le m^2$. Let x_h denote the sequence with any repetitions removed so that $F(x_h) - F(x_{h-1}) \not\approx 0$ (always keeping x_{-m^2} and x_{m^2}, say). Let

$$\Omega_h = \{\omega : x_{h-1} < X(\omega) \le x_h\}$$

and

$$\Lambda_h = \{\omega : x_{h-1} < Y(\omega) \le x_h\}.$$

We know ${}^{\#}[\Lambda_h]/{}^{\#}[\Omega] \approx {}^{\#}[\Omega_h]/{}^{\#}[\Omega]$ because the distributions are nearly the same. For each h we may choose an internal injection σ_m from part of Ω_h to part of Λ_h pairing up $\min[{}^{\#}[\Omega_h], {}^{\#}[\Lambda_h]]$ points. Doing this for all the finite number of pairs only leaves an infinitesimal proportion of points left over and we may take σ_m to be an arbitrary pairing of those points. We have $P[|X(\omega) - Y(\sigma_m(\omega))| \ge 2/m] \approx 0$ for each finite m. Extend the sequence σ_m to an internal sequence of internal transformations and apply Robinsonn's Sequential Lemma to find an infinite m satisfying the inequality $|X(\omega) - Y(\sigma_m(\omega))| < 2/m$ nearly surely. Let $\sigma = \sigma_m$ for such an infinite m.

Keisler and Hoover have a reshuffling theorem for very general stochastic processes. This result just shuffles one variable onto another.

(4.1.7) **INVERSION FORMULA:**

Let $F : {}^*\mathbb{R} \longrightarrow {}^*[0,1]$ *be internal, increasing, and let*

$$f(u) = \int_{-\infty}^{\infty} e^{iux}\, dF(x).$$

For example, F *could be the distribution of internal finite a.s. variable and* f *its characteristic function. If* F *is S-continuous at the two finite numbers* $a < b$, *then for any infinite* $v > 0$,

$$F(b) - F(a) \approx \frac{1}{2\pi} \int_{-v}^{v} \frac{e^{-iua} - e^{-iub}}{iu} f(u)du.$$

PROOF:

We need facts about the integral $I(\alpha,\beta) = \frac{1}{\pi} \int_{\alpha}^{\beta} \frac{\sin w}{w}\, dw$. $I(\alpha,\beta) \approx 1$ if α is negative infinite and β is positive infinite $(I(\alpha,\beta) \approx \frac{1}{2}$ if $\alpha \approx 0$ and β is positive infinite) and $I(\alpha,\beta) \approx 0$ if α and β are infinite of the same sign. First, we rewrite the integral term above using the integral formula for $f(u)$,

$$\frac{1}{2\pi} \int_{-v}^{v} \int_{-\infty}^{\infty} \left[\frac{e^{-iua} - e^{-iub}}{iu}\right] e^{iux}\, dF(x)du$$

$$= \int_{-\infty}^{\infty} \frac{1}{2\pi} \int_{-v}^{v} \frac{e^{iu(x-a)} - e^{iu(x-b)}}{iu}\, dudF(x)$$

$$= \int_{-\infty}^{\infty} \frac{1}{\pi} \int_{-v(x-b)}^{v(x-a)} \frac{\sin w}{w}\, dwdF(x),$$

by transfer of classical formulas. The inside dw-integral is infinitely near 1 for $a \ll x \ll b$ and since F is S-continuous at a and b we may ignore the cases $x \approx a$ and $x \approx b$; the net $dF(x)$-contribution is negligible. The inside dw-integral is infinitesimal when x is finitely outside the interval from a to b. Hence our original integral is within an infinitesimal of

$$\int_a^b dF(x) = F(b) - F(a).$$

(4.1.8) **EXERCISE:**

Suppose $f(u)$ and $g(u)$ are characteristic functions of the internal random variables X and Y defined on the uniform hyperfinite probability space Ω. If $f(u) \approx g(u)$ and both f and g are S-continuous, what is the relationship between X and Y?

(4.2) Joint Distributions, Laws and Independence

In this section we continue to let (Ω, P) be a *finite probability space. We let $\{X(t) : t \in \mathbb{T}\}$ be a family of random variables on Ω. The index set $\mathbb{T}$ and the family itself may be either internal or external.

(4.2.1) DEFINITIONS:

Let $\{X(t) : t \in \mathbb{T}\}$ *be a family of internal random variables. For each finite subset* $\{t_1, \cdots, t_m\} = T \subseteq \mathbb{T}$,

(a) *the joint distribution function of* $\{X(t) : t \in T \subseteq \mathbb{T}\}$ *is*

$$F_T(x_1, \cdots, x_m) = P[X(t_1) \le x_1 \ \& \cdots \& \ X(t_m) \le x_m],$$

(b) *the internal distribution measure of* $\{X(t) : t \in T\}$ *is*

$$\mu_T(A) = P[(X(t_1), \cdots, X(t_m)) \in A],$$

for an internal subset $A \subseteq {}^*\mathbb{R}^m$,

(c) *the joint characteristic function of* $\{X(t) : t \in T\}$ *is*

$$f_T(u) = E[\exp(iu \cdot X(T))]$$

where $u = (u_1, \cdots, u_m)$ *and* $u \cdot X(T) = \sum_{i=1}^{m} u_i X(t_i)$.

Let $\{Y(t) : t \in \mathbb{T}\}$ *be a family of measurable random variables on an arbitrary probability space. Let* $T \subseteq \mathbb{T}$

be finite. The definition of the distribution functions and characteristic functions of finite subfamilies of Y *are formally the same as above. The joint distribution measure of* (Y,T) *is the Borel measure given on Borel-measurable rectangles by:*

$$\upsilon_T(B_1 \times \cdots \times B_m) = P[Y(t_1) \in B_1 \ \& \cdots \& \ Y(t_m) \in B_m].$$

The measure μ_T is related to F_T by repeated Stieltjes integration. We could proceed in a manner similar to the last section . We would define $\tilde{F}_T = S\text{-}\lim_{y_j \downarrow x_j} F_T(x)$ and show that it is the standard distribution of the projected finite family if and only if the random variables are finite almot surely or if and only if μ_T is S-tite. Then the Borel measure $(\mu_T \circ st^{-1})$ could be related to the Lebesgue-Stieltjes integrals of $d\tilde{F}_T$. Finally, S-continuity and standard parts of f_T could be related to finiteness and the characteristic function of the projected random variables. We shall not carry this out for two reasons: first, it is similar to the ideas in section 4.1, second, the effect of the distributions is easier to work with, so (4.2.6) is the main fact we need.

Recall the definition of S-tite measure from (2.4.4).

(4.2.2) PROPOSITION:

If each X(t) *is finite a.s., then each* μ_T *is S-tite.*

PROOF:

Suppose that μ_T is not S-tite for some finite T. Then

there is an internal set of infinite points, $A \subseteq {}^*\mathbb{R}^m \setminus \mathcal{O}^m$, such that

$$P[(X(t_1),\cdots,X(t_m)) \in A] \gg 0.$$

Let $b = {}^*\inf[\max_j(|a_j| : a = (a_1,\cdots,a_m) \in A]$, a max-norm lower bound for A. This number b exists and is infinite because no finite point a' is in A and A is internal. The probability above is smaller than the following sum

$$\sum_{j=1}^{m} P[|X(t_j)| \geq b].$$

Hence, either an $X(t_j)$ is not finite a.s. or we have the contradiction that a finite sum of infinitesimals is not infinitesimal.

(4.2.3.M) DEFINITION:

Let $\{Y(t) : t \in \mathbb{T}\}$ *be a family of measurable random variables. The law of the family of random variables,* $\mathrm{Law}(Y,\mathbb{T})$, *is the collection of all the (finite dimensional real-valued joint) distributions from the family.*

If we select the Borel measure representation (cf. (4.2.1)) we can take

$$\mathrm{Law}(Y,\mathbb{T}) = \{\upsilon_T : T \text{ is a finite subset of } \mathbb{T}\}.$$

This notion automatically has a *transform for the internal *families* of random variables, but we usually do not wish to consider the whole *Law even in this case.

(4.2.3.I) **DEFINITION:**

Let $\{X(t) : t \in \mathbb{T}\}$ *be a family of internal random variables. The family (or indexing* $t \longrightarrow X(t)$) *may be external. The* S-*law of the family of random variables,* S-Law$(X,\mathbb{T})$, *is the collection of all the finite distributions from the family.*

Even if $X(t)$ is an internal family, S-Law$(X,\mathbb{T})$ is the external subset of *Law$(X,\mathbb{T})$ given by (cf. (4.2.1))

$$\text{S-Law}(X,\mathbb{T}) = \{\mu_T : T \text{ is a finite subset of } \mathbb{T}\}.$$

Notice that we do NOT allow T to be an unlimited *finite set. If the family $\{X(t) : t \in \mathbb{T}\}$ is internal,

$$^*\text{Law}(X,\mathbb{T}) = \{\mu_T : T \text{ is a } {}^*\text{finite subset of } \mathbb{T}\}.$$

(4.2.4) **DEFINITION:**

Let X *and* Z *be families of internal random variables indexed by* $\mathbb{T}$ *with internal distribution measures* μ_T^X *and* μ_T^Z, *respectively, for* T *finite in* $\mathbb{T}$. *Let* Y *be a family of measurable random variables with the same index set* $\mathbb{T}$ *and Borel distribution measures* ν_T *for* T *finite in* $\mathbb{T}$. *We shall say that the laws of* X *and* Z *are infinitely close and near-standard,*

$$\text{S-Law}(X,\mathbb{T}) \approx \text{S-Law}(Z,\mathbb{T}),$$

if all their finite dimensional distributions are S-*tite and pairwise have the same weak-star-standard-part,*

$$\mu_T^X \circ st^{-1} = \mu_T^Z \circ st^{-1}.$$

We shall say that the laws of X *and* Y *are infinitely close*

$$\text{S-Law}(X,\mathbb{T}) \approx \text{Law}(Y,\mathbb{T}),$$

if for each finite $T \subseteq \mathbb{T}$, μ_T^X *is* S-*tite and*

$$\mu_T^X \circ st^{-1} = \nu_T.$$

We studied the relationship between Borel measures and the inverse standard part in section (2.3). Now we may apply that to distributions.

(4.2.5) **PROPOSITION:**

If each X(t) *is finite a.s., then*

$$\text{S-Law}(X,\mathbb{T}) \approx \text{Law}(\tilde{X},\mathbb{T}).$$

PROOF:

By(4.2.2) each μ_T has finite carrier or is S-tite, so we only need to show that $\mu_T \circ st^{-1}$ is the distribution measure ν_T of $\tilde{X}$. Checking on simple intervals, we see

$$\nu_T[(-\infty,a_1] \times \cdots \times (-\infty,a_m]]$$

$$= P[\tilde{X}(t_1) \leq a_1 \ \& \cdots \& \ \tilde{X}(t_m) \leq a_m]$$

$$= \underset{k\to\infty}{\text{S-lim}}\ P[X(t_1) < a_1 + \frac{1}{k} \ \& \cdots \& \ X(t_m) < a_m + \frac{1}{k}]$$

$$= \mu_T \circ st^{-1}[(-\infty,a_1] \times \cdots \times (-\infty,a_m]].$$

Once the Borel measures ν_T and $\mu_T \circ st^{-1}$ agree on the intervals, they must agree on all Borel sets.

We could rephrase equal and infinitesimal laws by looking at characterizations in terms of distribution functions or characteristic functions, but the following is the most useful characterization for us:

(4.2.6) PROPOSITION:

Let $\{X(t) : t \in \mathbb{T}\}$ *be a family of internal random variables on a* **finite probability space and let* $\tilde{X}(t)$ *denote the projection of* $X(t)$ *on the hyperfinite space* (Ω,P). *The following are equivalent:*

(a) *each* $X(t)$ *is finite a.s. and*

$$\text{S-Law}(X,\mathbb{T}) \approx \text{Law}(\tilde{X},\mathbb{T}),$$

(b) *for each finite* $\{t_1, \cdots, t_m\} = T \subseteq \mathbb{T}$ *and each standard bounded continuous real-valued function* φ *of* m *real variables,*

$$E[\varphi(X(t_1), \cdots, X(t_m))] \approx E[\varphi(\tilde{X}(t_1), \cdots, \tilde{X}(t_m))].$$

PROOF:

(a) implies (b) is essentially our definition of close laws. If the $X(t)$ are finite a.s., then each *measure in the law is S-tite and (2.4.5) shows (b) since $X_t \in st^{-1}(B)$ if and only if $st\, X_t \in B$, so $\mu_T^X \circ st^{-1} = \mu_T^{\tilde{X}}$.

(b) implies (a) because we can approximate the indicator function I_a of each rectangle $A = (-\infty, a_1] \times \cdots \times (-\infty, a_m]$ by a sequence of continuous functions $0 \leq \varphi_n \leq 1$ with $\varphi_n \downarrow I_x$ and $\varphi_n(x) = 1$ if $x_j < a_j + \frac{1}{n}$. Condition (b) means that $\mu_T^X(st^{-1}(A)) = \mu_T^{\tilde{X}}(A)$.

(4.2.7) DEFINITION:

Let $X_1, \cdots, X_m$ *be a finite set of random variables. They are said to be independent if one of the following equivalent conditions holds:*

$$\text{(a)} \quad F_1(x_1)F_2(x_2)\cdots F_m(x_m) = F_{\{1,\cdots,m\}}(x_1, \cdots, x_m),$$

the joint distribution function is the product of the separate distributions.

(b) *The joint distribution measure is the product of the individual distribution measures.*

(c) *The characteristic functions multiply:*

$$f_{\{1,\cdots,m\}}(u_1,\cdots,u_m) = f_1(u_1)f_2(u_2)\cdots f_m(u_m).$$

(d) *For bounded continuous real-valued functions* $\varphi_1,\cdots,\varphi_m$

$$E[\varphi_1(X_1)]E[\varphi_2(X_2)]\cdots E[\varphi_m(X_m)] = E[\varphi_1(X_1)\varphi_2(X_2)\cdots\varphi_m(X_m)].$$

The m-fold extension of the inversion formula from the last section can be used to prove that (c) implies (a) and (b). The equivalence of (a) and (b) is through repeated Stieltjes integration. Condition (a) implies (d) by Fubini's theorem. Notice that (c) is essentially a complexified special case of (d), $e^{i\vartheta} = \cos\vartheta + i\sin\vartheta$.

(4.2.8) DEFINITION:

A family of random variables $\{X(t) : t \in \mathbb{T}\}$ *is said to be an independent family if each finite subfamily is independent.*

Our next result is a basic fact that we shall put to work in the next section.

(4.2.9) **PROPOSITION:**

Let (Ω, P) *be a* **finite probability space. If* X *and* Y *are independent internal random variables with distribution functions* F *and* G *and characteristic functions* f *and* g, *respectively, then the distribution function of* $Z = X + Y$ *is:*

$$P[X+Y \leq z] = \int_{-\infty}^{\infty} G(z-x)dF(x), \quad \textit{the convolution,}$$

and the characteristic function of Z *is:*

$$E[e^{iu(X+Y)}] = f(u)g(u), \quad \textit{the product.}$$

PROOF:

The characteristic function follows easily from (4.2.7)(d),

$$E[e^{iu(X+Y)}] = E[e^{iuX}\, e^{iuY}] = f(u)g(u).$$

Moreover this formulation easily extends to any finite number of independent variables: the characteristic function of the sum is the product of the characteristic functions.

The convolution formula is easy enough in the *finite setting because X and Y only take on *finite sets of values, hence $P[X+Y \leq z] = \sum_{x}[P[Y \leq z-x]P[X = x] : X = x]$.

(4.3) Some *Finite Independent Sums

Random variables (or indicator functions of events) that are functions of separate factors in a product of probability spaces are independent. For example, suppose $\Omega = \Omega_1 \times \Omega_2$ and $P = P_1 \times P_2$, while $X(\omega_1,\omega_2) = f(\omega_1)$ and $Y(\omega_1,\omega_2) = g(\omega_2)$. Then $P[X \le x \ \& \ Y \le y] = P[\{\omega_1 : f(\omega_1) \le x\} \times \{\omega_2 : g(\omega_2) \le y\}] = P_1[f \le x]P_2[g \le y] = P[X \le x]P[Y \le y]$, so X and Y are independent. In this section we give some examples of *finite extensions of this idea.

Thruout the section we let $\delta t = \frac{1}{n}$ for infinite $n = h!$ in $^*\mathbb{N}$,

$$\mathbb{T} = \{t \in {}^*\mathbb{R} : t = k\delta t,\ k \in {}^*\mathbb{N},\ 0 < t \le 1\}.$$

$$W = \{w \in {}^*\mathbb{N} : 1 \le w \le n^n\}$$

$$\Omega = \{\omega : \mathbb{T} \longrightarrow W\} = \text{internal functions from } \mathbb{T} \text{ into } W.$$

$\delta P(\omega) = \frac{1}{{}^\#[\Omega]}$, the uniformly weighted *finite probability on Ω.

(4.3.1) PROPOSITION:

Suppose $f(t,w)$ *is an internal function from* $\mathbb{T} \times W$ *into* $^*\mathbb{R}$. *The internal family* $X(t,\omega) = f(t,\omega_t)$ *is* **independent, that is, if* x_t *is an internal function from* $\mathbb{T}$ *into* $^*[0,\infty]$, *then*

$$P\{\omega : X(t,\omega) \le x_t \text{ for all } t\} = \Pi[P\{X(t) \le x_t\} : t \in \mathbb{T}].$$

In particular, we may take $x_t = \infty$ *for all but finitely many factors, so*

$$P[X(t_1) \leq x_1 \ \& \cdots \& \ X(t_m) \leq x_m] = \prod_{j=1}^{m} P[X(t_j) \leq x_j].$$

PROOF:

Each separate constraint $X(t,\omega) \leq x_t$ is only a condition on the $t^{\underline{th}}$ factor,

$$\{\omega : X(t,\omega) \leq x_t\} = \prod_{s<t} W \times \{w \in W : f(t,w) \leq x_t\} \times \prod_{s>t} W,$$

so that the net internal constraint is a *finite rectangle,

$$\{\omega : X(t,\omega) \leq x_t \text{ for all } t\} = \Pi[\{w : f(t,w) \leq x_t\} : t \in \mathbb{T}].$$

By *transform of the cardinality of a product of factors, P is the internal product of the uniform probability on W taken $^{\#}[\mathbb{T}]$-times. The product formula follows from this since the measure of a rectangle is the product of the measures of the factors.

(4.3.2) EXAMPLE: [Compare to (0.2.11).]

Let $\beta : W \longrightarrow \{-1,1\}$ be an internal function that equals -1 on one half of W and equals $+1$ on the other half. Then $\beta(\omega_t) = f(t,\omega_t)$ forms an independent family and

$$B(t,\omega) = \Sigma[\sqrt{\delta t}\ \beta(\omega_s) : 0 < s \leq t,\ s \in \mathbb{T}]$$

is a sum of independent factors. Hence, by (4.2.6) its characteristic function,

$$E[\exp(iuB(t))] = \{E[\exp(iu\sqrt{\delta t}\ \beta(w))]\}^{t/\delta t}.$$

We simply compute:

$$\begin{aligned} E[\exp(iu\sqrt{\delta t}\ \beta(w))] &= \tfrac{1}{2}[e^{iu\sqrt{\delta t}} + e^{-iu\sqrt{\delta t}}] \\ &= \cos(u\sqrt{\delta t}) \\ &= 1 - \frac{u^2}{2}\delta t + \delta t\eta], \quad \text{for} \quad \eta \approx 0. \end{aligned}$$

We thus obtain the characteristic function

$$E[\exp(iuB(t))] \approx [1 - \frac{u^2}{2}\delta t]^{t/\delta t} \approx e^{\frac{-u^2 t}{2}}.$$

By (4.1.3), the inversion formula and the classical computation that the charcteristic function of the normal law is $e^{\frac{-u^2 t}{2}}$,

$$\frac{1}{\sqrt{2\pi t}} \int_{-\infty}^{\infty} e^{iux}\, e^{\frac{-x^2}{2t}}\, dx = e^{\frac{-u^2 t}{2}},$$

we arrive at the conclusion of De Moivre's limit theorem (0.3.12) by a simpler if less direct route. This idea can be generalized to prove a more general central limit theorem.

(4.3.3) **EXAMPLE:** [Compare to (0.3.6).]

Let a be any standard positive real number. Define the internal function $\pi(w) = 1$ for $1 \leq w \leq \frac{an^n}{n}$ and $\pi(w) = 0$ otherwise. We know that $a\delta t - \frac{1}{n^n} \leq P[\pi(w) = 1] \leq a\delta t$ and since $\frac{n}{n^n}$ is infinitesimal, letting $p = P[\pi(w) = 1]/\delta t$, we have $p \approx a$. By (4.3.1), the family $f(t,\omega_t) = \pi(\omega_t)$ is independent and

$$J(t,\omega) = \Sigma[\pi(\omega_s) : 0 < s \leq t,\ s \in \mathbb{T}]$$

is a sum of independent random variables. By (4.2.6), its characteristic function,

$$E[\exp(iuJ(t))] = \{E[\exp(iu\pi(w))]\}^{t/\delta t}$$

and the inside term,

$$\begin{aligned} E[\exp(iu\pi(w))] &= e^{iu}\ p\delta t + e^{0}(1-p\delta t) \\ &= [1 + p(e^{iu}-1)\delta t]. \end{aligned}$$

We see that

$$E[\exp(iuJ(t))] \approx [1 + a(e^{iu}-1)\delta t]^{t/\delta t} \approx e^{a(e^{iu}-1)}.$$

Again we arrive at a result we have seen before, (0.3.7), by a simple indirect method once we know that the characteristic function of a Poisson distribution is

$$\sum_{k=0}^{\infty} e^{iuk} e^{-at} \frac{(at)^k}{k!} = e^{at(e^{iu}-1)} .$$

The ideas in the last two examples can be extended to give a general representation formula for "infinitely divisible laws" along the lines of the classical Lévy-Itô formula, but combining the continuous and discontinuous parts and using sums of infinitesimal Bernoulli trials instead of integrals with Poisson measures. We shall not give the details of that representation. We give one general result that we shall use in an example in the next chapter and sketch the representation for the Cauchy process in the closing exercise.

(4.3.4) **PROPOSITION:**

Let $X(t) = \Sigma[\delta X(s) : 0 < s \leq t, s \in \mathbb{T}]$, *where the* $\delta X(s)$ *are an internal* *independent *family with the same distribution*, $F(x)$. *Let* $\delta f(u) = E[\exp(iu\delta X)]$, *so that the characteristic function of* $X(t)$ *is* $f(t,u) = [\delta f(u)]^{t/\delta t}$. *If* $X(t)$ *is finite a.s. for finite* t, *then* $\delta f(u) = 1 + \delta t\psi(u)$ *with* $\psi(u)$ *finite and S-continuous for finite* u.

PROOF:

We know $f(1,u)$ is S-continuous by (4.1.3). By the ϵ-ϑ formulation of S-continuity, we know that there is a standard positive ϑ such that $|f(1,u)| > \frac{1}{2}$ for $|u| < \vartheta$. For each finite m, $\sqrt[m]{|f(1,u)|} = [|\delta f(u)|]^{\frac{n}{m}}$ exists and $\text{S-lim}|f(1,u)|^{\frac{1}{m}}$ equals either zero or one for all finite u according to whether $f(1,u) \approx 0$ or not. Hence the limit is one for $|u| < \vartheta$. Since the square absolute value of a charcteristic function is also a characteristic function, we may apply (4.1.3) to see that the S-limit is a standard characteristic function (it is S-conntinuous at zero). The S-limit is continuous, takes the value one and only can take at most one other value, zero; therefore it is identically one, so $f(1,u) \not\approx 0$ for any finite u.

We let

$$\psi(u) = \frac{[\delta f(u)-1]}{\delta t},$$

so that we have

$$f(1,u) = [\delta f(u)]^{1/\delta t} = [1+\delta t\psi(u)]^{1/\delta t}.$$

We know that when ψ is finite in ${}^*\mathbb{C}$,

$$[1+\delta t\psi]^{1/\delta t} \approx e^{\psi}.$$

Moreover, by Robinson's Sequential Lemma, this approximate identity continues to hold on an infinite disk $|\psi| \leq R$. We know that $\psi(0) = 0$ and that $\psi(u)$ is *continuous because $\delta f(u)$ is an internal charcteristic function. Since $f(1,u)$ is S-continuous and noninfinitesimal, it follows that $\psi(u)$ is finite and S-continuous for finite u.

First, suppose $\psi(u_1)$ lies outside $|\psi| \leq R$ for some finite u_1. The *arc $\psi[0,u_1]$ first crosses the boundary $|\psi| = R$ at $u_0 < u_1$ where $\mathrm{Re}[\psi(u)] > \log[\min(|f(1,x)| : 0 \leq x \leq u_0)] - 1$ $(\not\approx -\infty)$ for $0 \leq u \leq u_0$. This is because

$e^{\psi(u)} \approx f(1,u)$ for $|\psi(u)| \le R$ (and because $f(1,x) \not\approx 0$ for $x \le u_0$). When $u \approx v$ in $[0,u_0]$, we must have $\psi(u) - \psi(v) = (\eta + i2k\pi)$ for η infinitesimal and an integer k. Since the *arc has $Re[\psi]$ bounded below, it is easy to show that $k = 0$. This means that ψ is S-continuous on $[0,u_0]$ and it follows that u_1 cannot be finite. Similar reasoning shows that ψ is S-continuous for finite u and since $\psi(0) = 0$, we are done.

(4.3.5) **EXERCISE:**

Show that the terms $\delta X(t)$ in the last proposition are infinitesimal a.s. (Hint: (4.1.4).)

(4.3.6) **EXAMPLE:**

We may view $W = n^n$ as a *finite product. The set of functions from an m-element set into an $n^{[n/m]}$-element set has $n^{\frac{n}{m}\cdot m}$ elements whenever $m \le h$, where $n = h!$ as above. Any finite $m \le h$ is such that $\frac{m}{n}$ and $\frac{n}{n^{[n/m]}}$ are both infinitesimal. Let $\partial x = 1/\sqrt{m}$. Consider the values of x such that $x = k\partial x$, $k \ne 0$, k in $^*\mathbb{Z}$ and $|x| \le \frac{\sqrt{m}}{2}$. There are m such x's, so we may view all w in W as functions $w(x)$ with values in a set with $n^{[n/m]}$ points.

Since $n/n^{[n/m]}$ is infinitesimal for each x as above we may deine an internal *independent family of functions

$$\Gamma(x,w(x)) = \begin{cases} 1\ , & \text{probability } p(x) \\ 0\ , & \text{otherwise} \end{cases}$$

where $p(-x) = p(x) = \frac{1}{\pi}\frac{\partial x}{x^2}\delta t(1+\iota)$, with $\iota \approx 0$.

We let

$$\gamma(w) = \sum_x x\Gamma(x,w(x))$$

and

$$C(t,\omega) = \Sigma[\gamma(\omega_s) : 0 < s \leq t,\ s \in \mathbb{T}].$$

The characteristic function of $C(t,\omega)$ equals $[\delta f(u)]^{t/\delta t}$ where $\delta f(u)$ is the characteristic function of $\gamma(w)$,

$$\delta f(u) = \prod_{x\neq 0}\{1+p(x)[e^{iux}-1]$$

$$= \prod_{x>0}\{1+2p(x)[\cos(ux)-1] + p^2(x)[e^{iux}-1][e^{-ix}-1]\}$$

$$= \prod_{x>0}\{1+2p(x)[\cos(ux)-1][1-p(x)]\}.$$

(4.3.7) **EXERCISE:**

Show that $\delta f(u) = 1+\delta t\left[\sum_{x>0}\frac{1}{2}\frac{[\cos(ux)-1]}{x^2}\partial x+\eta\right]$, $\eta \approx 0$.

Then use the fact that $\frac{2}{\pi}\int_0^\infty \frac{[\cos(ux)-1]}{x^2}dx = -|u|$ to conclude that $[\delta f(u)]^{t/\delta t} \approx e^{-t|u|}$. The last expression is the characteristic function of the Cauchy distribution, $dF_t(x) = \frac{1}{\pi}\frac{t}{t^2+x^2}dx$.

(4.3.8) **EXERCISE:** [Compare to (0.2.11).]

The construction of (4.3.2) is certainly not unique. Define

$$\beta_1(w) = \begin{cases} -1, & 1 \leq w \leq (n^n)/2 \\ +1, & w > (n^n)/2 \end{cases}$$

and also define

$$\beta_2(w) = (-1)^w.$$

We may consider W as the product V^U, where $U = \{-1,1\}$ and $V = \{1,2,\cdots,n^{(\frac{n}{2})}\}$. Of course there are many *bijections of W onto V^U; fix one, $\upsilon : W \longrightarrow V^U$. Define

$$\beta_3(w) = \begin{cases} -1, & \upsilon(w)(-1) \leq n^{(n/2)}/2 \\ +1, & \upsilon(w)(-1) > n^{(n/2)}/2 \end{cases}$$

and

$$\beta_4(w) = \begin{cases} -1, & \upsilon(w)(+1) \leq n^{(n/2)}/2 \\ +1, & \upsilon(w)(+1) > n^{(n/2)}/2. \end{cases}$$

Each β_j defines an infinitesimal random walk:

$$B_j(t,\omega) = \Sigma[\beta_j(\omega_s)\sqrt{\delta t} : 0 < s \leq t, \ s \in \mathbb{T}].$$

Show that $\{\beta_1(\omega_s) : s \in \mathbb{T}\}$ and $\{\beta_2(\omega_s) : s \in \mathbb{T}\}$ are *independent families. Show how to select υ so that $\beta_1 = \beta_3$ and $\beta_2 = \beta_4$. Show how to select υ so that $\beta_1, \beta_2, \beta_3$ and β_4 are all independent.

What is $P[B_1(1) > 0 \mid B_2(1) > 0]$?

For each pair (i,j) show that there is a bijection $\sigma : W \longrightarrow W$ so that $B_i(t,\sigma(\omega)) = B_j(t,\omega)$.

(4.3.9) EXERCISE: [Compare to (0.3.5).]

Use the methods of (4.3.3, 6, 8) to construct four independent (approximate-Poisson) jump processes $J_k(t,\omega)$ with four (possibly) different finite a_k; $k = 1,2,3,4$.

CHAPTER 5: PATHS OF PROCESSES

This chapter studies the paths of processes over a hyperfinite evolution scheme. The idea is to find corresponding properties for stochastic processes defined on $[0,1]$ and internal processes on the infinitesimal time line $\mathbb{T}$.

Hyperfinite evolution will always be relative to the space

$$\Omega = W^{\mathbb{T}} = \{\omega : \mathbb{T} \longrightarrow W, \text{ internal}\}$$

where

$$\mathbb{T} = \{t \in {}^*\mathbb{R} : t = k\delta t,\ k \in {}^*\mathbb{N},\ 0 < t \leq 1\}$$

and

$$\delta t = \frac{1}{n} \text{ for some infinite } n = h! \text{ in } {}^*\mathbb{N}$$

and

$$W = \{k \in {}^*\mathbb{N} : 1 \leq k \leq n^n\}.$$

We take the uniform internal probability with weight function

$$\delta P(\omega) = \frac{1}{{}^{\#}[\Omega]}.$$

A function $X(t,\omega) : \mathbb{T} \times \Omega \longrightarrow {}^*\mathbb{R}$ can also be thought of as a function-valued random variable, $\omega \to X_\omega$, where the section $X_\omega(t) = f(t)$ with $f(t) = X(t,\omega)$. This approach requires us to consider some simple spaces of functions.

(5.1) Metric Lifting and Projecting

Let $(\mathbb{M},\rho)$ be a metric space entity, $\mathbb{M},\rho \in \mathfrak{X}$. The distance function or *metric*, $\rho : \mathbb{M} \times \mathbb{M} \longrightarrow [0,\infty)$ satisfies:

1. $\rho(x,y) = \rho(y,x)$; for x,y in $\mathbb{M}$.

2. $\rho(x,z) \leq \rho(x,y) + \rho(y,z)$; for x,y,z in $\mathbb{M}$.

3. $\rho(x,y) = 0$ if and only if $x = y$.

If ρ only satisfies 1 and 2 it is called a *semimetric*. [In that case the set of equivalence classes $x_\rho = \{y : \rho(x,y) = 0\}$ forms a metric space under ρ.] Our main examples of metric spaces are as follows:

(5.1.1) EXAMPLES:

(a) $\mathbb{M} = \mathbb{R}^d = \{(x_1,x_2,\cdots,x_d) : x_j \in \mathbb{R}\}$,

$$\rho(x,y) = \left[\sum_{j=1}^{d} (x_j-y_j)^2\right]^{\frac{1}{2}} = |x-y|.$$

(b) $\mathbb{M} = C[0,1]$ = continuous real-valued functions defined on $[0,1]$,

$$\rho(x,y) = \sup[|x(r)-y(r)| : 0 \leq r \leq 1],$$

the uniform norm.

(c) $\mathbb{M} = D[0,1]$ = the right-continuous real-valued functions with left limits on $[0,1]$,

$$(x,y) = k(x,y), \text{ the Kolmogorov metric.}$$

Example (c) is explained in greater detail below, especially in section 5.3.

Recall that a metric space $\mathbb{M}$ is called *complete* if every Cauchy sequence in $\mathbb{M}$ converges in $\mathbb{M}$ and is called *separable* if it has a countable dense subset. All three of the above examples are complete and separable. A topological space is said to be a *Polish space* if the topology is induced by some complete separable metric.

Some useful extensions of these examples are to consider $C(\mathbb{R})$ or $C([0,\infty))$, the continuous real-valued functions with domain $\mathbb{R}$ or $[0,,\infty)$. In these cases the metric is the metric of uniform convergence on compact subsets (no longer a norm). One can also extend the domain for $D[0,1]$ or allow the values for C or D spaces to lie in a complete separable metric space instead of $\mathbb{R}$. These spaces are still Polish spaces.

One slightly useful semimetric example is $L^0[\Omega]$, the mesureable functions with the semimetric

$$\rho(x,y) = \int \frac{|x(\omega)-y(\omega)|}{1+|x(\omega)-y(\omega)|} \, dP(\omega).$$

This semimetric measures convergence in probability—two functions are close if they are close except on a set of small measure.

(5.1.2) DEFINITION:

Let $(\mathbb{M},\rho) \in \mathfrak{X}$ *be a standard metric space. A point* y *in* ${}^*\mathbb{M}$ *is near-standard for* ρ *if there is a standard* x *in* ${}^\sigma\mathbb{M}$ *such that* $\rho(x,y) \approx 0$. *In this case we define* $st_\rho(y) = x$, *the standard part of* y. *The set of near standard points is denoted* $ns_\rho({}^*\mathbb{M})$.

For example, if $\mathbb{M} = \mathbb{R}^d$, then the near-standard points of

${}^*\mathbb{R}^d$ are just those $y = (y_1, y_2, \cdots, y_d)$ with each $y_j \in \mathcal{O}$ limited. In that case $x = st(y) = (st(y_1), st(y_2), \cdots, st(y_d))$ satisfies $|x-y| \approx 0$.

Another simple characterization of near-standardness is:

(5.1.3) **PROPOSITION:**

A function $y(r)$ *in* ${}^*C[0,1]$ *is near-standard for the uniform convergence metric if and only if* $y(r)$ *is finite and S-continuous, that is, if* $r \approx s$ *in* ${}^*[0,1]$, *then* $y(r) \approx y(s)$ *and both are finite.*

PROOF:

Let $x(r)$ be standard and suppose $\sup[|x(r)-y(r)| : 0 \lesssim r \lesssim 1] = \delta \approx 0$. Then we know $x(st(r))$ is finite and $x(st(r)) \approx x(r) \approx y(r)$ and $x(r) \approx x(s) \approx y(s)$. Thus, $y(s) \approx y(r)$ and both are finite. (See Appendix 1.)

Conversely, let x be the standard continuous function given by $x(r) = st\ y(r)$, for standard r. If r is standard, $x(r) \approx y(r)$ and otherwise letting $r = st(s)$ we have $x(s) \approx x(r)$ by S-continuity (see Appendix 1). Now, $x(r) \approx y(r)$, by construction, and $y(r) \approx y(s)$ by S-continuity. Thus $x(s) \approx y(s)$ for all s in ${}^*[0,1]$, so that $|x(s)-y(s)| < \epsilon$ for every standard positive ϵ. Therefore ${}^*\sup[|x(s)-y(s)| : 0 \lesssim s \lesssim 1] \approx 0$.

For a generalization of this result and its relation to the Ascoli-Arzela theorem, see Stroyan and Luxemburg [1976, (8.4.43)].

Other characterizations of near standard points in the various spaces of paths are given below. We will also need an internal generalization of this notion for hyperfinite processes, but give the standard case first.

Here is the basic idea. The infinitesimal random walk $B(t,\omega)$ we have seen above in (0.2.10), (0.3.12), etc., is almost surely finite and S-continuous. [We will prove this in the next section; we already know that it is finite by (0.3.12).] We could fill it in piecewise linearly in order to view the map $\omega \to B_\omega$ from Ω into ${}^*C[0,1]$ [where the section is the function $B_\omega(t) = B(t,\omega)$]. In this case then, for almost all ω, $st_\rho(B_\omega)$ exists and by the last result, $st_\rho(B_\omega)(r) = st(B(r,\omega))$.

Let $X : \Omega \longrightarrow {}^*\mathbb{M}$ be a function. We say that X is P-*almost surely near standard* if there is a set $\Lambda \subseteq \Omega$ with $P[\Lambda] = 0$ such that if $\omega \notin \Omega$, then $X(\omega) \in ns_\rho({}^*\mathbb{M})$. When X is almost surely near standard we define the *metric projection* of X by choosing any $a \in \mathbb{M}$ and letting $\tilde{X}(\omega) = st_\rho[X(\omega)]$, when $X(\omega)$ is near standard and $\tilde{X}(\omega) = a$ otherwise.

(5.1.4) THE METRIC PROJECTION THEOREM:

Let (Ω,P) be a *finite probability space and let $(\mathbb{M},\rho)$ be a standard complete separable metric space. If $X : \Omega \longrightarrow {}^*\mathbb{M}$ is internal and is almost surely near standard then the metric projection $\tilde{X}(\omega) = st_\rho(X(\omega))$ is P-measurable. Moreover, if $X(\omega)$ is near standard for $\omega \notin \Lambda$ with Λ internal and $P[\Lambda] = 0$, then $\tilde{X}$ has compact range.

PROOF:

Measurability is simple. Take b in $\mathbb{M}$ and ϵ positive.

The difference between $\{\omega \in \Omega : \rho(\tilde{X}(\omega),b) < \epsilon\}$ and $\bigcup_m \{\omega \in \Omega : \rho(X(\omega),b) < \epsilon - 1/m\}$ is contained in the null set Λ. The latter set is in the Loeb algebra since it is a countable union of internal sets.

The standard part of an internal set of near standard points is always compact, see Stroyan & Luxemburg [1976, (8.3.11)].

(5.1.5) DEFINITION:

Let $Y : \Omega \longrightarrow \mathbb{M}$ *be a function and let* $X : \Omega \longrightarrow {}^*\mathbb{M}$ *be an internal function. We say* X *is a lifting of* Y *if* $Y = st_\rho(X)$ *a.s.* $[P]$. *This depends on* ρ *and* P *up to null sets, so if needed we may say* X *is a* P*-lifting for the* ρ *metric.*

The result above shows that an a.s. near-standard internal function is a lifting of its projection.

(5.1.6) THE METRIC LIFTING THEOREM:

Let (Ω,P) *be a* **finite measure space, let* $(\mathbb{M},\rho)$ *be a standard complete separable metric space and let* $K \subseteq {}^*\mathbb{M}$ *be internal and S-dense,* $st_\rho(K) = \mathbb{M}$. *If a function* $Y : \Omega \longrightarrow \mathbb{M}$ *is* P*-measurable then it has a lifting with values in* K, $X : \Omega \longrightarrow K$.

PROOF:

This is very similar to the scalar case (1.3.9). Here is a slightly different approach. Let $\{z_k\}$ be a countable dense subset of $\mathbb{M}$. Since K is S-dense, for each z_k there is a

y_k in K with $st(y_k) = z_k$, so choose an associated sequence y_k. For each finite m in $\mathbb{N}$, let

$$L_k^m = \{x \in \mathbb{M} : k = \min[h : \rho(x,y_h) < \tfrac{1}{m}]\}$$

For m fixed the L_k^m are a Borel partition of $\mathbb{M}$.

Let $\Lambda_k^m = Y^{-1}(L_k^m)$. For each k choose an internal $\Omega_k^m \subseteq \Lambda_k^m$ with $P[\Lambda_k^m \setminus \Omega_k^m] < \frac{1}{m2^k}$. Extend y_k and Ω_k^m to internal sequences using countable comprehension (0.4), keeping the extension of y_k in K. Let $X^m(\omega) = y_k$ on Ω_k^m and $X^m(\omega) = a$ off the union of Ω_k^m. Then for each m there is an internal X^m with $P[\rho(X^m,Y) > \frac{1}{m}] \leq \frac{1}{m}$. Extend X^m to an internal sequence and choose an infinite n such that $P[\rho(X^m,X^n) > \frac{3}{m}] < \frac{2}{m}$ for $m < n$. Such an X^n lifts Y.

(5.2) Continuous Path Processes

Let $X(t,\omega) : \mathbb{T} \times \Omega \longrightarrow {}^*\mathbb{R}$ be an internal function. We may view this as a function from Ω into the internal space $F(\mathbb{T})$ of all internal functions from $\mathbb{T}$ into ${}^*\mathbb{R}$ by considering the section map $\omega \longrightarrow X_\omega$ where $(X_\omega)(t) = X(t,\omega)$. We call one of these sections, X_ω, the *path of* X at ω. We say X_ω is almost surely finite and S-continuous, if there is a set $\Lambda \subseteq \Omega$ with $P[\Lambda] = 0$ and $(X_\omega)(t) \approx (X_\omega)(s) \in \mathcal{O}$ for $t \approx s$ in $\mathbb{T}$ when $\omega \notin \Lambda$. The standard parts of finite S-continuous functions are standard continuous functions on $[0,1]$. Suppose that X_ω is finite and S-continuous and let $[r] = \max[t \in \mathbb{T} : t \leq r]$. The equation $\tilde{X}(r,\omega) = stX([r],\omega)$ defines a function on $[0,1]$. For each r, the set of numbers $\{stX(t,\omega) \mid t \approx r\}$ is only a singleton, so $stX(st^{-1}(r),\omega) = \tilde{X}(r,\omega)$ is well defined.

(5.2.1) DEFINITION:

Let $X : \mathbb{T} \times \Omega \longrightarrow {}^*\mathbb{R}$ *be internal and almost surely have finite S-continuous paths. The continuous path projection of* X *is the function*

$$\tilde{X} : [0,1] \times \Omega \longrightarrow \mathbb{R}$$

given by $\tilde{X}(r,\omega) = st[X(st^{-1}(r),\omega)]$, *when* X_ω *is finite and S-continuous, and* $\tilde{X}(r,\omega) = 0$, *otherwise.*

We know from Appendix 1 that the paths of $\tilde{X}$ are continuous, but moreover, the assignment of paths is measurable.

We may also describe the relation between X and $\tilde{X}$ by

$$\tilde{X}(st(t),\omega) = st\ X(t,\omega) \quad \text{a.s.}$$

(5.2.2) **THE S-CONTINUOUS PROJECTION THEOREM:**

Let X *and* $\tilde{X}$ *be as in (5.2.1). Then:*

(a) *The map* $\omega \longrightarrow \tilde{X}_\omega$ *of* Ω *into* $C[0,1]$ *is P-measurable.*

(b) *For each* r *in* $[0,1]$, *the random variable* $\tilde{X}^r$ *is P-measurable.*

(c) *The distributions of* $\tilde{X}$ *vary continuously, if* $\varphi(x_1,\cdots,x_m)$ *is a standard bounded continuous function of* m *variables and* $r_1 = st(t_1),\cdots, r_m = st(t_m)$, *then* $E[\varphi(X^{t_1},\cdots,X^{t_m})] \approx E[\varphi(\tilde{X}^{r_1},\cdots,\tilde{X}^{r_m})]$.

(c') *The map* $r \longrightarrow \tilde{X}^r$ *of* $[0,1]$ *into* $L^0(P)$ *with the convergence in probability semimetric, is continuous.*

(d) $\tilde{X} : [0,1] \times \Omega \longrightarrow \mathbb{R}$ *is* $(\mathrm{Borel}[0,1] \times \mathrm{Meas}(P))$-*measurable.*

PROOF:

Part (a) follows from the metric projection theorem (5.1.4) by extending $X(t,\omega)$ to be piecewise linear between the points of $\mathbb{T}$. Then we may identify $\tilde{X}$ and $st_\rho(X)$ where ρ is the uniform norm (up to a P null set).

We can argue on $X(t,\omega)$ for part (b) as follows. The evaluation map $x \rightarrow x(r)$ of $C[0,1]$ into $\mathbb{R}$ is continuous in the uniform norm, so $\{x : x(r) < a\}$ is open. By part (a),

$\{\omega : \tilde{X}_\omega(r) < a\} = (\tilde{X}_\omega)^{-1}(\{x : x(r) < a\})$ is measurable.

Part (c) is proved as follows. First, let $s_j \approx t_j$ for finitely many indices $1 \leq j \leq m$. Then

$$\{\omega : X(s_j,\omega) \not\approx X(t_j,\omega);\ 1 \leq j \leq m\} \subseteq \Lambda$$

where Λ is the null set off which X is S-continuous. Let φ be a bounded continuous standard function, and consider the internal probability:

$$P[\omega : |\varphi(X(t_1,\omega),\cdots,X(t_m,\omega))-\varphi(X(s_1,\omega),\cdots,X(s_m,\omega))| \geq \tfrac{\epsilon}{2}].$$

This probability is infinitesimal since it is internal and contained in the null set Λ, hence

$$|E[\varphi(X(t_1),\cdots,X(t_m))]-E[\varphi(X(s_1),\cdots,X(s_m))]| < \epsilon.$$

The internal set of tolerances,

$$T(\epsilon) = \{\vartheta \in {}^*\mathbb{R}^+ : |t_1-s_1| < \vartheta \ \&\ \cdots\ \&\ |t_m-s_m| < \vartheta \Rightarrow$$
$$|E[\varphi(X(t\text{'s}))] - E[\varphi(X(s\text{'s}))]|<\epsilon\}$$

contains the external set of all positive infinitesimals, thus it contains a standard positive ϑ. Finally, if $q_j = st(s_j)$ and $r_j = st(t_j)$ and $|s_j-t_j| < \vartheta/2$, then $E[\varphi(\tilde{X}(q\text{'s}))] \approx E[\varphi(X(s\text{'s}))]$ and $E[\varphi(X(t\text{'s}))] \approx E[\varphi(\tilde{X}(r\text{'s}))]$, so $|E[\varphi(\tilde{X}(q\text{'s}))] - E[\varphi(X(r\text{'s}))]| \leq \epsilon$. Compare this with (4.2.3), but note the change in notation, $\tilde{X}(t) = st(X(t))$ with nonstandard time t in (4.2.3), while now we index $\tilde{X}$ with standard instants $r \approx t$.

Part (c') is a special case of part (c) since the semimetric of convergence in probability is given by $E[\varphi(X,Y)]$ where φ is the standard bounded continuous functon:

$$\varphi(x,y) = \frac{|x-y|}{1+|x-y|}.$$

We prove part (d) by showing that $\tilde{X}$ is a uniform limit of step functions. This is based on part (c). Let $\vartheta > 0$ and let $X_\vartheta(t,\omega) = X([k\vartheta],\omega)$ for $k\vartheta \leq t < (k+1)\vartheta$ where $[k\vartheta] = \min[s \in \mathbb{T} : s \geq k\vartheta]$ Also let $m_\vartheta(\omega) = \max_t |X(t,\omega)-X_\vartheta(t,\omega)|$.

If $\omega \notin \Lambda$ and $\vartheta \approx 0$, then $m_\vartheta(\omega) \approx 0$. Since $P[\Lambda] = 0$, m_ϑ is infinitesimal almost surely and thus also nearly surely by (1.3.10). If ϵ and η are standard positive tolerances $P[m_\vartheta \geq \epsilon] < \eta$, so there must be a standard ϑ with this property (consider the internal set of ϑ's that work). This means m_ϑ and $\tilde{m}_\vartheta$ tend to zero in probability, so for a sequence of ϑ's, $\tilde{m}_\vartheta \longrightarrow 0$ a.s.

When ϑ is standard and positive we let $\tilde{X}_\vartheta(r,\omega) = \mathrm{st}(X_\vartheta(t,\omega))$, for $k\vartheta \leq r,t < (k+1)\vartheta$, so $|\tilde{X}_\vartheta(r,\omega)-\tilde{X}(r,\omega)| \leq \tilde{m}_\vartheta(\omega) \longrightarrow 0$ a.s. P. Each $\tilde{X}_\vartheta$ is (Borel$\times$ P)-measurable. This proves (d).

The next result shows that Anderson's random walk [see (0.2.10), (0.3.12), (4.3.2)] has S-continuous paths. That, together with all the machinery we have built up, shows that it is infinitely close to a conventional Brownian motion. Later we will give a general S-continuity theorem for martingales.

(5.2.3) **THEOREM:**

Anderson's infinitesimal random walk

$$B(t,\omega) = \Sigma[\sqrt{\delta t}\ \beta(\omega_s) : 0 < s \leq t, \text{ step } \delta t]$$

almost surely has S-continuous paths. Its projection, $\tilde{B} : [0,1] \times \Omega \longrightarrow \mathbb{R}$, *has continuous paths a.s. and has stationary independent normally distributed increments, that is,*

(a) (stationary) *the distribution of* $[\tilde{B}(r+\epsilon)-\tilde{B}(r)]$ *depends only on* ϵ.

(b) (independent increments) *if* $0 \leq r_0 < r_1 < \cdots < r_m \leq 1$, *then the family* $\{\tilde{B}(r_j)-\tilde{B}(r_{j-1})\}$ *is independent.*

(c) (normal) $$P[\tilde{B}(r) \leq a] = \frac{1}{\sqrt{2\pi r}} \int_{-\infty}^{a} e^{-\frac{x^2}{2r}}\, dx.$$

PROOF:

To prove S-continuity we begin by estimating violation of an ϵ-ϑ condition with $\epsilon = \frac{2}{m}$ and $\vartheta = \frac{1}{n}$ where m and n are finite positive integers. Define the internal set

$$\Omega_n^m = \left\{\omega : (\exists k)\left[\max_{\frac{k}{n}\leq t\leq\frac{k+1}{n}} B(t,\omega) - \min_{\frac{k}{n}\leq t\leq\frac{k+1}{n}} B(t,\omega)\right] > \frac{2}{m}\right\}.$$

Since the differences between max and min on $\frac{1}{n}$-time intervals at worst occur for separate sample sequences ω,

$$P[\Omega_n^m] \le nP\left[\left(\max_{0\le t\le\frac{1}{n}} B(t,\omega) - \min_{0\le t\le\frac{1}{n}} B(t,\omega)\right) > \frac{2}{m}\right].$$

If the difference between a max and min for ω is more than $\frac{2}{m}$, then B_ω either goes up by at least half that or down by at least half, starting from zero; the last probability is

$$\le nP\left[\max_{0\le t\le\frac{1}{n}} |B(t)| > \frac{1}{m}\right]$$

$$\le nP\left[\max_{0\le t\le\frac{1}{n}} B(t) > \frac{1}{m}\right] + nP\left[\min_{0\le t\le\frac{1}{n}} B(t) < -\frac{1}{m}\right].$$

Any sample ω up to $t < \frac{1}{n}$ is followed by an equal number of sample choices making $B(\frac{1}{n}) - B(t)$ positive and an equal number negative, so the above probabilities are

$$\le 2nP[B(\tfrac{1}{n}) > \tfrac{1}{m}] + 2nP[B(\tfrac{1}{n}) < -\tfrac{1}{m}]$$

$$= 4nP[B(\tfrac{1}{n}) > \tfrac{1}{m}]$$

$$\approx \frac{4n}{\sqrt{2\pi}} \int_{\frac{1}{m}}^{\infty} e^{-\frac{x^2 n}{2}} \sqrt{n}\, dx\,, \quad \text{by (0.3.12) or (4.3.2).}$$

Finally a simple calculation shows that

$$P[\Omega_n^m] < 2ne^{-\frac{\sqrt{n}}{m}}, \quad \text{if} \quad n > 4m^2,$$

so that $\operatorname*{S-lim}_{n\to\infty} P[\Omega_n^m] = 0$ for each fixed m. The S-discontinuous paths are contained in the samples from

$$\Lambda = \bigcup_m \bigcap_n \Omega_n^m$$

and

$$P[\Lambda] = \sup(\inf P[\Omega_n^m]) = 0.$$

This proves that $B(t)$ is a.s. S-continuous. Since $B(\delta t) \approx 0$, an S-continuous path is finite as well.

Continuity of $\tilde{B}$ now follows off Λ as in (5.2.2).

The stationarity comes simply from the defining sum for B; $B(r+\epsilon) - B(r)$ has the same form as the sum defining $B(\epsilon)$.

Independence of the increments follows from (4.2.3) and (4.3.2).

(5.2.4) **REMARK ON WIENER MEASURE:**

One definition of Wiener measure is: the unique Borel measure W on $C[0,1]$ satisfying:

(a) $$W\{x : x(r) \leq a\} = \frac{1}{\sqrt{2\pi r}} \int_{-\infty}^{a} e^{-\frac{x^2}{2r}} dx$$

and

(b) if $0 \leq r_0 < r_1 < \cdots < r_m \leq 1$, then the random variables $x \longrightarrow (x(r_j) - x(r_{j-1}))$ form an independent family.

One can show that W is given by

$$W(E) = P[\{\omega : \tilde{B}_\omega \in E\}]$$

where E is a Borel subset of $C[0,1]$. Roughly the idea is to associate $B(t,\omega)$ with piecewise linear paths and transfer the measure P into $^*C[0,1]$, so the formula above is

$$W(E) = P[st_\rho^{-1}(E)]$$

analogous to the construction of (2.3.4).

(5.2.5) DEFINITION:

Two functions $Y : [0,1] \times \Omega \longrightarrow \mathbb{R}$ *and* $Z : [0,1] \times \Omega \longrightarrow \mathbb{R}$ *are* P-*indistinguishable if there is a set* Λ, $P[\Lambda] = 0$, *such that when* $\omega \notin \Lambda$, *then* $Y(r,\omega) = Z(r,\omega)$ *for all* r *in* $[0,1]$, *that is, if their full paths agree almost surely.*

Let $Y : [0,1] \times \Omega \longrightarrow \mathbb{R}$ *be a function and let* $X : \mathbb{T} \times \Omega \longrightarrow {}^*\mathbb{R}$ *be internal and a.s. have finite* S-*continuous paths. Then* X *is an* S-*continuous path lifting of* Y *provided that the projection* $\tilde{X}$ *of (5.2.1) and* Y *are indistinguishable.*

Suppose almost all paths of Y are continuous on $[0,1]$. The projection theorem (5.2.2) tells us that Y must have several measurability properties if it has a lifting, the question is: How little can we assume about Y and still have a lifting?

(5.2.6) DEFINITION:

A function $Y : [0,1] \times \Omega \longrightarrow \mathbb{R}$ *is called a stochastic process on* $[0,1]$ *if for each* r *in* $[0,1]$ *the section* Y^r *is* P-*measurable.*

The answer to the above question is that we will assume Y

is a stochastic process.

(5.2.7) **THE S-CONTINUOUS PATH LIFTING THEOREM:**

If a stochastic process $Y : [0,1] \times \Omega \longrightarrow \mathbb{R}$ *a.s. has continuous paths, then it has an S-continuous path lifting.*

PROOF:

Assume that Y_ω is continuous except for ω in Λ with $P[\Lambda] = 0$. We want to show that $\omega \longrightarrow Y_\omega$ is P-measurable as a map into $C[0,1]$. That is, if U is open in $C[0,1]$ with the uniform norm, then $\{\omega \in \Omega \setminus \Lambda : Y_\omega \in U\}$ is P-measurable. This will allow us to apply the metric lifting theorem.

We know, since each Y^r is measurable, that $\{\omega : a \le Y(r,\omega) \le b\}$ is P-measurable. Define the one-dimensional cylinder set in $C[0,1]$ by

$$\mathcal{C}(a,b;r) = \{x : a \le x(r) \le b\}.$$

This set is closed in $C[0,1]$ since the evaluation is continuous. Also, $(Y^r)^{-1}[a,b] = Y_\omega^{-1}[\mathcal{C}(a,b;r)]$, so Y_ω is measurable with respect to cylinders. The closed ball in $C[0,1]$,

$$\{y : \text{for all } r, |x(r)-y(r)| \le \epsilon\} = \bigcap_{q \in \mathbb{Q}} \mathcal{C}(x(q)-\epsilon, x(q)+\epsilon; q)$$

by continuity. Hence inverse images of closed balls in $C[0,1]$ are measurable. Finally, separability of $C[0,1]$ (rational coefficient polynomials are dense by Weierstrass' approximation theorem) means that open sets are countable unions of closed balls, so Y_ω is measurable.

Let $K \subseteq {}^*C[0,1]$ be the set of piecewise linear continuous functions with corners at points of $\mathbb{T}$, so K is internal, S-dense for the uniform norm and in 1-1 correspondence with internal functions defined on $\mathbb{T}$. By (5.1.6), Y_ω has a lifting with values in K and the restriction to $\mathbb{T}$ is an S-continuous path lifting by (5.1.3).

(5.2.8) **EXERCISE:**

Wiener measure is unique, yet we had considerable choice in constructing $B(t)$. Show that $\tilde{B}_1(r)$ and $\tilde{B}_2(r)$ from (4.3.8) are distinguishable. You should construct similar examples based on (4.3.9) when you read about decent path projection in the next section.

(5.3) Decent Path Processes

We shall say that a function $x : [0,1] \longrightarrow \mathbb{R}$ is a *decent path* if it is continuous from the right and has left limits. We denote the space of decent paths by

$$D[0,1] = \{x : [0,1] \longrightarrow \mathbb{R} \mid \lim_{q \uparrow r} x(q) \text{ exists and } \lim_{s \downarrow r} x(s) = x(r)\}.$$

("Decent" is a free translation of the Alsatian word *cadlag*, which sometimes slips into North American articles. Cf. Dellacherie and Meyer [1978, p. 90].)

Skorohod introduced a number of topologies on $D[0,1]$. Later, Kolmogorov gave a metric that induces the most interesting of those topologies and makes $D[0,1]$ a complete separable metric space. Kolmogorov's metric is given as follows.

A strictly increasing Lipschitz function ρ *from* $[0,1]$ *onto* $[0,1]$ *is called a time deformation of* $[0,1]$. The set of all time deformations of $[0,1]$ is denoted $\Lambda[0,1]$. The measure of the amount of deformation for ρ in $\Lambda[0,1]$ is:

$$\delta(\rho) = \sup \left| \log \left[\frac{\rho(r)-\rho(s)}{r-s} \right] \right| ,$$

so that small deformations have secant lines of slope near one. Notice that $\delta(\rho^{-1}) = \delta(\rho)$ and if $\delta(\rho_m)$ tends to zero, then ρ_n tends uniformly to the identity map.

(5.3.1) DEFINITION:

The Kolmogorov metric (for Skorohod's topology) on $D[0,1]$ *is given by* $k(x,y) =$

$$\inf\{\epsilon : (\exists\rho\in\Lambda[0,1])[\sup_r |x(r)-y(\rho(r))|<\epsilon \ \&\ \delta(\rho)<\epsilon]\}.$$

Of course the uniform metric is also defined on $D[0,1]$ since $\sup_r |x(r)-y(r)|$ is finite when x and y belong to $D[0,1]$.

The instant $r = 1$ plays a somewhat artificial role in the theory of $D[0,1]$; for example, jumps at 1 cannot be shifted. This technical annoyance disappears in the study of $D[0,\infty)$, but at the expense of other difficulties with the metric. The infinitesimal analysis of $D[0,\infty)$, when r is finite, is just like that of $D[0,1]$, when r is less than 1.

We begin with some examples to illustrate how the metric works. The first one shows that we can shift jumps an infinitesimal amount—the uniform metric would not allow that. This will make our infinitesimal Bernoulli processes like the approximate Poisson process $J(t)$ from (0.3.7) near standard.

(5.3.2) EXAMPLE:

Define a sequence of functions by:

$$x_m(r) = I[\tfrac{1}{2} - \tfrac{1}{m}, 1](r) = \begin{cases} 0, & 0 \le r < \frac{1}{2} - \frac{1}{m} \\ 1, & \frac{1}{2} - \frac{1}{m} \le r \le 1 \end{cases}.$$

This sequence does not converge in the uniform norm, in fact,

$\sup|x_m(r)-x_n(r)| = 1$ for $m \neq n$. Consider the time deformations

$$\rho_m(r) = \begin{cases} \frac{m}{(m-2)} r , & 0 \leq r \leq \frac{1}{2} - \frac{1}{m} \\ \frac{m}{(m+2)} r + \frac{2}{(m+2)} , & \frac{1}{2} - \frac{1}{m} \leq r \leq 1 \end{cases} .$$

With these deformations we have $x_m(\rho_m^{-1}(r)) = x_\infty(r) = I[\frac{1}{2} ,1]$. The figures show the functions and deformations:

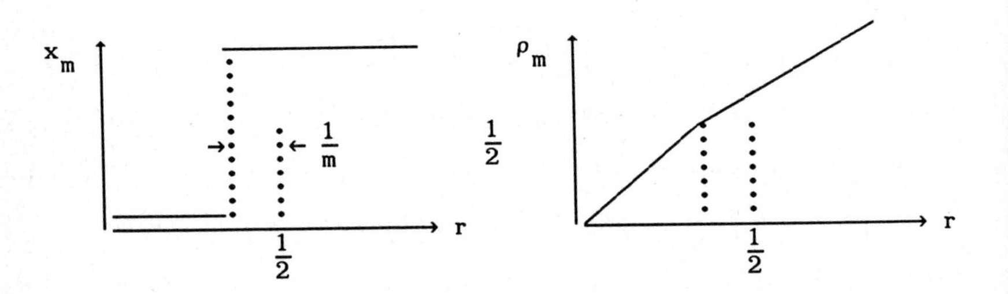

We see that $\delta(\rho_n) \approx 0$ when n is infinite, since the extreme slope, $\frac{n}{n-2} \approx 1$. Therefore, $k(x_n,x_\infty) \approx 0$ and $x_m \xrightarrow{k} x_\infty$.

The next example shows how the measure of deformation can prevent undesirable convergence.

(5.3.3) EXAMPLE:

The sequence $x_m = I[\frac{1}{2}, \frac{1}{2} + \frac{1}{m}]$ does not converge. In order to make $|x_n(r)-x_m(\rho(r))| < 1$ we need a time deformation like the one in the following figure:

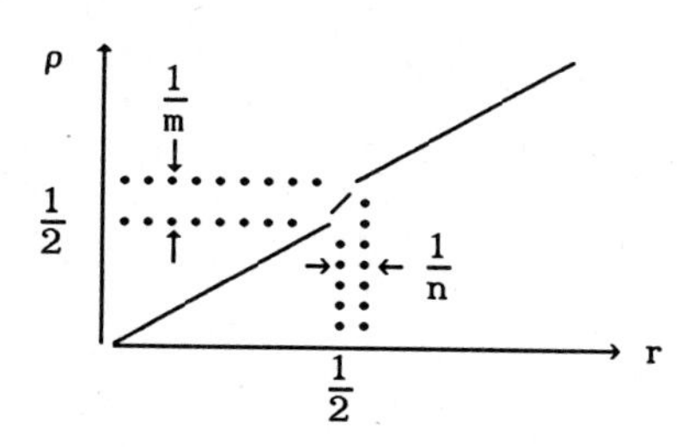

If we take $n = m^m$ and m infinite, then the short segment has infinite slope. This prevents x_m from forming a Cauchy sequence. We don't want it to converge either, because it is tending toward the indecent path $I\{\frac{1}{2}\}$.

Our next task is to say which internal functions y in $^*D[0,1]$ are a Kolmogorov infinitesimal from some standard path x in $D[0,1]$. After we have done that, we want to see how to view functons on $\mathbb{T}$ as near functions in $^*D[0,1]$. Lifting and projecting will be done between internal processes on $\mathbb{T}$ and standard ones on $[0,1]$ similar to the continuous case of the last section.

By transform of the definition of k, if $k(x,y) \approx 0$, there is an internal increasing function ρ with secant slopes infinitely near 1 such that for all r in $^*[0,1]$, $x(r) \approx y(\rho(r))$. That is, up to an infinitesimal time deformation x and y are *uniformly* close.

When x is standard in $D[0,1]$, it cannot have two standard postive jumps in infinitesimal time—being standard, such jumps occur at standard times. By the last paragraph and this, we see that finite jumps of y are finitely separated when $k(x,y) \approx 0$. The next result gives three ways to say that noninfinitesimal jumps are finitely separated.

(5.3.4) PROPOSITION & DEFINITION:

Let y *be an internal function in* $^*D[0,1]$. *We say* y *has* S-*separated jumps if it satisfies the following equivalent conditions:*

(a) *For every positive infinitesimal* δ, *there is an internal* **finite sequence* $0 = r_0 < r_1 < \cdots < r_n = 1$ *with* $(r_j - r_{j-1}) > \delta$, *for* $1 \le j \le n$, *satisfying* $y(r) \approx y(r_j)$ *whenever* $r_j \le r < r_{j+1}$, *for* $0 \le j < n$.

(b) *For every standard* r *in* $(0,1)$, *there exists* $s \approx r$ *such that if* $r_1 \approx r_2 \approx r$ *and* $r_1 < r_2 < s$, *then* $y(r_1) \approx y(r_2)$ *and if* $s_1 \approx r$ *and* $s < s_1$, *then* $y(s) \approx y(s_1)$. *Also, if* δ *is a positive infinitesimal,* $y(0) \approx y(\delta)$ *and* $y(1-\delta) \approx y(1-\iota)$ *for any positive infinitesimal* ι.

(c) *For every standard positive* ϵ, *there exists a standard positive* ϑ *and a sequence* $0 = r_0 < r_1 < \cdots < r_m = 1$ *with* $(r_j - r_{j-1}) > \vartheta$, *for* $1 \le j \le m$, *and* $|y(r) - y(r_j)| < \epsilon$ *when* $r_j \le r < r_{j+1}$, *for* $0 \le j < m$.

PROOF (of equivalence):

(a) ⇒ (b): If (b) fails, there are three times $s_1 < s_2 < s_3$ with $s_1 \approx s_2 \approx s_3$ and $y(s_1) \not\approx y(s_2)$ and $y(s_2) \not\approx y(s_3)$. Let $\delta = s_3 - s_1 \approx 0$. Condition (a) cannot hold for this δ. Failure of (b) at an endpoint implies not (a) by similar reasoning.

(b) ⇒ (c): If (b) holds and ϵ is standard and positive, let $r_0 = 0$ and $s_{j+1} = \text{st}\ ^*\inf\{r : |y(r)-y(r_j)| \geq \epsilon\} \wedge 1$. Use condition (b) to find $r_{j+1} \approx s_{j+1}$ such that whenever $q_1 \approx q_2 \approx r_{j+1}$ and $q_1 < q_2 < r_{j+1}$, then $y(q_1) \approx y(q_2)$ while whenever $r_{j+1} < t$ and $t \approx r_{j+1}$, then $y(r_{j+1}) \approx y(t)$. Condition (b) implies that $\vartheta = \min(r_j - r_{j-1}) \not\approx 0$.

(c) $\Rightarrow$ (a): If (a) fails, there is an infinitesimal δ such that no sequence of δ-separated points has infinitesimal right y variation. Let $\epsilon = \frac{1}{2}\,\text{st}(^*\inf\{\max[|y(r)-y(r_j)| : r_j \leq r < r_{j+1}, \text{etc.}] : r_j - r_{j-1} > \delta\})$. This internal inf is bounded below by all infinitesimals, hence $\epsilon > 0$. Condition (c) fails for this epsilon.

Notice that an S-continuous internal function satisfies our condition of S-separated jumps. Perhaps we should say y 'has S-separated jumps, if it has any, and otherwise is S-continuous'. Besides having separated jumps, a near-standard function $y \in \text{st}_k^{-1}(x)$ must take finite values, and in particular only jump finite amounts. This is clear from the *uniform approximation*

(5.3.5) $$x(r) \approx y(\rho(r)), \quad \textit{for all} \quad r,$$

where ρ *is an infinitesimal time deformation,* $\delta(\rho) \approx 0$. This approximation also makes it clear that

(5.3.6) $$x(r) = \operatorname*{S-lim}_{s \downarrow r} y(s)$$

meaning, for every standard positive ϵ, there is a standard positive ϑ, such that if $r << s < r+\vartheta$, then $|y(s)-x(r)| < \epsilon$. Condition (b) is an infinitesimal way to say left and right S-limits exist.

Our discussion of finiteness and separation of jumps proves one implication of the next result.

(5.3.7) PROPOSITION:

Let y *be an internal function in* $^*D[0,1]$. *Then* y *is near-standard for Kolmogorov's metric if and only if* y *takes finite values and has S-separated jumps. Moreover, these are the only internal functions* y *such that for every standard positive* ϵ, *there is a standard* x *in* $D[0,1]$ *with* $k(x,y) < \epsilon$.

PROOF:

First we show that an internal finite S-separated function can be approximated within ϵ by standard functions. Let ϵ be standard and positive and take a sequence $0 = r_0 < r_1 < \cdots < r_m = 1$ with $r_j - r_{j-1} > \vartheta \not\approx 0$ and $|y(r_j)-y(r)| < \frac{\epsilon}{2}$, for $r_j \leq r < r_{j+1}$, from (5.3.4)(c). Let $s_j = st(r_j)$ and define a step function $x_\epsilon(t) = st(y(r_j))$ for $s_j \leq t < s_{j+1}$ when $j < m$ and $x_\epsilon(1) = st(y(1))$. Since the s_j are separated by $\vartheta \not\approx 0$, the piecewise linear function ρ with corners at s_j satisfying $\rho(s_j) = r_j$ has $\delta(\rho) \approx 0$,

$$\frac{\rho(r)-\rho(s)}{r-s} \approx 1.$$

Also, $|x_\epsilon(r)-y(\rho(r))| < \epsilon$, so $k(x_\epsilon,y) < \epsilon$.

We finish the proof by showing that $D[0,1]$ is complete in

Kolmogorov's metric. If x_m is a sequence of standard functions with $k(y,x_m) < 2^{-m-1}$, then $k(x_m,x_{m+n}) < 2^{-m}$, so the x_m's form a Cauchy sequence. If we knew $x_m \longrightarrow x_\infty$ in D[0,1] then $k(x_\infty,x_n) \approx 0$ for infinite n (in the extended standard sequence) and also $k(x_n,y) < 2^{-n-1}$ (for sufficiently small infinite n) so $k(y,x_\infty) \approx 0$. In other words, any y approximated by a standard sequence is near-standard. (See Stroyan & Luxemburg [1976, (8.4.28 & 29].)

Here is the proof of completeness. Since $k(x_m,x_{m+1}) < 2^{-m}$ there is a time deformation ρ_m such that

$$\sup|x_m(r)-x_{m+1}(\rho_m(r))| < 2^{-m}$$

and

$$\delta(\rho_m) < 2^{-m}.$$

For each fixed m the sequence of repeated compositions

$$\rho_m, \rho_{m+1} \circ \rho_m, \rho_{m+2} \circ \rho_{m+1} \circ \rho_m, \cdots, \rho_{m+n} \circ \cdots \circ \rho_m$$

(n compositions) forms a uniformly convergent sequence of functions; in fact,

$$\delta(\rho_{m+n} \circ \cdots \circ \rho_m) \le \sum_{k=m}^{m+n} 2^{-k} < 2^{-m+1}$$

so that the iterated compositions remain S-continuous when n is infinite. Let $\sigma_m = \lim_{n\to\infty} \rho_{m+n} \circ \cdots \circ \rho_m$ and notice that the above shows $\delta(\sigma_m) \le 2^{-m+1}$. Also, $\sigma_m = \sigma_{m+1}\rho_m$, so $\sup_r|x_m(\sigma_m^{-1}(r))-x_{m+1}(\sigma_{m+1}^{-1}(r))| = \sup_r|x_m(r)-x_{m+1}(\rho_m(r))| < 2^{-m}$ and $x_m \circ \sigma_m^{-1}$ is uniformly convergent, say to x_∞. Since

$\delta(\sigma_m) \longrightarrow 0$, $k(x_m, x_\infty) \longrightarrow 0$ once we know $x_\infty \in D[0,1]$, that is, has left and right limits. This follows easily from condition (b) of (5.3.4). (Notice that condition (c) is inherited directly from an approximating sequence, so x_∞ has separated jumps even if it didn't start out near such a y.)

(5.3.8) EXAMPLE:

Consider the internal process

$$J(t,\omega) = \sum[\pi(\omega_s) : 0 < s \leq t, \text{ step } \delta t]$$

given in example (4.3.3) and (0.3.7). We can extend $J(t,\omega)$ to a step function on ${}^*[0,1]$ by letting $J(r,\omega) = J(t,\omega)$, for $t \leq r < t+\delta t$. Then $J(r)$ is finite and has S-separated jumps if and only if $J(t)$ does, while $J(r)$ is in ${}^*D[0,1]$. This will mean we can form the Kolmogorov standard part $st_k(J_\omega)$. Later we want to describe the standard part directly in terms of t.

To show that jumps of $J(t)$ are separated, define

$$\tau_k(\omega) = \min[1 \ \& \ t \in \mathbb{T} : J(t,\omega) \geq k].$$

The quantity τ_1 is the random time you wait for the first jump of J. For each t in $\mathbb{T}$,

$$(5.3.9) \quad P[\tau_1 > t] = \text{"independently draw } \frac{t}{\delta t} \text{ zeros"}$$

$$= [1-p\delta t]^{(t/\delta t)}$$

$$\approx e^{-pt} \approx e^{-at}.$$

An especially important feature of this waiting time is that there is no "aging":

$$(5.3.10) \qquad P[\tau_1 > t+s \mid \tau_1 > s]$$
$$= \text{"independently draw } \frac{t}{\delta t} \text{ more zeros"}$$
$$= [1-p\delta t]^{\frac{t}{\delta t}}$$
$$= P[\tau_1 > t].$$

Note: For nonempty internal sets the probability of Λ given Λ', $P[\Lambda|\Lambda'] = P[\Lambda \cap \Lambda']/P[\Lambda']$.

The next jump is just like the last:

$$P[\tau_2 > t+s \mid \tau_1 = s]$$
$$= \text{"independently draw } \frac{t}{\delta t} \text{ zeros"}$$
$$= P[\tau_1 > t].$$

Finally, by (5.3.9), for sufficiently large infinitesimal Δt,

$$(5.3.11) \qquad \frac{1}{\Delta t} P[\tau_1 \leq \Delta t] \approx a$$

or equivalently,

$$\underset{\Delta t \downarrow 0}{\text{S-lim}}\, P[\tau_1 \leq \Delta t]/\Delta t = a,$$

for small finite time the *rate* of jumps is approximately a.

Let $\Delta t = \frac{1}{m}$, for some finite natural number m. The set Λ_m of sample sequences ω with two or more jumps during a Δt

interval and $J(1)$ finite consists of the following samples summarized in the table below.

First, consider $\{\omega : \tau_1(\omega) \leq \Delta t\} = [\tau_1 \leq \Delta t]$. Denote the probability $P[J(\Delta t) \geq 1]$ by $p(\Delta t)$. Second, we may eliminate paths with a jump right before 1, $[\tau_1 > 1-\Delta t \ \& \ J(1) \geq 1]$. The probability of a jump after $1-\Delta t$ is $p(\Delta t)$, independent of $J(1-\Delta t)$ and the probability that $J(1-\Delta t) = 0$ is approximately $e^{-a(1-\Delta t)}$. Thus the probability of our second piece of the sample with indecent jumps is $\lesssim p(\Delta t)$. Next we suppose the first jump is o.k., but the second one is too close to the first: $[\Delta t < \tau_1 \leq 1-\Delta t \ \& \ \tau_2 \leq \tau_1+\Delta t \ \& \ J(1) \geq 2]$. This probability is

$$\sum_{\substack{\Delta t<s<1-\Delta t \\ \text{step } \delta t}} P[\tau_2 \leq \tau_1+\Delta t \mid \tau_1 = s]P[\tau_1 = s]$$

$$= p(\Delta t) \sum_{\Delta t<s<1-\Delta t} P[\tau_1 = s] \leq p(\Delta t)P[J(1-\Delta t) \geq 1]$$

$$\lesssim p(\Delta t)[1-e^{-a}].$$

Our final example before the table is the case where τ_1 and τ_2 are Δt-separated, but τ_3 is too close to τ_2: $[\Delta t < \tau_1 < \tau_2-\Delta t < 1-2\Delta t \ \& \ \tau_3 \leq \tau_2+\Delta t \ \& \ J(1) \geq 3]$. This probability is

$$\sum_{\Delta t<s<1-2\Delta t} P[\tau_3 \le \tau_2+\Delta t \ \& \ \tau_2>\tau_1+\Delta t|\tau_1 = s]P[\tau_1 = s]$$

$$= \sum_{[\Delta t<s<1-2\Delta t]} \sum_{[s+\Delta t<t<1-\Delta t]} P[\tau_3 \le t_2+\Delta t|\tau_2=t]P[\tau_2=t|\tau_1=s]P[\tau_1=s]$$

$$= p(\Delta t) \sum_s \sum_t P[\tau_2=t \mid \tau_1=s]P[\tau_1=s].$$

The inside term,

$$\sum_{s+\Delta t<t<1-\Delta t} P[\tau_2=t|\tau_1 = s] = \sum_{\Delta t<t<1-\Delta t-s} P[\tau_1=t]$$

$$= P[J(\Delta t) = 0 \ \& \ J(1-\Delta t-s) \ge 1] \le P[J(\Delta t) = 0 \ \& \ J(1-2\Delta t) \ge 1]$$

$$\le (1-e^{-a}).$$

Again, the outside term,

$$\sum_{\Delta t<t<1-2\Delta t} P[\tau_1 = s] \lesssim (1-e^{-a}).$$

So the whole probability is $\lesssim p(\Delta t)[1-e^{-a}]^2$.

	BAD SAMPLE	PROBABILITY
(0):	$[\tau_1 \le \Delta t]$	$\lesssim p(\Delta t)$
(1):	$[\tau_1 > 1-\Delta t \ \& \ J(1) \ge 1]$	$\lesssim p(\Delta t)$
(2):	$[\Delta t < \tau_1 \le 1-\Delta t \ \& \ \tau_2 \le \tau_1+\Delta t]$	$\lesssim p(\Delta t)[1-e^{-a}]$
(3):	$[\Delta t < \tau_1 < \tau_2-\Delta t < 1-2\Delta t \ \& \ \tau_3 \le \tau_2+\Delta t]$	$\lesssim p(\Delta t)[1-e^{-a}]^2$

$$(k+1):\quad [\mathop{\&}_{j=1}^{k} \tau_j - \Delta t > \tau_{j-1} \;\&\; \tau_k \leq 1-\Delta t$$

$$\& \;\tau_{k+1} \leq \tau_k + \Delta t \;\&\; J(1) \geq k+1] \lessapprox p(\Delta t)[1-e^{-a}]^k.$$

The $(k+1)^{st}$ estimate is obtained in the same manner as the $(3)^{rd}$ was above.

$$\sum_{t_1}\sum_{t_2}\cdots\sum_{t_k} P[\tau_{k+1} \leq \tau_k + t \,|\, \tau_k = t_k] P[\tau_k = t_k \,|\, \tau_{k-1} = t_{k-1}]$$

$$\cdots P[\tau_2 = t_2 \,|\, \tau_1 = t_1] P[\tau_1 = t_1].$$

The innermost term has probability $p(\Delta t)$ while each $\sum_{t_j} P[\tau_{j+1} = t_{j+1} \,|\, \tau_j = t_j]$ is less than $(1-e^{-a})$.

Thus the total probability of two or more jumps within Δt, $P[\Lambda_m] \leq p(\Delta t)[e^a+1]$. We know from (5.3.11) $p(\Delta t)$ is asymptotic to $(\Delta t)a$ as Δt finitely tends to zero. Since J has finite S-separated paths except on $\cap \Lambda_m = \Lambda$, we see that J satisfies the next definition with $\delta t = \Delta t$.

(5.3.12) DEFINITION:

Let $X : \mathbb{T} \times \Omega \longrightarrow {}^*\mathbb{R}$ *be internal and let* $\Delta t \in \mathbb{T}$. *The* Δt*-sample of* X *is the internal function on* ${}^*[0,1] \times \Omega$ *given by*

$$X^{\Delta t}(r,\omega) = \begin{cases} X(\Delta t,\omega) & , \text{ for } 0 \leq r < 2\Delta t \\ X(k\Delta t,\omega) & , \text{ for } k\Delta t \leq r < \min[1,(k+1)\Delta t] \\ X(1,\omega) & , \text{ for } r = 1. \end{cases}$$

We say that X *a.s. has a decent path sample or D-sample*

if there is a set Λ *with* $P[\Lambda] = 0$ *and an infinitesimal* $\Delta t \geq \delta t$ *in* $\mathbb{T}$ *such that the* Δt*-sample paths* $X_\omega^{\Delta t}$ *are Kolmogorov-metric near-standard for* $\omega \notin \Lambda$.

To be specific we will refer to the sample mesh by saying X has a *Δt-decent path sample.* The function $X_\omega^{\Delta t}$ is completely determined by its values on

$$\mathbb{T}_\Delta = \{t \in \mathbb{T} \mid t = k\cdot\Delta t, \text{ for some } k \in \mathbb{N}^+\} \cup \{1\}.$$

Notice that indecent paths [like the extension of the sequence in (5.3.3)] *can* have D-samples provided oscillations or close jumps are confined. For example, if $X(t,\omega) = 1$ when $t = \frac{1}{2}$ and zero for all other t, independent of ω, then a Δt-sample which avoids $\frac{1}{2}$ has $X^{\Delta t} \equiv 0$. Sampling will be needed later to assure that *martingales have decent projections, see (5.3.25).

(5.3.13) **DEFINITION:**

Let X *be internal and a.s. have a* D*-sample* $X^{\Delta t}$. *The function* $\tilde{X}(r,\omega) = st_k(X_\omega^{\Delta t})(r)$ *from* $[0,1] \times \Omega$ *into* $\mathbb{R}$ *is the decent path projection of the sample of* X.

As remarked above in (5.3.6), we know

$$\tilde{X}(r) = \underset{t\downarrow r}{\text{S-lim}}\, X^{\Delta t}(t) = \underset{\mathbb{T}\ t\downarrow r}{\text{S-lim}}\, X(t) \qquad \text{a.s. } P.$$

Also note $\tilde{X}(1) = st\, X(1)$.

(5.3.14) DEFINITION:

Let $x : {}^*[0,1] \longrightarrow {}^*\mathbb{R}$ *be internal and let* Δt *be a positive infinitesimal. The forward* $\frac{1}{h}$*-average of* x *is*

$$x_h(r) = \begin{cases} h\Sigma[x(s)\Delta t : r \leq s < r + \frac{1}{h},\ s \in \mathbb{T}_\Delta],\ 0 \leq r \leq 1 - \frac{1}{h} \\ x_h(1 - \frac{1}{h}), \quad 1 - \frac{1}{h} \leq r < 1, \\ x(1), \quad r = 1 \end{cases}$$

abbreviated

$$x_h(r) = h \sum_{\substack{s=r \\ \text{step } \Delta t}}^{r+1/h} x(s)\Delta t.$$

Our next result says in detail that forward averaging is almost a continuous map from $D[0,1]$ into $C[0,1]$. We make use of this in the projection theorem for decent paths by reducing it to limits of continuous paths.

(5.3.15) LEMMA:

Let $x, y \in {}^*D[0,1]$ *be Kolmogorov near-standard and satisfy* $k(x,y) \approx 0$. *Let* h *be a finite natural number and* Δt *a positive infinitesimal and let* x_h *and* y_h *be the forward averages. Then*

(a) x_h *is finite and S-continuous on* ${}^*[0,1)$.

(b) $x_h(r) \approx y_h(r)$ *for all* r, $0 \leq r \leq 1$, x_h *and* y_h *differ by a uniform metric infinitesimal.*

(c) $st_k(x)(r) = \lim_{h\to\infty} st\ x_h(r)$.

(d) $st_k(x_h(r)) = h\int_r^{r+1/h} st_k(x)(s)ds$ *for standard* $r \leq 1 - \frac{1}{h}$.

PROOF:

(a) Averages of finite functions are finite. If $q \approx r$ then $x_h(q)$ and $x_h(r)$ differ at most by $2|r-q|$ times the finite bound for x, so $x_h(q) \approx x_h(r)$.

(b) Let $x(r) \approx y(\rho(r))$ for an infinitesimal time deformation. Then $x_h(r) - [y(\rho(r))]_h \approx 0$ by the triangle inequality. We complete the proof by showing that $[y(\rho(r))]_h \approx y_h(r)$. Let $\epsilon >> 0$ be a standard positive tolerance and apply condition (c) of (5.3.4) to obtain $\vartheta >> 0$ and a sequence $0 = r_0 < r_1 < \cdots < r_m = 1$ with $r_j - r_{j-1} > \vartheta$ and $|y(r)-y(r_j)| < \frac{\epsilon}{3}$ if $r \in [r_j, r_{j+1})$. Define $\rho_j = \rho^{-1}(r_j)$, so $|y(\rho(\sigma))-y(\rho(\rho_j))| < \frac{\epsilon}{3}$ when $\sigma \in [\rho_j, \rho_{j+1})$. When $r \in [r_j \vee \rho_j, r_{j+1} \wedge \rho_{j+1})$, then $|y(r)-y(\rho(r))| < \epsilon$. Since ρ is an infinitesimal time deformation, $\sum_{j=1}^{m} [(r_j \vee \rho_j)-(r_j \wedge \rho_j)] \approx 0$ and thus $|y_h(r)-y \circ \rho_h(r)| \lesssim \epsilon$. Since ϵ is an arbitrary standard tolerance we see that $y_h(r) \approx y \circ \rho_h(r)$.

(c) Since $st_k(x)(r) = \text{S-}\lim_{s \downarrow r} x(s)$, for each standard $\epsilon >> 0$, there exist s_1, s_2, $r \approx s_1 << s_2$ such that

$$|st_k(x)(r)-x(s)| < \epsilon \quad \text{for} \quad s_1 \leq s \leq s_2.$$

For any finite h such that $r + \frac{1}{h} \leq s_2$,

$$|st_k(x)(r)-x_h(r)| \lesssim \epsilon.$$

(d) The standard path $st_k(x)$ is Riemann integrable, so

this is just the standard infinitesimal characterization of Riemann integrals (see Stroyan & Luxemburg [1976]) plus the triangle inequality and a simple estimate with time deformation as in the proof of (b) above.

(5.3.16) THE DECENT PATH PROJECTION THEOREM:

Let X *have a* Δt*-decent path sample for some infinitesimal* $\Delta t \geq \delta t$ *in* $\mathbb{T}$ *and let* $\tilde{X}$ *be the decent path projection of* X. *Then:*

(a) *The map* $\omega \longrightarrow \tilde{X}_\omega$ *of* Ω *into* $D[0,1]$ *is* P-*measurable.*

(b) *For each* r *in* $[0,1]$, *the random variable* $\tilde{X}^r$ *is* P-*measurable.*

(c) $\tilde{X} : [0,1] \times \Omega \longrightarrow \mathbb{R}$ *is* (Borel[0,1] × Meas(P))-*measurable.*

(d) *The distributions of* $\tilde{X}$ *are right continuous, if* $\varphi(x_1, \cdots, x_m)$ *is a standard bounded continuous function of* m *variables and* r_j, $1 \leq j \leq m$, *are in* $[0,1)$, *there is an* $\eta \approx 0$ *such that* $r_j \approx t_j > r_j + \eta$ *for* $1 \leq j \leq m$ *implies* $E[\varphi(X^{t_1}, \cdots, X^{t_m})] \approx E[\varphi(\tilde{X}^{r_1}, \cdots, \tilde{X}^{r_m})]$.

PROOF:

Let $X^{\Delta t}$ be as in (5.3.12), let h be a finite natural number and let $X_h(r,\omega)$ be the forward $\frac{1}{h}$-average for $X^{\Delta t}$. (Notice that X_h only depends on the sample of X itself.) By (5.3.15) each $X_h(r,\omega)$ is finite and S-continuous off Λ, so, except at $r = 1$, $\tilde{X}_h$ is a continuous path projection as in (5.2.2). Therefore $\tilde{X}_h$ is

(Borel × P)-measurable for each h. Since $\tilde{X}(r,\omega) = \lim_{h\to\infty} \tilde{X}_h(r,\omega)$ for each ω outside Λ and each r, $\tilde{X}$ is (Borel × P)-measurable. This proves (c) and (a) and (b) follow from (c).

(d) We know $S\text{-}\lim_{t_j \downarrow r_j} X^{t_j} = \tilde{X}^{r_j}$ off a certain P-null set Λ, hence for $\omega \notin \Lambda$ there exists a positive infinitesimal $\vartheta(\omega)$ such that $\tilde{X}(r_j,\omega) \approx X(t_j,\omega)$ if $r_j+\vartheta(\omega) < t_j \approx r_j$ for $1 \le j \le m$. For each standard $\epsilon > 0$, there is an internal $\Lambda_\epsilon \supseteq \Lambda$ with $P[\Lambda_\epsilon] < \epsilon$; call

$$\vartheta(\epsilon) = \max\{\vartheta(\omega) : \omega \notin \Lambda_\epsilon\} \approx 0,$$

and take an infinitesimal ϑ greater than a countable family of $\vartheta(\epsilon)$, say $\vartheta(\frac{1}{p})$. Then the set $\Lambda' = \cap \Lambda_{1/p}$ has zero probability, and for $\omega \in \Lambda'$

$$r_j+\vartheta < t_j \approx r_j, \quad 1 \le j \le m \quad \text{imply} \quad \tilde{X}(r,\omega) \approx X(t,\omega);$$

in other words, the measurable projection of $\varphi(X^{t_1},\cdots,X^{t_m})$ equals $\varphi(\tilde{X}^{r_1},\cdots,\tilde{X}^{r_m})$ a.s. Since $\varphi(X^{t_1},\cdots,X^{t_m}) \in SL^1(P)$, the integral of its projection is infinitely close to its sum.

(5.3.17) **EXERCISE:**

Let $J(t,\omega)$ be the Bernouilli process infinitesimally far from a Poisson as in (0.3.7). Prove that J is continuous in distribution (from both sides). Give an example of an internal separated jump process with a left discontinuity in distribution.

Our next example needs the following standard result. We include the (*-transformed) proof for the reader's convenience.

(5.3.18) **SKOROHOD'S LEMMA:**

Let $X(t) = \Sigma[\delta X(s) : 0 < s \le t, \text{ step } \delta t]$, $X(0) = 0$, *where* $\{\delta X(s) : s \in \mathbb{T}\}$ *is a* **independent family. Then*

$$P\left[\max_{0 \le s \le t} |X(s)| > 2\epsilon\right] \le \frac{P[|X(t)| > \epsilon]}{1 - \max_{0 \le s \le t}\{P[|X(t)-X(s)| > \epsilon]\}}.$$

PROOF:

Let $\tau(\omega) = \min[1 \;\&\; t : |X(t,\omega)| > 2\epsilon]$.

$$\begin{aligned}
P[|X(t)| > \epsilon] &\ge P[|X(t)| > \epsilon \;\&\; \max_{s \le t}|X(s)| > 2\epsilon] \\
&\ge \sum_{s \le t} P[|X(t)| > \epsilon \;\&\; \tau = s] \\
&\ge \sum_{s \le t} P[|X(t)-X(s)| \le \epsilon \;\&\; \tau = s] \\
&\ge \sum_{s \le t} P[|X(t)-X(s)| \le \epsilon]P[\tau = s] \\
&\ge \{1 - \max_{s \le t} P[|X(t)-X(s)| > \epsilon]\} \sum_{s \le t} P[\tau = s] \\
&\ge \{1 - \max_{s \le t} P[|X(t)-X(s)|>\epsilon]\}P[\max_{s \le t}|X(s)| > 2\epsilon].
\end{aligned}$$

(5.3.19) **EXAMPLES:**

Let $X(t) = \Sigma[\delta X(s) : 0 < s \le t, \text{ step } \delta t]$, $X(0) = 0$, be a sum of *independent $\delta(X)$'s all with the same distribution. Let $\delta f(u) = E[\exp(iu\delta X)]$ be the characteristic function of the infinitesimal increment, cf. (4.3.5). If $\delta f(u) = 1 + \delta t\ \psi(u)$ for a finite S-continuous function $\psi(u)$, then the characteristic function of $X(t)$ for $t \gg 0$, $f(t,u) = [1+\delta t\psi(u)]^{t/\delta t} \approx e^{t\psi(u)}$ is S-continuous and $X(t)$ is finite a.s. by (4.1.3). Proposition (4.3.4) is the converse of the last remark. These processes are infinitely close to stationary independent increment standard processes. We have seen the examples of Brownian motion, the Poisson process and the Cauchy process (4.3.6-7) above. Observe that the latter has no moments, e.g., $X(t) \notin SL^1(\Omega)$. Deterministic drift $X(t) = ct$ arises this way by taking $\delta X(s) = c\delta t$ for all s with probability one. Also, a deterministic process is independent of anything else and $f(t,u) = e^{iuct}$ in this case.

The following results show that internal 'versions' have decent paths. We only prove the easy part to indicate how Skohorod's lemma works. More 'general' sampling results with weaker hypotheses and conclusions are proved later.

(5.3.20) LEMMA:

Let $X(t) = \Sigma[\delta X(s) : 0 < s \leq t,\ \text{step } \delta t]$, $X(0) = 0$, *be a sum of* **independent identically distributed infinitesimal increments* $\delta X(s)$ *such that* $X(t)$ *is finite a.s. for* $0 < t \leq 1$, *or equivalently,* $\psi(u)$ *as above is finite and S-continuous. Then the paths of* X *are finite a.s.*

PROOF:

It won't do to have $X(t,\omega)$ finite if $X(t+\delta t,\omega)$ is infinite: $\mathbb{T}$ is uncountable so sets of measure zero could build up. We estimate the easy term in Skorohod's lemma using (4.1.4):

$$P[|X(s)| > \epsilon] \leq \alpha\epsilon \int_0^{1/\epsilon} [1-\text{Re}[1+\delta t\psi(v)]^{s/\delta t}]dv$$
$$\leq \alpha\epsilon \int_0^{1/\epsilon} (e^{|\psi(v)|}-1)dv.$$

Since $\psi(v) \approx 0$, for $v \approx 0$, when ϵ is infinite the probability above is infinitesimal. Therefore $\max_{0\leq t\leq 1} \{P[|X(t)| > b]\} \approx 0$ whenever b is infinite. By (5.3.18) and the identical distribution of the $\delta X(s)$,

$$P\left[\max_{t\leq 1}|X(t)| > 2\epsilon\right] \leq \frac{P[|X(1)| > \epsilon]}{1- \max_{s\leq 1} P[|X(s)| > \epsilon]}.$$

So $P[\max_{t\leq 1}|X(t)| > 2n] \approx 0$ whenever n is infinite. For finite natural numbers $m > 0$, let

$$\Lambda_m = \{\omega : \max_{t\leq 1}|X(t)| > 2m\} \quad \text{and} \quad \Lambda_0 = \cap \Lambda_m.$$

The paths of x are finite off Λ_0 and $P[\Lambda_0] = 0$. This proves the lemma.

Next, we show that the paths of X a.s. are S-continuous at the endpoints. This time we use Skorohod's lemma with small epsilon. Let $\epsilon >> 0$ be standard.

$$P\left[\max_{s\leq 1/n} |X(s)| > 2\epsilon\right] \leq \frac{P[|X(1/n)| > \epsilon]}{1- \max_{s\leq 1/n} P[|X(s)| > \epsilon]}$$

since the distribution of $X(t) - X(s)$ is the same as $X(t-s)$.

Using (4.1.4) again, when $s \leq \frac{1}{n}$ we see

$$P[|X(s)| > \epsilon] \leq \alpha\epsilon\int_0^{1/\epsilon} \{1-\mathrm{Re}[1+\delta t\psi(v)]^{s/\delta t}\}dv,$$

so that

$$\max_{0\leq s\leq 1/n} P[|X(s)| > \epsilon] = O(1/n).$$

In particular, $\max_{s\leq 1/n} P[|X(s)| > \epsilon] \approx 0$ whenever n is infinite. By the inequalities above and the $\epsilon-\vartheta$ formulation of S-limits, $P[\max_{0\leq s\leq 1/m} |X(s)| > 2\epsilon]$ tends to zero as m tends finitely toward infinitey. This means the paths of X are S-continuous at zero a.s. A similar argument yields a.s. continuity at one.

In order to make a sum of stationary independent infinitesimal increments compatible with the measurable theory of time evolution (and notational conventions) in the remaining chapters we shall suppose that $X(0) = 0$ and

$$X(t) = \sum[\delta X(s) : 0 < s < t, \text{ step } \delta t]$$

where there is a fixed internal function $\gamma : W \longrightarrow {}^*\mathbb{R}$ and

$$\delta X(t) = \gamma(\omega_{t+\delta t}).$$

We must select γ carefully so that $\tilde{X}$ is finite a.s., of course. In chapter 4 we made the vague claim that all the classical laws can be represented this way, but we have explicitly shown that Brownian motions, Poisson processes and

Cauchy process (4.3.6) do arise this way. The latter kind are not integrable and so they offer certain technical challenges which we will sidestep by showing that such an X is a "semimartingale." (See (7.3.3).) There we also prove a general semimartingale lifting theorem. The paths of semimartingales in general must be sampled for infinitesimals $\Delta t > \delta t$ in order to have separated jumps. Our next undefended claim says that a 'natural classical object' has a 'nice internal version'. We shall not defend the claim because we must work with Δt-samples for other reasons below.

(5.3.21) **PROPOSITION:**

Let $X(0) = 0$ *and* $X(t) = \Sigma[\gamma(\omega_{s+\delta t}) : 0 < s < t,$ step $\delta t]$ *be a sum of* **independent identically distributed infinitesimal increments* $\delta X(t,\omega) = \gamma(\omega_{t+\delta t})$, *for a single internal* $\gamma : W \longrightarrow {}^*\mathbb{R}$. *Assume that* $X(t)$ *is finite a.s. on* $\mathbb{T}$. *Then* X *has a* δt*-decent path sample and its decent path projection* $\tilde{X}(r)$ *is a stationary independent increment process on* $[0,1]$, *that is, if* $0 \leq r_0 < r_1 < \cdots < r_1 \leq 1$, $\{\tilde{X}(r_j)-\tilde{X}(r_{j-1}) : 1 \leq j \leq m\}$ *is an independent family and the distribution of* $[\tilde{X}(r+s)-\tilde{X}(r)]$ *only depends on* s, $0 \leq r < r+s \leq 1$.

The reader wishing to explore this claim is referred to Gnedenko and Kolmogorov [1954, 1968] for classical laws and Gihman and Skorohod [1974] or Ethier and Kurtz [1980] for technical help with $D[0,1]$. This claim is not used except as motivation for introducing semimartingales, where different estimates play the role of Skorohod's lemma.

(5.3.22) **DEFINITION:**

Let $Y : [0,1] \times \Omega \longrightarrow \mathbb{R}$ *be a function and let* $X : \mathbb{T} \times \Omega \longrightarrow {}^*\mathbb{R}$ *be internal and a.s. have a* Δt*-decent path sample. We say* X *is a* Δt*-decent path lifting or D-lifting of* Y *if the D-projection* $\tilde{X}$ *and* Y *are indistinguishable.*

If we show that Y has a Δt-lifting, then the restriction of that lifting to multiples of another infinitesimal $\nabla t = k \cdot \Delta t$, $k \in \mathbb{N}$, is a ∇t-lifting with the same projection. This seems a little silly, but we'll see later that we sometimes need to make $\Delta t > \delta t$ in order to maintain some side condition on the lifting. The Projection Theorem says that the projection works for every infinitesimal $\Delta t \geq \delta t$.

Recall the Doob's definition (5.2.6); Y is a stochastic process if each section, Y^r, is P-measurable.

(5.3.23) **THE DECENT PATH LIFTING THEOREM:**

If a stochastic process $Y : [0,1] \times \Omega \longrightarrow \mathbb{R}$ *a.s. has decent paths in* $D[0,1]$*, then for each infinitesimal* $\Delta t \geq \delta t$ *in* $\mathbb{T}$*,* Y *has a* Δt*-decent path lifting.*

PROOF:

Assume $Y_\omega \in D[0,1]$ except for ω in Λ where $P[\Lambda] = 0$. We want to show that the map $\omega \longrightarrow Y_\omega$ is P-measurable as a map into $D[0,1]$. What we have to work with are the facts: (1) each Y^r is P-measurable and (2) evaluation $y \longrightarrow y(r)$

from $D[0,1]$ into $\mathbb{R}$ is Borel measurable. The second fact can be proved by forward integral averaging in a manner analogous to (5.3.16). Evaluation of each average is continuous and the evaluated averages tend to the evaluation.

Let $\{q_m\}$ enumerate the rationals in $[0,1]$ with $q_1 = 1$. For a fixed x in $D[0,1]$ and $\epsilon > 0$ define the set

$$\mathcal{D}(m,\epsilon;x) = \{y \in D[0,1] : \exists\ \rho \in \Lambda[0,1], \delta(\rho) < \epsilon$$

$$\& \ |x(\rho(q_j))-y(q_j)| < \epsilon, \text{ for } 1 \le j \le m\}.$$

Let V be the set of points $\xi = (\xi_1,\cdots,\xi_m)$ in $\mathbb{R}^m$ such that $\xi_j = x(\rho(q_j))$ for some ρ with $\delta(\rho) < \epsilon$. Let U be the set of η such that $|\eta_j-\xi_j| < \epsilon$, for $1 \le j \le m$. The set U is open and since evaluation is measurable,

$$\{\omega : Y(q_j,\omega) \in U\} \quad \text{is} \quad P\text{-measurable}.$$

In other words, $Y_\omega^{-1}(\mathcal{D}(m,\epsilon;x))$ is P-measurable for each m, ϵ and x. The lemma following the proof shows that open balls in the Kolmogorov metric are in the sigma algebra generated by the $\mathcal{D}$-sets. Since $D[0,1]$ is separable (step functions with rational values are dense), Y_ω is measurable.

Let $K \subseteq {}^*D[0,1]$ be the set of *right continuous internal step functions with jumps at the points of $\mathbb{T}$ (if anywhere). The set is internal, S-dense for the Kolmogorov metric and in 1-1 correspondence with internal functions defined on $\mathbb{T}$. By (5.1.6), Y_ω has a metric lifting with values in K. The restriction of the lifting to $\mathbb{T}$ is a δt-decent path lifting.

The effect of the next technical lemma is that the sigma algebra generated by $\mathcal{D}$-sets is the Borel algebra of $D[0,1]$. The lemma finishes the proof of the lifting theorem.

(5.3.24) **LEMMA:**

$$\{y \in D[0,1] : k(y,x) < \epsilon\} = \bigcup_{\vartheta}[\bigcap_{m} \mathcal{D}(m,\epsilon-\vartheta;x) : \vartheta \in \mathbb{Q} \cap (0,\epsilon)].$$

PROOF:

If $k(y,x) < \epsilon$, then $k(y,x) < \epsilon-\vartheta$ for some positive rational ϑ, so there is a time deformation ρ with $\delta(\rho) < \epsilon-\vartheta$ and $|y(r)-x(\rho(r))| < \epsilon-\vartheta$ for all r in $[0,1]$, hence in particular for rational r.

Now we show the other inclusion. Fix ϑ and suppose $y \in \bigcap_{m} \mathcal{D}(m,\epsilon-\vartheta;x)$. For each m choose a time deformation ρ_m so that $\delta(\rho_m) < \epsilon-\vartheta$ and $|x(\rho_m(q_j))-y(q_j)| < \epsilon-\vartheta$, for $1 \leq j \leq m$. Since any ρ_n in the *extension of the ρ_m sequence satisfies $\delta(\rho_n) < \epsilon-\vartheta$, ρ_n is Lipschitz, and the infinitesimal hull $\hat{\rho} = \hat{\rho}_n$, (0.3.22), is a standard time deformation with $\delta(\hat{\rho}) \leq \epsilon-\vartheta$. Also, we know $|x(\rho_n(q_j))-y(q_j)| < \epsilon-\vartheta$ for all standard rationals, in fact, for $1 \leq j \leq n$, n infinite. For each of these points either $|x(\hat{\rho}(q_j))-y(q_j)| \leq \epsilon-\vartheta$ or $|x(\hat{\rho}(q_j)^-)-y(q_j)| \leq \epsilon-\vartheta$, so $\sup|x(\hat{\rho}(r))-y(r)| \leq \epsilon-\vartheta$ and $k(x,y) \leq \epsilon-\vartheta$. This proves the inclusion.

Internal processes that almost surely have left and right S-limits always have decent path samples.

(5.3.25) **THE S-LIMIT LEMMA:**

Let X *be an internal process and suppose there is a set* $\Lambda \subseteq \Omega$ *with* $P[\Lambda] = 0$ *such that if* $\omega \notin \Lambda$, *then for all* $r \in [0,1]$, $S\text{-}\lim_{\mathbb{T}_\Delta\; t\uparrow r} X(t,\omega)$ *and* $S\text{-}\lim_{\mathbb{T}_\Delta\; t\downarrow r} X(t,\omega)$ *exist.* *Then there is a positive infinitesimal* $\nabla t = k\cdot\Delta t$ *such that* X *has a* ∇t*-decent path sample with projection* $\tilde{X}(r,\omega) = S\text{-}\lim_{\mathbb{T}_\Delta\; t\downarrow r} X(t,\omega)$.

PROOF:

Let $Y(r,\omega) = S\text{-}\lim X(t,\omega)$ for $\omega \in \Lambda_1$, for a Loeb null set $\Lambda_1 \supseteq \Lambda$, and zero otherwise. Let $Z(t,\omega)$ be a δt-decent path lifting of Y. Since Z and X have the same right S-limits, for each $m \in \mathbb{N}$, there is a $\Delta_m \in \mathbb{T}_\Delta$ such that $\Delta_m \approx \frac{1}{m}$ and $P[(\max_{1\leq k\leq m} |X(k\Delta_m) - Z(k\Delta_m)| \vee |X(1) - Z(1)|) > \frac{1}{m}] < \frac{1}{m}$. Extend Δ_m to an internal sequence with values in $\mathbb{T}_\Delta$ using comprehension. The probability inequality above remains true up to some infinite n_1 while $\Delta_k \approx \frac{1}{k}$ remains true up to a smaller infinite n, by Robinson's sequential lemma. Let $\nabla t = \Delta_n$. Since Z has a δt-decent path sample, it has a ∇t-decent path sample. We know $X(t) \approx Z(t)$ for $t = k\cdot\nabla t$, so X has decent paths for a sample along multiples of ∇t. This proves the lemma.

Our next topic is lifting (continuous or) decent path processes with a side condition of uniform integrability.

(5.3.26) **DEFINITION:**

A family of random variables, $\{X_\alpha : \alpha \in A\} \subseteq L^1(P)$ *is called uniformly integrable provided that for every* $\epsilon > 0$, *there exists* M *in* $\mathbb{N}$ *such that for all* $m \geq M$ *and all* α, $E[|X_\alpha(\omega)| I_{\{|X_\alpha|>m\}}(\omega)] < \epsilon$. *(The indicator* $I_{\{|X_\alpha|>m\}}(\omega) = 1$ *if* $|X_\alpha(\omega)| > m$ *and* $= 0$, *if* $|X_\alpha(\omega)| \leq m$.)

Uniform integrability means that the truncations $X_\alpha^m = X_\alpha I_{\{|X_\alpha| \leq m\}}$ converge in L^1-norm to X_α as m tends to infinity, uniformly in α. It is not hard to show that uniform integrability of a family $\{X_\alpha\}$ is equivalent to the two conditions:

(a) *the* L^1*-norms of the family are uniformly bounded,* there is a b such that for all α, $E[|X_\alpha|] < b$, and

(b) *the family is uniformly absolutely continuous,* for every $\epsilon > 0$ there exists $\vartheta > 0$ such that for every measurable set $\Lambda \subseteq \Omega$, if $P[\Lambda] < \vartheta$, then, for all α, $E[|X_\alpha| I_\Lambda] < \epsilon$.

If $\{Y_\alpha : \alpha \in A\}$ is an internal family of random variables and each Y_α is S-integrable, Lemma (1.6.2) implies that the family $\{\tilde{Y}_\alpha : \alpha \in A\}$ is uniformly integrable.

We want to show a correspondence between the family of real-instant-sections of a stochastic process and the time sections of a corresponding internal process. We say that a *stochastic process* $X : [0,1] \times \Omega \longrightarrow \mathbb{R}$ *is uniformly integrable* if the family of sections $\{X^r : r \in [0,1]\}$ is uniformly integrable.

(5.3.27) THE S-INTEGRABLE DECENT PATH LIFTING THEOREM:

A stochastic process $Y : [0,1] \times \Omega \longrightarrow \mathbb{R}$ *is uniformly integrable and a.s. has decent paths if and only if there is an infinitesimal* Δt *in* $\mathbb{T}$ *and a* Δt*-decent path lifting* X *for* Y *such that whenever* $t = k\Delta t$ *in* $\mathbb{T}$, *the section* X^t *is S-integrable.*

PROOF:

Suppose there is such an X. For each r in $[0,1]$ there is a $t = k\Delta t \approx r$ such that $\tilde{X}(r) = stX(t) = Y(r)$, a.s. Lemma (1.6.2) implies that the whole family $\{stX(t) : t \in \mathbb{T}\}$ is uniformly integrable, hence Y is.

Conversely, suppose Y is uniformly integrable and a.s. has decent paths. The family of sections and left-limits-sections,

$$\{Y(r) : r \in [0,1]\} \cup \{Y(r^-) = \lim_{s \uparrow r} Y(s) : r \in [0,1]\},$$

continues to be uniformly integrable by Fatou's lemma,

$$E\Big[|Y(r^-)| I_{\{|Y(r^-)|>m\}}\Big] = E\Big[\lim_{n\to\infty} |Y(r-\tfrac{1}{n})| I_{\{|Y(r-\frac{1}{n})|>m\}}\Big]$$

$$\le \liminf E\Big[|Y(r-\tfrac{1}{n})| I_{\{|Y(r-\frac{1}{n})|>m\}}\Big].$$

Let $M(\epsilon)$ be a function such that if $m \ge M(\epsilon)$, then

$$E[|Y(r)|I_{\{|Y(r)|\geq m\}}] < \frac{\epsilon}{3}.$$

Let $Z : \mathbb{T} \times \Omega \longrightarrow {}^*\mathbb{R}$ be a δt-decent path lifting of Y and define truncations for each n in ${}^*\mathbb{N}$,

$$Z_n(t,\omega) = \begin{cases} Z(t,\omega) & , \quad \text{if} \quad |Z(t,\omega)| \leq n \\ n \ \mathrm{sgn}[Z(t,\omega)], & \quad \text{if} \quad |Z(t,\omega)| \geq n \end{cases}$$

while for each finite n,

$$Y_n(r,\omega) = \begin{cases} Y(r,\omega) & , \quad \text{if} \quad |Y(r,\omega)| \leq n \\ n \ \mathrm{sgn}[Y(r,\omega)], & \quad \text{if} \quad |Y(r,\omega)| \geq n. \end{cases}$$

These truncations continue to have decent paths in the appropriate senses and Z_n lifts Y_n for finite n.

For any t, $P[st(Z_n(t)) = Y_n(st(t))$ or $st(Z_n(t)) = Y_n(st(t))^-)] = 1$, so that

$$E[|Z_n(t)|I_{\{|Z_n(t)|>m+1\}}]$$

$$\lesssim \begin{cases} E[|Y_n(st(t))|I_{\{|Y_n(s(t))| \ > m \ \& \ Z_n(t) \ \approx \ Y_n(st(t))\}}] \\ + \ E[|Y_n(st(t)^-)|I_{\{|Y_n(st(t)^-)| \ > m \ \& \ Z_n(t) \ \approx \ Y_n(st(t)^-)\}}] \end{cases}$$

$$\lesssim E[|Y(st(t))|I_{\{|Y(st(t))|>m\}} + Y(st(t)^-)|I_{\{|Y(st(t)^-)|>m\}}].$$

Hence we see that the internal sets

$$I_p = \{n \in {}^*\mathbb{N} : n>p \ \& \ (\forall m>M(\tfrac{1}{p}))(\forall t)E[|Z_n(t)|I_{\{|Z_n(t)|>m\}}] < \tfrac{1}{p}\}$$

contain all finite n's past p when p is finite. Let $n \in \cap[I_p : p \in \mathbb{N}]$. The countable intersection is nonempty by saturation (0.4). The process $X = Z_n$ is an S-integrable decent path lifting of Y.

(5.3.28) EXERCISE:

If Y is an a.s. continuous process and X is a D-lifting of Y, show that X is a.s. S-continuous. Logically, we could have done without section (5.2) and derived those results from this exercise. Re-state (5.3.26) for an a.s. continuous uniformly integrable process.

(5.4) Lebesgue and Borel Path Processes

We continue to work on the hyperfinite evolution scheme defined at the beginning of the chapter, $\Omega = W^{\mathbb{T}}$, $\mathbb{T} = \{\delta t, 2\delta t, \cdots, 1\}$, P, etc. The main results of this section are extensions of Theorems (2.1.4) and (2.2.6) to the time variable of a stochastic process. Let λ denote Lebesgue measure on $[0,1]$. Let δt denote internal counting measure on $\mathbb{T}$ multiplied by the constant δt and let $\boldsymbol{\delta t}$ denote its hyperfinite extension. Let $\boldsymbol{\delta t} \times \mathbf{P}$ denote the (conventional) complete product of the (ordinary) measures $\boldsymbol{\delta t}$ and $\mathbf{P}$. Let $\delta t \otimes \delta P$ denote the internal measure with weight function $\delta t \otimes \delta P(t,\omega) = \delta t \cdot \delta P(\omega)$ and let $\boldsymbol{\delta t \otimes \delta P}$ denote its hyperfinite extension. (See Chapter 3.)

Our first result is the analogue of the theorem for one variable that says standard parts of internal sets are closed. Thruout the section we need the half-standard-part function

$$s : \mathbb{T} \times \Omega \longrightarrow [0,1] \times \Omega, \quad s(t,\omega) = (st(t),\omega).$$

(5.4.1) LEMMA:

Let $\mathcal{E} \subseteq \mathbb{T} \times \Omega$ *be an internal set and* $s(t,\omega) = (st(t),\omega)$. *The set* $s(\mathcal{E})$ *is* $(\text{Borel}[0,1] \times \text{Loeb}(\Omega))$ *-measurable.*

PROOF:

Let $\{I_m\}$ be an enumeration of the standard open rational endpoint intervals in $\mathbb{R}$. Let $\mathcal{E}_m = \{\omega : \exists t \in {}^*I_m$ with $(t,\omega) \in \mathcal{E}\}$ and let

$$\mathcal{E}^m = ([0,1] \times \mathcal{E}_m \cup (I_m^c \times \mathcal{E}_m^c),$$

where A^c denotes the complement of A. Each $\mathcal{E}_m$ is internal, so $\mathcal{E}^m$ is (Borel $\times$ Loeb)-measurable. This construction gives

$$s(\mathcal{E}) = \bigcap_m \mathcal{E}^m$$

for the following reasons. Let $(r,\omega) \in s(\mathcal{E})$ and fix m. If $r \in I_m$, then since it is a standard open interval, the $(t,\omega) \in \mathcal{E}$ which makes $s(t,\omega) = (r,\omega)$ has $t \in {}^*I_m$ and $\omega \in \mathcal{E}_m$. If $r \notin I_m$ but some $t \in {}^*I_m$ satisfies $(t,\omega) \in \mathcal{E}$, we still have $(r,\omega) \in [0,1] \times \mathcal{E}_m$. In the final case where $r \notin I_m$ and no $t \in {}^*I_m$ has $(t,\omega) \in \mathcal{E}$, then $(r,\omega) \in I_m^c \times \mathcal{E}_m^c$. For the opposite inclusion, suppose $(r,\omega) \in \cap\mathcal{E}^m$ and for the subsequence with $r \in I_m$ and $\cap I_m = \{r\}$, choose $t_m \in {}^*I_m$ such that $(t_m,\omega) \in \mathcal{E}$. Next, select a sub-subsequence t_k so that $|r-t_k|$ is monotone decreasing. Extend t_k to an internal sequence and consider the internal set of indices

$$\{n \in {}^*\mathbb{N}: (\forall k \in {}^*\mathbb{N})[0 < k \leq n \Rightarrow [(t_k,\omega) \in \mathcal{E}$$
$$\& \ |r-t_k| < |r-t_{k-1}|]]\}.$$

An infinite n from this set satisfies $(r,\omega) = (st(t_n),\omega)$. This proves the lemma.

Let $(\mathbb{M},\rho)$ be a metric space entity, $\mathbb{M},\rho \in \mathfrak{X}$. We state our results for functions $X : [0,1] \times \Omega \longrightarrow \mathbb{M}$, since it is no harder than for real values.

(5.4.2) DEFINITION:

Given a function $X : [0,1] \times \Omega \longrightarrow \mathbb{M}$, *an internal function* $Y : \mathbb{T} \times \Omega \longrightarrow {}^{*}\mathbb{M}$ *such that* $X(st(t),\omega) = st_{\rho}Y(t,\omega)$ *a.e.* $\delta t \times \mathbf{P}$ *is called a Lebesgue lifting of* X.

Implicit in the definition of Lebesgue lifting is a weak notion of almost S-continuity. If $st(t) = st(s) = r$ then $Y(t,\omega) \approx Y(s,\omega) \approx X(r,\omega)$, at least a.s. in the product space.

(5.4.3) THEOREM:

Let $(\mathbb{M},\rho)$ *be a complete separable metric space entity. Let* K *be an S-dense internal subset of* ${}^{*}\mathbb{M}$, $st_{\rho}(K) = \mathbb{M}$. *A function* $X(r,\omega) : [0,1] \times \Omega \longrightarrow \mathbb{M}$ *is* $(\lambda\times\mathbf{P})$*-measurable if and only if there is an internal function* $Y(t,\omega) : \mathbb{T} \times \Omega \longrightarrow K$ *such that* $X(st(t),\omega) = st_{\rho}(Y(t,\omega))$ *a.s.* $\delta t \times \mathbf{P}$, *that is, if and only if has a Lebesgue lifting in* K.

The proof of this result, given below, is based on the two variable extension of Theorem (2.1.2) as follows. Notice that if $\mathcal{S} \subseteq [0,1] \times \Omega$, then $\{(t,\omega) \in \mathbb{T} \times \Omega : (st(t),\omega) \in \mathcal{S}\} = s^{-1}(\mathcal{S})$, where s is the half-standard-part map from above.

(5.4.4) **PROPOSITION:**

For each set $\mathcal{S} \subseteq [0,1] \times \Omega$, *the following conditions are equivalent:*

(a) $\mathcal{S}$ *is* $(\lambda \times \mathbf{P})$*-measurable.*

(b) $s^{-1}(\mathcal{S})$ *is* $(\delta t \times \mathbf{P})$*-measurable.*

(c) $s^{-1}(\mathcal{S})$ *is* $\delta t \otimes \delta \mathbf{P}$*-measurable.*

In this case, the map $s(t,\omega) = (st(t),\omega)$ *is measure preserving,* $(\delta t \times \mathbf{P}) s^{-1}(\mathcal{S}) = \lambda \times \mathbf{P}(\mathcal{S})$.

PROOF:

If $\mathcal{S} = [a,b] \times \Lambda$, where Λ is internal, then $s^{-1}(\mathcal{S}) = \bigcap_m [a-\frac{1}{m}, b+\frac{1}{m}] \times \Lambda$, so $s^{-1}(\mathcal{S})$ is measurable and $\delta t \times \mathbf{P}(s^{-1}(\mathcal{S})) = \lambda \times \mathbf{P}(\mathcal{S})$. The set

$$\sum = \{\mathcal{S} \subseteq [0,1] \times \Omega : \lambda \times \mathbf{P}(s^{-1}(\mathcal{S}))\}$$

is a $(\lambda \times \mathbf{P})$-complete sigma algebra containing Borel × Loeb. Since both measures are complete this shows that (a) implies (b).

Condition (b) implies (c) by (3.1.2).

$(c \Rightarrow a)$: Suppose $s^{-1}(\mathcal{S})$ is $\delta t \otimes \delta \mathbf{P}$-measurable, so there are chains of internal sets $\mathcal{C}_1 \subseteq \mathcal{C}_2 \subseteq \cdots$ and $\mathcal{D}_1 \subseteq \mathcal{D}_2 \subseteq \cdots$ such that $\mathcal{C} = \cup \mathcal{C}_m \subseteq s^{-1}(\mathcal{S})$, $\mathcal{D} = \cup \mathcal{D}_m \subseteq s^{-1}(\mathcal{S}^c)$, and $\delta t \otimes \delta \mathbf{P}(\mathcal{C}) + \delta t \otimes \delta \mathbf{P}(\mathcal{D}) = 1$. We know that $s(\mathcal{C}_m)$ and $s(\mathcal{D}_m)$ are measurable. Also,

$$\begin{aligned} 1 &= \delta t \otimes \delta \mathbf{P}(\mathcal{C}) + \delta t \otimes \delta \mathbf{P}(\mathcal{D}) \\ &\leq \delta t \otimes \delta \mathbf{P}[s^{-1}(s(\mathcal{C})] + \delta t \otimes \delta \mathbf{P}[s^{-1}(s(\mathcal{D}))]. \end{aligned}$$

From the beginning of our proof [before (a) implies (b)] and (3.1.2) we know that $\lambda \times P[s(\mathcal{C})] = \delta t \otimes \delta P[s^{-1}(s(\mathcal{C}))]$ and the same for $\mathcal{D}$. Hence,

$$1 \leq \lambda \times P[s(\mathcal{C})] + \lambda \times P[s(\mathcal{D})] \leq \lambda \times P[\mathcal{S}] + \lambda \times P[\mathcal{S}^c]$$

and $\mathcal{S}$ is measurable. Again, the first part of our proof shows s is measure preserving. This proves (5.4.4).

PROOF of (5.4.3):

Let X be $(\lambda \times P)$-measurable and define

$$Z(t,\omega) = X(st(t),\omega).$$

By (5.4.4), Z is $\delta t \otimes \delta P$-measurable. By (5.1.6), Z has a metric lifting Y, $st_\rho(Y) = Z$ a.s.

Conversely, suppose $st_\rho[Y(t,\omega)] = X(st(t),\omega)$ except for (t,ω) in $\mathcal{N}$ a $(\delta t \times P)$-null set. Since Y is internal, $st_\rho[Y]$ is $(\delta t \otimes \delta P)$-measurable. Let U be open in $\mathcal{M}$.

$$\begin{aligned}\{(t,\omega) : st_\rho[Y(t,\omega)] \in U\}\setminus\mathcal{N} &= \{(t,\omega) : X(st(t),\omega) \in U\}\setminus\mathcal{N} \\ &\subseteq s^{-1}\{(r,\omega) : X(r,\omega) \in U\} \\ &= \{(t,\omega) : X(st(t),\omega) \in U\} \\ &\subseteq \{(t,\omega) : st_\rho[Y(t,\omega)] \in U\} \cup \mathcal{N}.\end{aligned}$$

By completeness and (5.4.4), X is $(\lambda \times P)$-measurable.

Notice that we have neither assumed nor concluded that X

is a stochastic process in the sense of (5.2.6).

For reference, here is the analog of (2.1.4)(b) in the $(\lambda \times P)$-setting. We state a slight extension for vector values. Extensions of this result are useful in the study of stochastic processes.

(5.4.5) S-INTEGRABLE LEBESGUE LIFTING THEOREM:

Let d *be an integer, let* $\| \ \|$ *denote the norm on* $\mathbb{R}^d$, *and let* $p \geq 1$ *be real. A function* $X : [0,1] \times \Omega \longrightarrow \mathbb{R}^d$ *has* $\|X\| \in L^p(\lambda \times P)$ *if and only if there is an internal* $Y : \mathbb{T} \times \Omega \longrightarrow {}^*\mathbb{R}^d$ *such that* $\|Y\|^p$ *is* $(\delta t \otimes \delta P)$*-S-integrable and* $st[Y(t,\omega)] = X(st(t),\omega)$ *a.s., that is, if and only if* X *has an* SL^p*-Lebesgue-lifting.*

We leave the proof as an exercise. If the vector values bother you, first show the scalar case and then use the max norm on $\mathbb{R}^d$, $\|(x^1,\cdots,x^d)\| = \max|x^j|$.

(5.4.6) PROPOSITION:

If X *has a* Δt*-decent path lifting* Y, *then* Y *is a Lebesgue lifting for the internal measure* $\Delta t \times \delta P$ *of counting on* $\{t \in \mathbb{T} : t = k\Delta t,\ k \in {}^*\mathbb{N}\}$ *times* Δt *crossed with* δP.

PROOF:

By extending Y to be $Y(k\Delta t)$ for points of t between $k\Delta t$ and $(k+1)\Delta t$, we may assume that Y is a δt-lifting. We wish to show that $X(st(t),\omega) = st\ Y(t,\omega)$ a.s. $\Lambda t \otimes \delta P$.

Since X has a decent path lifting, it is (Borel × P)-measurable, so $X(st(t),\omega) = st\ Z(t,\omega)$ a.s., for some internal Z. Also, for a.a. ω, $\lim_{m\to\infty} st\ Y(t+\frac{1}{m}) = X(st(t))$. It follows that, except for a null set, for every p in $\mathbb{N}$ there is an m_p in $\mathbb{N}$ such that if m is finite and $m > m_p$, then $|Y(t+\frac{1}{m},\omega) - Z(t,\omega)| < \frac{1}{p}$. Therefore, the internal sets

$$\{n \in {}^*\mathbb{N} : \forall m \in {}^*\mathbb{N}, m_p \le m \le n \Rightarrow (\delta t \times P)[|Y(t+\tfrac{1}{m})-Z(t)| \ge \tfrac{1}{p}] \le \tfrac{1}{p}\}$$

contains an infinite n_p. Choose $n \in \cap\mathbb{N}[m_p,n_p]$, then $Y(t+\frac{1}{n})$ and hence $Y(t)$ is a Lebesgue lifting of X.

(5.4.7) **PROPOSITION:**

For each set $\mathcal{B} \subseteq [0,1] \times \Omega$, *the following conditions are equivalent:*

(a) $\mathcal{B}$ *is* (Borel[0,1] × Loeb(Ω))-*measurable.*

(b) $s^{-1}(\mathcal{B})$ *is* (Loeb($\mathbb{T}$) × Loeb(Ω))-*measurable.*

(c) $s^{-1}(\mathcal{B})$ *is* Loeb($\mathbb{T} \times \Omega$)-*measurable.*

PROOF:

(a) implies (b): If $\mathcal{B} = B \times \Lambda$, for $B \in$ Borel and $\Lambda \in$ Loeb, then Henson's theorem (2.2.6) says $s^{-1}(\mathcal{B}) = st^{-1}(B) \times \Lambda \in$ Loeb($\mathbb{T}$) × Loeb(Ω). Finite disjoint unions of such rectangles also have this property. A simple application of the Monotone Class Lemma (3.3.4) shows that every $\mathcal{B} \in$ Borel × Loeb has $s^{-1}(\mathcal{B}) \in$ Loeb × Loeb.

(b) implies (c) is trivial because Loeb($\mathbb{T} \times \Omega$) ⊃ Loeb($\mathbb{T}$) × Loeb(Ω).

(c) implies (a): We know that $s^{-1}(\mathfrak{B})$ and $[s^{-1}(\mathfrak{B})]^c$ are derived from internal subsets of $\mathbb{T} \times \Omega$ by Souslin operations. (See (2.2.1) and the following remark.) Let

$$s^{-1}(\mathfrak{B}) = \bigcup_{\sigma} \bigcap_{n} \mathcal{I}_{\sigma|n}$$

and

$$[s^{-1}(\mathfrak{B})]^c = \bigcup_{\sigma} \bigcap_{n} \mathcal{J}_{\sigma|n}$$

for internal $\mathcal{I}$'s and $\mathcal{J}$'s. By (5.4.7), $s(\mathcal{I}_{\sigma|n})$ and $s(\mathcal{J}_{\sigma|n})$ are Borel × Loeb sets and since

$$\mathfrak{B} = s(s^{-1}(\mathfrak{B})) = s(\bigcup_{\sigma} \bigcap_{n} \mathcal{I}_{\sigma|n}) = \bigcup_{\sigma} \bigcap_{n} s(\bigcap_{m=1}^{n} \mathcal{I}_{\sigma|m})$$

and

$$\mathfrak{B}^c = s[s^{-1}(\mathfrak{B})]^c = s(\bigcup \bigcap \mathcal{J}_{\sigma|n}) = \bigcup \bigcap s(\bigcap_{m=1}^{n} \mathcal{J}_{\sigma|m}),$$

it suffices to show that s commutes with decreasing intersections of internal sets (compare (2.2.4)) and that the analog of Lusin's Separation Theorem (2.2.3) holds for Borel × Loeb. The proof of the first fact is similar to that of (2.2.4). The necessary separation theorem says: *if* $\mathfrak{B}$ *and* $\mathfrak{B}^c$ *are analytic over the family of compact × internal rectangles, then* $\mathfrak{B}$ *is* (Borel[0,1] × Loeb(Ω))*-measurable.*

Let $F = [0,1] \times \Omega$ and let $\mathcal{F}$ be the family $\mathcal{K} \cdot \mathcal{E}$ of all disjoint finite unions of rectangles $K \times \Lambda$, where $K \in \mathcal{K}$ is a compact subset of $[0,1]$ and $\Lambda \in \mathcal{E}$ is an internal event $\Lambda \subseteq \Omega$. We claim that $\mathcal{K} \cdot \mathcal{E}$ is a semicompact paving of $[0,1] \times \Omega$. If $\{K_n \times \Lambda_n\}$ has the finite intersection property, then both $\{K_n\}$ and $\{\Lambda_n\}$ have the property. By compactness on $[0,1]$, $\cap K_n \neq \emptyset$ and by saturation on Ω, $\cap \Lambda_n \neq \emptyset$. Therefore, $\cap(K_n \times \Lambda_n) \neq \emptyset$. Theorem (2.2.3) now gives the separation result stated above.

The reason for proving Proposition (5.4.7) is to give the two-variable analog for the following function version of (2.2.8) (which says that $B \in \text{Borel}[0,1]$ if and only if $st^{-1}(B) \in \text{Loeb}(\mathbb{T})$). Suppose $f : [0,1] \longrightarrow \mathbb{R}$ is Borel measurable. Then $g : \mathbb{T} \longrightarrow \mathbb{R}$ given by $g(t) = f(st(t))$ is Loeb measurable by (2.2.8). Also, g is infinitesimally S-continuous (but usually not internal). Conversely, if g is Loeb measurable and $t \approx s$ implies $g(t) = g(s)$, then $f(r) = g(st^{-1}(r))$ is well defined and Borel measurable by (2.2.6). We did not give this version of (2.2.6) in Chapter 2, but instead showed how to replace g with an internal h that was S-continuous and satisfied $f(st(t)) = st(h(t))$ almost everywhere with respect to various hyperfinite measures (see section 2.1 for Lebesgue-like measures and 2.3 for Borel-like ones). The corresponding internal path lifting is postponed until Chapter 7.

(5.4.8) **PROPOSITION:**

Let $G : [0,1] \times \Omega \longrightarrow \mathbb{R}$ *be any function. The following are equivalent conditions:*

(a) $G(r,\omega)$ *is* $(\text{Borel}[0,1] \times \text{Loeb}(\Omega))$*-measurable.*

(b) $G(st(t),\omega)$ *is* $(\text{Loeb}(\mathbb{T}) \times \text{Loeb}(\Omega))$*-measurable.*

(c) $G(st(t),\omega)$ *is* $\text{Loeb}(\mathbb{T} \times \Omega)$*-measurable.*

PROOF:

Apply (5.4.8) to the sets $\mathcal{B} = \{(r,\omega) : G(r,\omega) < a\}$.

The above result can be extended to (Borel × P)-measurable functions by the following

(5.4.9) **PROPOSITION:**

A (Borel[0,1] × Meas(P))-measurable function, H, is indistinguishable from a (Borel[0,1] × Loeb(Ω))-measurable function. That is, there exists a (Borel[0,1] × Loeb(Ω))-measurable G and a Loeb(Ω)-null set Λ such that if $\omega \notin \Lambda$, then

$$H(r,\omega) = G(r,\omega).$$

PROOF:

By the Monotone Class Lemma (3.3.4) and (1.2.13) we see that if $\mathcal{S} \in$ (Borel[0,1] × Meas(P)), then there is a Loeb(Ω)-null set Λ and a $\mathcal{T} \in$ (Borel[0,1] × Loeb(Ω)) such that if $\omega \notin \Lambda$, then the sections agree, $\mathcal{S}_\omega = \mathcal{T}_\omega$.

Let H_m be a sequence of simple (Borel × P)-measurable functions such that $H_m \longrightarrow H$ everywhere. (See the proof of (1.3.9) for an example of constructing H_m's.) By the first paragraph there are null sets Λ_m and simple (Borel × Loeb)-measurable functions G_m such that $G_m = H_m$ off Λ_m for all r. Let $G = \limsup G_m$. Off $\cup\Lambda_m$, $G = H$ for all r.

(5.5) Beyond $[0,1]$ and Scalar Values

This section briefly describes extensions of the results of this chapter. It can be skipped without loss of continuity. There are two kinds of extensions: the range and the domain. Extending the path spaces to have values in a complete separable metric space presents no serious difficulties. Near-standard paths in ${}^*C([0,1],\mathbb{M})$ or ${}^*D([0,1],\mathbb{M})$ take values near-standard in ${}^*\mathbb{M}$, rather than just finite values. Things are a little more abstract, but the ideas are pretty much the same. Extending the domain for continuous functions from $[0,1]$ to $[0,\infty)$, $\mathbb{R}$ or even $\mathbb{R}^d$ presents no problem either. Conditions on x in ${}^*C([0,\infty),\mathbb{M})$ look the same on $x(t)$ except t is finite rather than in ${}^*[0,1]$. The extension $D[0,\infty)$ or $D([0,\infty),\mathbb{M})$ is more technical.

(5.5.1) DEFINITION:

Let $(\mathbb{M},\rho)$ be a complete separable metric space. Let $C([0,\infty),\mathbb{M}) = \{x : x$ is a continuous function from $[0,\infty)$ into $\mathbb{M}\}$. A metric for measuring uniform convergence on compact subsets of $[0,\infty)$ is for $x,y \in C([0,\infty),\mathbb{M})$,

$$c(x,y) = \int_0^\infty c(x,y;u)e^{-u}du$$

where

$$c(x,y;u) = \min\{1, \max[\rho(x(t),y(t)) : 0 \le t \le u]\}.$$

We shall call c the compact convergence metric.

(5.5.2) **EXERCISE:**

Let $x, y \in {}^*C([0,1],\mathbb{M})$ be internal *continuous functions with values in ${}^*\mathbb{M}$. Show that $c(x,y) \approx 0$, which we write as $x \overset{c}{\approx} y$, if and only if $x(t) \overset{\rho}{\approx} y(t)$ for all finite positive t in ${}^*[0,\infty)$. ($x(t) \overset{\rho}{\approx} y(t)$ means $\rho(x(t),y(t)) \approx 0$.) For $\mathbb{M} = \mathbb{R}$, give an example of an internal x with $x(t) \approx 0$ for all finite t, but $x(T) = 1$ for some infinite T. Why is your x in ${}^*C[0,\infty)$?

The notion of S-continuity is nearly the same as before, x is *S-continuous* at b in ${}^*[0,\infty)$ if $t \approx b$ implies $x(t) \overset{\rho}{\approx} x(b)$.

(5.5.3) **PROPOSITION:**

A function x *in* ${}^*C([0,\infty),\mathbb{M})$ *is compact convergence near-standard if and only if* $x(t) \in ns_\rho({}^*\mathbb{M})$ *for each finite* t *and* x *is S-continuous at each finite* t. *In this case,* $st_c(x)$ *is given externally by* $t \longrightarrow st_\rho(x(t))$.

This result plays the same role as (5.1.3) does for $C[0,1]$. The next result is the extension of (5.2.2). First we need to extend the hyperfinite evolution scheme.

(5.5.4) **FRAMEWORK FOR HYPERFINITE EVOLUTION OVER** $[0,\infty)$:

Let

$$\mathbb{T} = \{t \in {}^*\mathbb{R} : t = k\cdot\delta t,\ k \in {}^*\mathbb{N},\ t \le \tfrac{1}{\delta t}\}$$

where

$$\delta t = \frac{1}{n}, \quad \textit{for some infinite} \quad n = h! \quad \textit{in} \quad {}^*\mathbb{N}.$$

Let

$$\Omega = W^{\mathbb{T}} = \{\omega : \omega \textit{ is an internal map from } \mathbb{T} \textit{ into } W\}$$

where

$$W = \{k \in {}^*\mathbb{N} : 1 \le k \le n^n\}.$$

Let

$$P = \textit{uniform internal counting measure on } \Omega,$$

$$\delta P(\omega) = \frac{1}{{}^{\#}[\Omega]}.$$

If $x : \mathbb{T} \longrightarrow {}^*\mathbb{M}$ is S-continuous and takes ρ-near-standard values in ${}^*\mathbb{M}$ for finite t, then

$$r \longrightarrow st_{\rho}[x(st^{-1}(r))]$$

is well-defined as a map from $[0,\infty)$ into $\mathbb{M}$. Denote this map by $\tilde{x}(r)$, the *continuous path projection of* x.

(5.5.5) EXERCISE:

Show that the map just described is a continuous (standard) function.

(5.5.6) PROPOSITION:

Let $X : \mathbb{T} \times \Omega \longrightarrow {}^*\mathbb{M}$ *be an internal function. Suppose there is a set* $\Lambda \subseteq \Omega$ *with* $P(\Lambda) = 0$ *such that if* $\omega \notin \Lambda$, *then* $X_{\omega}(t)$ *is* ρ*-near-standard and*

S-continuous for finite t *in* $\mathbb{T}$. *Then the continuous path projection* $\tilde{X}(r,\omega) = st_\rho[X(st^{-1}(r),\omega)]$ *is* $(\text{Borel}[0,\infty)\times\text{Meas}(P))$*-measurable as a function from* $[0,\infty) \times \Omega$ *into* $\mathbb{M}$.

To prove the extension of the S-continuous lifting theroem (5.2.7), we need to know the fact that separability of the range $\mathbb{M}$ makes $C([0,\infty),\mathbb{M})$ separable. Completeness of $\mathbb{M}$ also makes $C([0,\infty),\mathbb{M})$ complete.

(5.5.7) PROPOSITION:

Let $X : [0,\infty) \times \Omega \longrightarrow \mathbb{M}$ *be a stochastic process and suppose that for a.a.* ω, $X_\omega \in C([0,\infty),\mathbb{M})$. *Then there is an a.s. S-continuous internal* $Y \in \mathbb{T} \times \Omega \longrightarrow {}^*\mathbb{M}$ *such that* $\tilde{Y}$ *and* X *are indistinguishable. The internal process* Y *is called an S-continuous path lifting of* X, $\tilde{Y}(r,\omega) = st_\rho[Y(st^{-1}(r),\omega)]$ *a.s.* $\mathbf{P}$.

We now sketch the ideas in extending the decent paths to $[0,\infty)$. More details can be found in Ethier and Kurtz [1980].

(5.5.8) DEFINITION:

Let $(\mathbb{M},\rho)$ *be a complete separable metric space. Let* $D([0,\infty),\mathbb{M}) = \{x : [0,\infty) \longrightarrow \mathbb{M} \mid \lim_{q\uparrow r} x(q)$ *exists &* $\lim_{s\downarrow r} x(s) = x(r)\}$, *the space of right continuous with left limit functions taking values in* $\mathbb{M}$. *The Kolmogorov metric (for Skorohod's topology) is defined as follows.*

A time deformation on $[0,\infty)$ *is a strictly increasing Lipschitz continuous function* τ *mapping* $[0,\infty)$ *onto* $[0,\infty)$. *Let* $\Delta[0,\infty) = \{\tau \mid \tau$ *is a time deformation of* $[0,\infty)\}$. *The measure of deformation for* τ *in* $\Delta[0,\infty)$ *is* $\delta(\tau) = \text{ess sup}[|\log \tau'(r)| \mid r \geq 0]$. *For* $x,y \in D([0,\infty),\mathbb{M})$, $\tau \in \Delta[0,\infty)$ *and* $u \geq 0$ *define*

$$k(x,y;\tau,u) = \min\{1, \sup_r[\rho[x(\tau(r) \wedge u),y(r\wedge u)] \vee \delta(\tau)],$$

corresponding roughly to Kolmogorov's metric for $[0,u]$. *Finally, Kolmogorov's metric for* $[0,\infty)$ *is*

$$k(x,y) = \inf_{\tau\in\Delta[0,\infty)} \int_0^\infty k(x,y;\tau,u)e^{-u}du.$$

The space $D([0,\infty),\mathbb{M})$ is a complete separable metric space because we have assumed that $\mathbb{M}$ is.

Here is the Ascoli-type compactness theorem for $D[0,\infty)$.

(5.5.9) **PROPOSITION:**

Let y *be an internal function in* ${}^*D([0,\infty),\mathbb{M})$. *Then* y *is near-standard for Kolmogorov's metric if and only if*

(a) *for each finite* r *in* ${}^*[0,\infty)$, $y(r)$ *is near-standard in* ${}^*\mathbb{M}$, *and*

(b) *for each standard positive* u *and each standard positive* ϵ, *there exists a standard positive*

ϑ and an increasing sequence $0 = r_0 < r_1 < \cdots < r_k = u$ with $r_{j+1} - r_j > \vartheta$ such that $\rho[y(r), y(r_j)] < \epsilon$, when $r_j \le r < r_{j+1}$.

PROOF ($\Leftarrow$):

For each m in $\mathbb{N}$, let $\epsilon = \frac{1}{2m}$ and $u = \log(2m)$ and choose an associated ϑ and $r^m_{j+1} - r^m_j > \vartheta$, $0 \le j \le k(m)$. Define a standard step funcion

$$x_m(r) = st_\rho(y(r_j)), \quad \text{for} \quad s_j \le r < s_{j+1},$$

where $s_j = st(r_j)$ for $0 \le j \le k(m)$. Since the r^m_j's are finitely separated, the following internal piecewise linear function τ satisfies $\delta(\tau) \approx 0$. The map τ is piecewise linear between the corners s_j,

$$\tau(s_j) = r_j, \qquad 0 \le j \le k(m),$$

and

$$\tau(r) = r - s_{k(m)} + r_{k(m)}, \qquad \text{for} \quad r \ge s_{k(m)}.$$

We also know

$$\rho(y(\tau(r)), x_m(r)) < \frac{1}{2m}, \qquad \text{for} \quad 0 \le r < u$$

and

$$\int_0^\infty k(y, x_m; \tau, u) e^{-u} du < \frac{1}{2m}.$$

This proves that y is "pre-near-standard" in the terminology of Stroyan & Luxemburg [1976, 8.4], so by completeness of D, y is near-standard.

We shall not give a proof of the converse implication.

(5.5.10) **DEFINITION:**

Let $X : \mathbb{T} \times \Omega \longrightarrow {}^*\mathbb{M}$ *be internal. We say* X *a.s. has a* Δt*-decent path sample or* D*-sample over* $[0,\infty)$ *if there is a set* $\Lambda \subseteq \Omega$ *with* $P(\Lambda) = 0$ *and an infinitesimal* $\Delta t \geq \delta t$ *in* $\mathbb{T}$ *such that the extended step paths*

$$X^{\Delta t}(r,\omega) = \begin{cases} X(\Delta t,\omega) & , \quad for \quad 0 \leq r < 2\Delta t \\ X(k\Delta t,\omega) & , \quad for \quad k\Delta t \leq r < (k+1)\Delta t \\ X(1/\delta t,\omega), & \quad for \quad r \geq 1/\delta t \end{cases}$$

are Kolmogorov-near-standard in ${}^*D([0,\infty),\mathbb{M})$ *when* $\omega \notin \Lambda$. *Then* $\tilde{X}(r,\omega) = st_k[X^{\Delta t}(r,\omega)]$ *is called the decent path projection of the sample of* X.

(5.5.11) **PROPOSITION:**

If an internal function $X : \mathbb{T} \times \Omega \longrightarrow {}^*\mathbb{M}$ *has a* D*-sample, then its decent path projection* $\tilde{X}(r,\omega)$ *is* $(\text{Borel}[0,\infty)\times\text{Meas}(P))$*-measurable.*

The lifting half for decent paths is as follows.

(5.5.12) **PROPOSITION:**

If a stochastic process $Y : [0,\infty) \times \Omega \longrightarrow \mathbb{M}$ *a.s. has decent paths then for every infinitesimal* $\Delta t \geq \delta t$ *in* $\mathbb{T}$ *there is an internal* $X : \mathbb{T} \times \Omega \longrightarrow {}^*\mathbb{M}$ *such that* X *has a* Δt*-decent path sample and* $\tilde{X}$ *is indistinguishalbe from* Y. *Such an* X *is called a* Δt*-decent path lifting of* Y.

See the proof of (6.7.1) below.

(5.5.13) PROPOSITIONS:

A stochastic process $Y : [0,\infty) \times \Omega \longrightarrow \mathbb{M}$ *is uniformly integrable (on compact intervals* $[0,m]$*) and a.s. has decent paths if and only if for some infinitesimal* Δt *in* $\mathbb{T}$*, there is a* Δt*-decent path lifting* $X : \mathbb{T} \times \Omega \longrightarrow {}^*\mathbb{M}$ *of* Y *such that for each (finite)* $s = k\Delta t$*,* X^s *is S-integrable.*

Finally, the analogs of (5.4.8 & 9) say the following two things. First, a [Borel$[0,\infty)$×Meas(**P**)]-measurable function, H, is indistinguishable from some [Borel$[0,\infty)$×Loeb(Ω))-measurable function G. Second, if $\mathbb{M}$ is a complete separable metric space, we have the following

(5.5.14) PROPOSITION:

Let $G : [0,\infty) \times \Omega \longrightarrow \mathbb{M}$ *be any function. The following are equivalent conditions where* $K(t,\omega) = G(st(t),\omega)$ *is defined for finite times* $t \in \mathcal{O} \cap \mathbb{T}$ *and* $K(t,\omega)$ *equals any fixed* $a \in \mathbb{M}$ *for infinite* t.

(a) $G(r,\omega)$ *is* [Borel$[0,\infty)$×Loeb(Ω)]*-measurable.*

(b) $K(t,\omega)$ *is* [Loeb($\mathbb{T}$)×Loeb(Ω)]*-measurable.*

(c) $K(t,\omega)$ *is* Loeb($\mathbb{T}$×Ω)*-measurable.*

CHAPTER 6: HYPERFINITE EVOLUTION

In this chapter we begin the study of time evolution of stochastic processes. The basic internal framework is the same as in Chapter 5 except that now we add zero to the internal time line, $\mathbb{T}$. (This is done for technical reasons involving previsible processes.) Let $h \in {}^*\mathbb{N}$ be infinite, $n = h!$, $\delta t = \frac{1}{n}$, and

$$\mathbb{T} = \{t \in {}^*\mathbb{R} \mid (\exists k \in {}^*\mathbb{N})[t = k\delta t,\ 0 \leq k \leq n]\}$$

$$W = \{k \in {}^*\mathbb{N} \mid 1 \leq k \leq n^n\}$$

$$\Omega = W^{\mathbb{T}} = \{\omega : \mathbb{T} \longrightarrow W \mid \omega \textit{ internal}\}$$

$$P = \text{uniform internal counting measure on } \Omega, \quad \text{with weight function } \delta P(\omega) = \frac{1}{{}^{\#}[\Omega]}.$$

One contemporary general way to describe time evolution of random processes is to have a time-indexed family of sigma algebras called a "filtration." The idea of this is that "everything random up to time t" is measurable with respect to the sigma algebra indexed by t. There is an obvious combinatorial way to restrict an internal random process on our internal Ω to the internal times $t \in \mathbb{T}$, but since infinitely many $s \in \mathbb{T}$ have the same standard part, $st(s) = st(t)$, it may not be obvious what this means "in standard terms." We begin the chapter with a look at the difference between determining an event at a time t in $\mathbb{T}$ and determining an event during the instant r in $[0,1]$. All the t's in $st^{-1}(r)$ may be involved during the instant.

The remainder of the chapter gives the basic 'time respecting lifting and projection theorems'. That is, it shows which time properties of internal processes correspond to filtration properties of their standard parts and vice versa.

(6.1) Events Determined at Times and Instants

A partial schematic 'picture' of Ω in case $W = \{-1,1\}$ is the tree of choices in Figure (6.6.1). A particular ω is one branch.

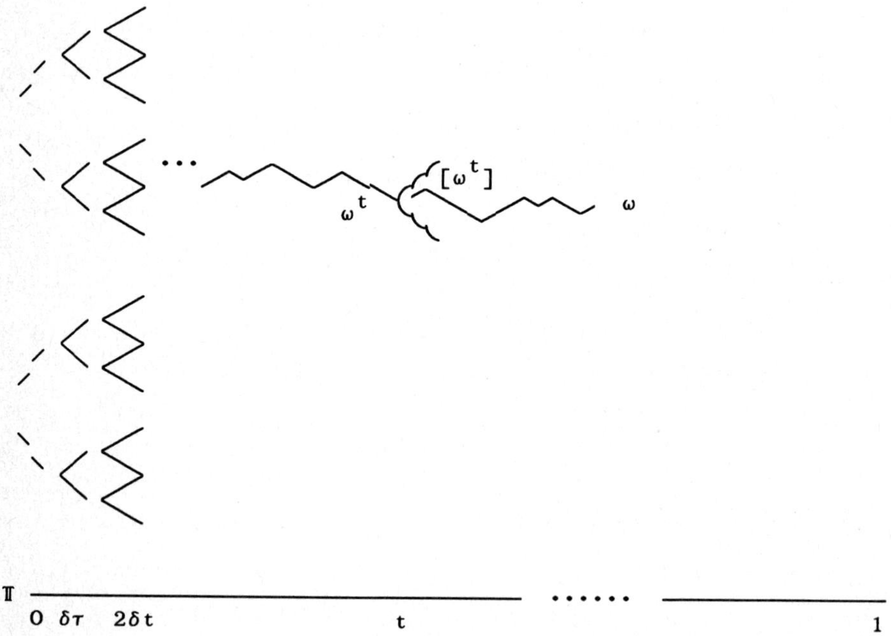

$[\omega^t]$ ω^t ω

$\mathbb{T}$ ......
0 $\delta\tau$ $2\delta t$ t 1

Fig. (6.1.1)

(6.1.2) **NOTATION:**

For $\upsilon \in \Omega$ *and* $t \in \mathbb{T}$ *let*

$$\upsilon^t = (\upsilon_0, \upsilon_{\delta t}, \cdots, \upsilon_t) = \upsilon|_{\mathbb{T}[0,t]}$$

the restricted sequence of choices in $\mathbb{T}$ *from* 0 *to* t.

Let

$$[\upsilon^t] = \{\omega \in \Omega : \omega^t = \upsilon^t\},$$

the internal set of all samples that agree with υ *up to time* t. *For any* $\Lambda \subseteq \Omega$, *let* $[\Lambda]^t = \cup\{[\lambda^t] : \lambda \in \Lambda\}$ $= \{\omega \in \Omega \mid \exists \lambda \in \Lambda$ *with* $\omega^t = \lambda^t\}$.

A schematic υ^t and $[\upsilon^t]$ are shown in Figure (6.1.1).

If Λ is an internal subset of Ω, then $[\Lambda^t]$ is also internal by the internal definition principle.

(6.1.3) **DEFINITION:**

An internal event determined at time t *in* $\mathbb{T}$ *is an internal subset* $\Lambda \subseteq \Omega$ *such that* $[\Lambda]^t = \Lambda$.

A measurable event determined at time t *in* $\mathbb{T}$ *is a P-measurable subset* $\Lambda \subseteq \Omega$ *such that* $[\Lambda]^t = \Lambda$. *Let* $\mathcal{M}(t)$ *denote the sigma algebra of measurable events determined at* t.

The reader should recall (1.5.6) thru (1.5.11) which show that $\mathcal{M}(t)$ is a sigma algebra and describe it in terms of internal sets determined at time t.

(6.1.4) **DEFINITION:**

Let $r \in [0,1]$ *and define*

(a) $\omega \overset{r}{\sim} \upsilon$ *if and only if* $\omega^t = \upsilon^t$ *for all* t *in* $\mathbb{T}$ *with* $t \approx r$.

For $\Lambda \subseteq \Omega$, *let* $(\Lambda)^r = \{\omega \in \Omega \mid \exists\lambda \in \Lambda \text{ with } \lambda \overset{r}{\sim} \omega\}$. *A measurable event determined during the instant* r *is a* P*-measurable set* $\Lambda \subseteq \Omega$ *such that* $(\Lambda)^r = \Lambda$. *Let* $\mathcal{F}(r)$ *denote the sigma algebra of measurable events determined during* r. *The family* $\{\mathcal{F}(r) \mid r \in [0,1]\}$ *is called the progressive filtration.*

(b) $\omega \underset{r}{\sim} \upsilon$ *if and only if* $\omega^t = \upsilon^t$ *for all* t *in* $\mathbb{T}$ *with* $t \ll r$ $(t < r$ & $t \not\approx r)$, *when* $r > 0$, *while* $\omega \underset{0}{\sim} \upsilon$ *means* $\omega_0 = \upsilon_0$. *For* $\Lambda \subseteq \Omega$, *let* $(\Lambda)_r = \{\omega \in \Omega \mid \exists\lambda \in \Lambda \text{ with } \lambda \underset{r}{\sim} \omega\}$.

A measurable event determined before the instant r *is a* P*-measurable set* $\Lambda \subseteq \Omega$ *such that* $(\Lambda)_r = \Lambda$. *Let* $\mathcal{D}(r)$ *denote the sigma algebra of measurable events determined before* r. *The family* $\{\mathcal{D}(r) \mid r \in [0,1]\}$ *is called the previsible filtration.*

Notice that the times $t \in \mathbb{T}$ index the sigma algebras $\mathcal{A}(t)$, but that real instants $r \in [0,1]$ index $\mathcal{F}(r)$ and $\mathcal{D}(r)$.

This is a lot of jargon, so here are some simple

(6.1.5) **FILTRATION FACTS:**

(a) $\omega \overset{r}{\sim} \upsilon$ *if and only if* $\omega^t = \upsilon^t$ *for some* $t \gg r$.

(b) $\omega \underset{r}{\sim} \upsilon$ *if and only if* $\omega^t = \upsilon^t$ *for some* $t \approx r$.

(c) *If* E *is internal and in* $\mathcal{F}(r)$, *then* E *is determined at some* $t \approx r$, $E \in \mathcal{M}(t)$.

(d) *If* E *is internal and in* $\mathcal{D}(r)$, $r > 0$, *then* E *was determined at some* $t \ll r$, $E \in \mathcal{M}(t)$.

(e) *For* $t' \ll r \approx t$, $t', t \in \mathbb{T}$, $r \in [0,1]$, *the sigma algebras below are nested as shown:*

$$\mathcal{M}(t') \subset \mathcal{D}(r) \subset \mathcal{M}(t) \subset \bigcup_{s \approx r} \mathcal{M}(s) \subset \mathcal{F}(r) = \bigcap_{s \gg r} \mathcal{M}(s) = \bigcap_{u \gg r} \mathcal{F}(u).$$

PROOF:

(a) $\{t \mid \omega^t = \upsilon^t\}$ is internal and contains all $t \approx r$, an external set [cf. (0.3.8)]. Thus it contains a $t \gg r$.

(b) $\{t \mid \omega^t = \upsilon^t\}$ is internal and contains all $t \gg r$, an external set [cf. (0.3.8)]. Thus it contains a $t \approx r$.

(c) We can easily verify that $(E)^r = E$ implies $[E]^t = E$ for all $t \gg r$. Since E is internal, $\{t \mid [E]^t = E\}$ is also internal and thus contains a $t \approx r$.

(d) We observe that $(E)_r = E$ implies $[E]^t = E$ when $t \gtrsim r$. Since E is internal, $\{t \mid [E]^t = E\}$ also contains a $t \ll r$.

(e) <u>First inclusion</u>: Let $[M]^{t'} = M$ for M P-measurable. If $\omega \in M$ and $\upsilon \underset{r}{\sim} \omega$ then $\upsilon^{t'} = \omega^{t'}$, so $\upsilon \in M$. Thus $M \in \mathcal{D}(r)$.

<u>Strictness of first inclusion</u>: Let t'' satisfy $t' \ll t'' \ll r$, then $[\omega^{t''}] \in \mathcal{D}(r) \setminus \mathcal{M}(t')$.

<u>Second inclusion</u>: If $(D)_r = D$, $\omega \in D$ and $\upsilon^t = \omega^t$, then $\upsilon \underset{r}{\sim} \omega$ by (b) and $\upsilon \in D$ or $[D]^t = D$.

Strictness of second inclusion: $([\omega^t])_r \neq [\omega^t]$ because $\upsilon_s = \omega_s$ for $0 \leq s \leq t-\delta t$ and $\upsilon_t \neq \omega_t$ still satisfies $\upsilon \underset{r}{\sim} \omega$.

Third inclusion: Trivial, but the union is also a sigma algebra by $\aleph_1$-saturation. If $M_j \in \mathcal{M}(s_j)$, for $s_j \approx r$ then there is an $s_\infty \approx r$ such that $s_j < s_\infty$, for example, take $s_\infty \in \bigcap_j \mathbb{T}[s_j, r+\frac{1}{j}]$.

Strictness of third inclusion: Since $[\omega^{s+\delta t}] \neq [\omega^s]$, the union is increasing.

Fourth inclusion: If $[M]^s = M$ when $s \approx r$ and $\omega \in M$ and $\upsilon \overset{r}{\sim} \omega$, then $\upsilon^s = \omega^s$, so $(M)^r = M$.

Strictness of fourth inclusion: $(\omega)^r \notin \cup\mathcal{M}(s)$.

Fifth and sixth inclusion and equalities: If $(M)^r = M$, then $[M]^s = M$ for any $s >> r$, hence $\mathcal{F}(r) \subseteq \mathcal{M}(s)$ in this instance. This fact means that the last two intersections are equal by virtue of the third inclusion applied to different r's and s's. If $M \in \bigcap_{s>>r} \mathcal{M}(s)$, $\omega \in M$ and $\upsilon \overset{r}{\sim} \omega$, then, by (a), $\upsilon^s = \omega^s$, for some $s >> r$ and $\upsilon \in M$, so $(M)^r = M$.

The next result is a set lifting lemma like Lemma (1.2.13), but adding time.

(6.1.6) PROPOSITION:

(a) *If* $F \in \mathcal{F}(r)$, *then there is an internal set* $E \in \mathcal{M}(t)$ *determined at a time* $t \approx r$ *such that* $P[F \triangledown E] = 0$.

(b) *If* $D \in \mathscr{D}(r)$, $r > 0$, *then for every standard positive* $\epsilon >> 0$, *there is an internal set* $E \in \mathscr{M}(t)$ *determined at a time* $t << r$ *such that* $P[D \nabla E] < \epsilon$. *If* $D \in \mathscr{D}(0)$, *there is an internal* $E \in \mathscr{D}(0)$ *such that* $P[D \nabla E] = 0$

PROOF:

(a) Let $F \in \mathscr{F}(r)$. For each finite natural number, let C_m^1 and G_m^1 be internal sets such that $C_m^1 \subseteq F \subseteq G_m^1$ and $P[G_m^1 \setminus C_m^1] < \frac{1}{m}$, cf. (1.1.3). Let $t_m = [r + \frac{1}{m}]$ $= \max[t \in \mathbb{T} \mid t \leq r + \frac{1}{m}]$. Define $C_m = [C_m^1]^{t_m}$. Since $t_m >> r$, $C_m = [C_m^1]^{t_m} \subseteq (C_m^1)^r \subseteq (F)^r = F$. Define $G_m = \Omega \setminus [\Omega \setminus G_m^1]^{t_m}$. Again, since $t_m >> r$, $[\Omega \setminus G_m^1]^{t_m} \subseteq (\Omega \setminus G_m^1)^r \subseteq (\Omega \setminus F)^r = (\Omega \setminus F)$, so

$$C_m \subseteq F \subseteq G_m.$$

C_m amd G_m are determined at t_m and $P[G_m \setminus C_m] < \frac{1}{m}$. Making dependent choices, we may select sequences $C_1 \subseteq C_2 \subseteq \cdots \subseteq F$, $G_1 \supseteq G_2 \supseteq \cdots \supseteq F$, G_m determined at t_m and $P[G_m \setminus C_m] < \frac{1}{m}$. By countable comprehension, we may extend (C_m, G_m, t_m) to an internal sequence, (0.4.3). Choose an infinite n so that the G and C sequences are monotone up to n, $r \leq t_n \leq r + \frac{1}{n}$ and $P[G_n \setminus C_n] < \frac{1}{n}$. We may take $E = G_n$ or C_n because

$$\cup C_m \subseteq C_n \subseteq G_n \subseteq \cap G_m$$

and

$$\cup C_m \subseteq F \subseteq \cap G_m.$$

(b) Let $D \in \mathcal{D}(r)$ and let $C \subseteq D$ be internal and such that $P[D \setminus C] < \epsilon$. Let $t_m = [r - \frac{1}{m}] = \max[t \in \mathbb{T} : t \leq r - \frac{1}{m}]$ and $C_m = [C]^{t_m}$.

First, $\bigcap_m C_m = (C)_r$. Let $\upsilon \underset{r}{\sim} \omega \in C$, so $\upsilon^t = \omega^t$ for all $t \ll r$ and $\upsilon^{t_m} = \omega^{t_m}$ in particular. This shows one inclusion; for the other suppose $\upsilon^{t_m} = \omega^{t_m}$ for every finite m. Then clearly $\upsilon \underset{r}{\sim} \omega$ and $\upsilon \in (C)_r$.

Next,

$$C \subseteq \bigcap_m C_m = (C)_r \subseteq (D)_r = D.$$

We know $\operatorname{S-lim}_{m \to \infty} P[C_m] \leq P[D]$, so there exists finite m for which $|P[C_m] - P[D]| < \epsilon$. We may let $E = C_m$.

If $D \in \mathcal{D}(0)$, then for every m in $\mathbb{N}$ there is an internal $C_m \subseteq D$ such that $P[D \setminus C_m] < \frac{1}{m}$. Since the relation $\underset{0}{\approx}$ is internal, $\omega \underset{0}{\approx} \upsilon$ if and only if $\omega_0 = \upsilon_0$, $(C_m)_0$ is internal and $C_m \subseteq (C_m)_0 \subseteq D$. Now we may form an increasing chain of internal sets determined at 0 and contained in D. A sufficiently small infinite $C_n = E$.

The next lemma is useful below.

(6.1.7) **LEMMA:**

(a) *A random variable* $X : \Omega \longrightarrow \mathbb{R}$ *is measurable with respect to the completion of* $\mathcal{F}(r)$ *if and only if there is a time* $t \approx r$ *and an* $\mathcal{M}(t)$*-measurable internal random variable* Y *such that* $\tilde{Y} = X$ *a.s.* P.

(b) *If* X *is* $\mathcal{F}(r)$*-measurable and bounded,* $|X| \leq c$, *then there is a* $t \approx r$ *and an* $\mathcal{M}(t)$*-measurable internal* Y *with* $|Y| \lesssim c$ *such that* $\tilde{Y} = X$ *a.s.* P.

(c) *If* X *is* $\mathcal{F}(r)$*-measurable and* P*-integrable, then there is a* $t \approx r$ *and an* S*-integrable, internal* Y *determined at time* t *such that* $E[|\tilde{Y}-X|] = 0$.

PROOF:

(a) Suppose $X = \tilde{Y}$ a.s. for Y and t as stated. Then $\tilde{Y}$ is P-measurable and if $\omega \overset{r}{\sim} \upsilon$, then $\omega^t = \upsilon^t$, so $Y(\omega) = Y(\upsilon)$, by $\mathcal{M}(t)$-measurability. This means $\tilde{Y}$ is $\mathcal{F}(r)$-measurable. Since $X = \tilde{Y}$ a.s., X is measurable with respect to the completion of $\mathcal{F}(r)$.

Next, suppose X is measurable with respect to the completion and for each m and k in $\mathbb{N}$ with $-2^{2m}-1 \leq k \leq 2^{2m}$ define

$$\Phi_m^{-2^{2m}-1} = \left\{\omega : X(\omega) < -2^{-m}\right\}$$

$$\Phi_m^k = \left\{\omega : \frac{k}{2^m} \leq X(\omega) < \frac{k+1}{2^m}\right\}$$

$$\Phi_m^{2^{2m}} = \{\omega : 2^m \leq X(\omega)\}.$$

For each m, the Φ_m^k form a partition of Ω and there are sets with $P[\Lambda_m^k \nabla \Phi_m^k] = 0$ and $\Lambda_m^k \in \mathcal{F}(r)$. Apply (6.1.6) to obtain internal sets Ω_m^k such that $P[\Lambda_m^k \nabla \Omega_m^k] = 0$. By refining the Ω_m^k to an internal family of nonoverlapping sets and then discarding the original overlaps, whose external measure was zero, $P[\Lambda_m^h \cap \Lambda_m^k] = 0$, we may assume that the Ω_m^k are disjoint and all determined at the maximal time, $t_m \approx r$, determining the originals. Let

$$Y_m(\omega) = \sum\left[\frac{k}{2^m} \quad I_m^k(\omega) \ ; \ |k| \le 2^{2m}\right]$$

where I_m^k is the indicator function of Ω_m^k.

Extend the sequence Y_m to an internal sequence of internal random variables determined at time t_m and observe that for large enough finite m there must be an infinite n_m such that whenever $m \le j \le n_m$, $|t_j - t_m| < \frac{1}{m}$ and $P[|Y_m(\omega) - Y_j(\omega)| \ge \frac{1}{2^m}] < \epsilon$. By saturation, there is an n in $\bigcap_m [m, n_m]$. Let $Y = Y_n$ for that n. The idea is similar to (1.3.9). This proves part (a).

Notice that if X is bounded, some of the Λ_m^k-sets will be empty and we can replace the corresponding Ω_m^k-sets by the empty set to make $|Y| \lesssim c$. This proves part (b).

The proof of part (c) follows the lines of (1.4.10). For each m, truncate X to make an approximation $|X^m| \le m$. We know by dominated convergence that $X^m \longrightarrow X$ in L^1-norm. Lift X^m using part (b) to a bounded Y^m. Boundedness and part (b) assure us that $E[|\tilde{Y}^m - X^m|] = 0$, so that Y^m is an S-Cauchy sequence, for every finite p in $\mathbb{N}$, there exists an $m_p \ge p$ such that for all finite $j,k \ge m_p$, $E[|Y^j - Y^k|] < \frac{1}{p}$. Extend the sequence of pairs $\{(Y^m, t_m)\}$ to an internal sequence and observe that the internal set

$$\{n \in {}^*\mathbb{N} : \forall j,k \in {}^*\mathbb{N}, m_p \le j \le k \le n \Rightarrow E[|Y^j - Y^k|] < \frac{1}{p} \ \& \ |t_n - r| < \frac{1}{n}\}$$

contains an infinite n_p whenever p is in $\mathbb{N}$. The countable intersection $\bigcap_p [m_p, n_p] \cap {}^*\mathbb{N}$ is nonempty by saturation (0.4.2). The necessary lifting is Y^n for an infinite n in the intersection.

(6.1.8) **LEMMA:**

A function $X : \Omega \longrightarrow \mathbb{R}$ *is measurable with respect to the completion of* $\mathfrak{D}(r)$ *if and only if for every* $\epsilon >> 0$ *there is an internal random variable* Y *determined at time* t *where* $t << r$ *such that* $P[|X - \tilde{Y}| > \epsilon] < \epsilon$.

PROOF:

Define sets Φ_m^k as in the proof of (6.1.7)(a) except now there exist $\Lambda_m^k \in \mathfrak{D}(r)$ such that $P[\Phi_m^k \triangledown \Lambda_m^k] = 0$. Let Ω_m^k be disjoint internal sets determined at times $t_m^k << r$ such that $\sum_{|k| \le 2^{2m}} P[\Lambda_m^k \triangledown \Omega_m^k] < \frac{\epsilon}{2}$. Let $t_m = \max[t_m^k] << r$, so all $\Omega_m^k \in \mathcal{M}(t_m) \subset \mathfrak{D}(r)$. The internal variable

$$Y(\omega) = \sum \frac{k}{2^m} I_{\Omega_m^k}(\omega)$$

works whenever m large enough so that $\epsilon < \frac{1}{2^m}$ and $P[\cup\{\Lambda_m^k : |k| < 2^{2m}\}] > 1 - \frac{\epsilon}{2}$.

For the converse, choose a sequence Y_m of internal variables determined before r such that $P[|X - \tilde{Y}| > \frac{1}{m}] < \frac{1}{m}$. Then each $\tilde{Y}_m$ is $\mathfrak{D}(r)$-measurable and $\tilde{Y}_m \longrightarrow X$ in probability.

A subsequence $\tilde{Y}_{m_k} \longrightarrow X$ a.s. so X is measurable with respect to the completion of $\mathcal{D}(r)$.

(6.1.9) EXERCISE:

Prove (6.1.7)(a) and (6.1.8) in the case where $X : \Omega \longrightarrow \mathbb{M}$ takes values in a complete separable metric space, $\mathbb{M}$.

(6.1.10) EXERCISE:

A function $X : \Omega \longrightarrow \mathbb{R}$ is measurable with respect to the completion of $\mathcal{D}(r)$ if and only if for each $t \approx r$, there is an internal Y determined at time t such that $\tilde{Y} = X$ a.s.

(6.1.11) EXERCISE:

The filtration facts (6.1.5) can be completed as follows:

(a) Let $t_m \ll r$ be a sequence with S-lim $t_m = r$. The smallest complete sigma algebra containing $\cup\mathcal{M}(t_m)$,

$$\overline{\Sigma\mathcal{M}(t_m)} \supseteq \mathcal{D}(r).$$

(b) The following holds.

$$\mathcal{D}(r) = \bigcap_{t\approx r} \mathcal{M}(t).$$

(c) The smallest complete sigma algebra containing $\bigcup_{t\approx r} \mathcal{M}(t)$,

$$\overline{\sum_{t\approx r} \mathcal{M}(t)} \supseteq \mathcal{F}(r).$$

(6.1.12) **EXERCISE:**

If $D \in \mathcal{D}(r)$, for $0 < r$, and if $t \approx r$, then there exists an internal $E \in \mathcal{M}(t)$ such that $P[D \nabla E] = 0$. (First approximate D by a sequence E_m from (6.1.6)(b), then with any internal F, $P[D \nabla F] = 0$, and compare F with E_m as m becomes infinite.)

(6.2) Progressive and Previsible Measurability

This section examines some basic properties of the hyperfinite filtrations $\mathcal{D}(r)$ and $\mathcal{F}(r)$ indexed by standard instants $r \in [0,1]$. In particular, these filtrations enjoy a special "completeness" property that adapted *does* imply progressively measurable, see (6.2.2).

Let $\mathbb{M}$ denote a Polish (complete separable metric) space, for example, $\mathbb{M} = \mathbb{R}^d$. Recall the definitions of $\mathcal{F}(r)$, $\overset{r}{\sim}$, $\mathcal{D}(r)$ and $\underset{r}{\sim}$ from (6.1.4) and the definition of sections, X^r, from (3.1.3).

(6.2.1) DEFINITION:

(a) *A function* $X : [0,1] \times \Omega \longrightarrow \mathbb{M}$ *is said to be* $\mathcal{F}$*-adapted if each section* X^r *is* $\mathcal{F}(r)$*-measurable; in other words, if for each* $r \in [0,1]$, X^r *is* P*-measurable and whenever* $\omega \overset{r}{\sim} \upsilon$, *then* $X(r,\upsilon) = X(r,\omega)$. *If* X *is* [Borel[0,1] × Loeb(Ω)]*-measurable and* $\mathcal{F}$*-adapted, then we say* X *is progressively measurable.*

(b) *A function* $X : [0,1] \times \Omega \longrightarrow \mathbb{M}$ *is said to be* $\mathcal{D}$*-adapted if each section* X^r *is* $\mathcal{D}(r)$*-measurable. If* X *is* [Borel[0,1] × Loeb(Ω)]*-measurable and* $\mathcal{D}$*-adapted, we say* X *is previsibly measurable.*

The next result is not true for general filtrations on probability spaces. In that case one usually changes the definition of "progressively measurable" to the conclusion of our next result. The reader may review product measures in section (3.3) above.

(6.2.2) THE PROGRESSIVE AND PREVISIBLE MEASURABILITY THEOREM:

(a) *If* X *is progressively measurable, then for each* $r \in [0,1]$, *the restriction of* X *to* $[0,r] \times \Omega$ *is* $[\mathrm{Borel}[0,r] \times \mathcal{F}(r)]$*-measurable.*

(b) *If* X *is previsibly measurable, then for each* $r \in [0,1]$, *the restriction of* X *to* $[0,r] \times \Omega$ *is* $[\mathrm{Borel}[0,r] \times \mathcal{D}(r)]$*-measurable.*

DISCUSSION:

Notice that the hypotheses in both (a) and (b) imply that X is a stochastic process in the weak sense that X^r is measurable for each r, since X is $[\mathrm{Borel}[0,1] \times \mathrm{Loeb}(\Omega)]$-measurable. Since X is a stochastic process we will show that for each Borel set B, the set

$$\mathcal{S} = \{(s,\omega) \in [0,1] \times \Omega : s < r \ \& \ X(s,\omega) \in B\}$$

is in the product of the Borel algebra and $\mathcal{D}(r)$. Then in case (a), $\{(s,\omega) : s \leq r \ \& \ X(s,\omega) \in B\}$ has the claimed property because $\{\omega : X(r,\omega) \in B\} \in \mathcal{F}(r)$ and $\mathcal{F}(r) \supset \mathcal{D}(r)$, while in case (b), $\{\omega : X(r,\omega) \in B\} \in \mathcal{D}(r)$.

If $X(s,\omega) \in B$ and $s \ll t$ for t in $\mathbb{T}$, then whenever $\omega^t = \upsilon^t$, we know $\omega \overset{s}{\approx} \upsilon$ and $\omega \underset{s}{\sim} \upsilon$, so that the sections $\mathcal{S}^s$ are determined at time t, $[\mathcal{S}^s]^t = \mathcal{S}^s$.

The last two observations lead us to formulate the

(6.2.3) **LEMMA:**

If $\mathcal{S}$ *is* [Borel[0,1] × Loeb(Ω)]*-measurable and if for each instant* r *in* [0,1], *the section* $\mathcal{S}^r$ *is determined at time* t *in* $\mathbb{T}$ *then* $\mathcal{S}$ *is* [Borel × $\mathcal{M}(t)$]*-measurable.*

PROOF THAT (6.2.3) IMPLIES (6.2.2):

By the preceding discussion it suffices to show that

$$\mathcal{S} = \{(s,\omega) \in [0,1] \times \Omega : s < r \ \&\ X(s,\omega) \in B\}$$

is in Borel × $\mathcal{D}(r)$. Keep the same fixed r and B. Choose a rational q and t in $\mathbb{T}$ so that $q \ll t \ll r$. The set $\mathcal{S} \cap \{[0,q) \times \Omega\}$ is (Borel × $\mathcal{M}(t)$)-measurable by the lemma and (Borel × $\mathcal{D}(r)$)-measurable by (6.1.5). Finally, $\mathcal{S} = \cup\{\mathcal{S} \cap [0,q) \times \Omega : q \in [0,r) \cap \mathbb{Q}\}$, a countable union of sets in Borel × $\mathcal{D}(r)$. This reduces (6.2.2) to proving the lemma (6.2.3).

PROOF OF (6.2.3):

Let $\Omega_t = W^{\mathbb{T}[0,t)}$ and $\Omega_t^c = W^{\mathbb{T}[t,1]}$ so that we may identify Ω with $\Omega_t \times \Omega_t^c$ under the correspondence $(\omega_0,\omega_{\delta t},\cdots,\omega_1) \longleftrightarrow ((\omega_0,\cdots,\omega_{t-\delta t});\ (\omega_t,\cdots,\omega_1))$. Let P_t and P_t^c be uniform counting measures on Ω_t and Ω_t^c, respectively. So $P = P_t \otimes P_t^c$ as an internal measures. There is a natural association between Meas(P_t) and $\mathcal{M}(t)$. Consider the class of sets

$$\{\mathcal{T} \in \text{Borel} \times \text{Meas}(P_t) : \mathcal{T} \times \Omega_t^c \in \text{Borel} \times \mathcal{M}(t)\}.$$

This is a sigma algebra because the second factor does not interfere with set algebra operations. It contains rectangles $S \times \Lambda_t$, since Meas(P) is an extension of $\text{Meas}(P_t) \times \text{Meas}(P_t^c)$ see (3.1.2). Therefore this class equals Borel $\times\ \mathcal{M}(P_t)$.

Let $\lambda \in \Omega_t^c$ and let $\mathcal{S}(\lambda) = \{(r,\upsilon) \in [0,1] \times \Omega_t :$ $(r,\upsilon,\lambda) \in \mathcal{S}\}$ be the section over Ω_t^c. Since each r-section of $\mathcal{S}$ is t-determined, whenever $(r,\omega) \in \mathcal{S}$, then $(r,[\omega^t]) \subseteq \mathcal{S}$, or $\mathcal{S} = \mathcal{T} \times \Omega_t^c$ for some set $\mathcal{T} \subseteq [0,1] \times \Omega_t$. The class of sets

$$\{\mathcal{S} \in \text{Borel} \times \text{Meas}(P) : \mathcal{S}(\lambda) \in \text{Borel} \times \text{Meas}(P_t) \text{ for } P_t^c\text{-a.e. } \lambda\}$$

is a sigma algebra because countable unions of null sets are null and taking sections commutes with set algebra operations. If $\mathcal{S} = S \times \Lambda$ for $S \in$ Borel and $\Lambda \in$ Meas(P), then $\mathcal{S}(\lambda) = S \times \Lambda_\lambda \in \text{Borel} \times \text{Meas}(P_t)$ by (3.1.4). Thus, this class is all of Borel $\times$ Meas(P). The important conclusion for us is that $\mathcal{T} \in \text{Borel} \times \text{Meas}(P_t)$.

By the first paragraph of the proof, $\mathcal{T} \times \Omega_t^c \in \text{Borel} \times \mathcal{M}(t)$ and $\mathcal{S} = \mathcal{T} \times \Omega_t^c$ so (6.2.3) and (6.2.2) are proved.

We let λ denote Lebesgue measure and say a set is $(\lambda \times P)$-measurable if it is in the complete product sigma algebra described in (3.1.2) for the measures λ and P. Alternately, $\mathcal{S} \subseteq [0,1] \times \Omega$ is $(\lambda \times P)$-measurable if and only if there exists $\mathcal{T}$ in (Lebesgue $\times$ Meas(P)) such that $(\lambda \times P)(\mathcal{S} \nabla \mathcal{T}) = 0$. Theorem (5.4.3) says that $\mathcal{S}$ is $(\lambda \times P)$-measurable if and only if $s^{-1}(\mathcal{S})$ is $\delta t \otimes P$-measurable. We know from Theorem (5.4.8) that $\mathcal{B} \in [\text{Borel}[0,1] \times \text{Loeb}(\Omega)]$ if and only if $s^{-1}(\mathcal{B}) \in \text{Loeb}(\mathbb{T}\times\Omega)$, so we could say that $(\lambda \times P)$-

measurable sets are those in the $\mu = [\delta t \otimes \delta P \circ s^{-1}]$-completion of $[\text{Borel}[0,1] \times \text{Loeb}(\Omega)]$. Every internal bounded measure, π, on $\mathbb{T} \times \Omega$ induces a $[\text{Borel}[0,1] \times \text{Loeb}(\Omega)]$-measure, μ, by the equation $\mu = \boldsymbol{\pi} \circ s^{-1}$. Moreover, $\mathcal{S} \subseteq [0,1] \times \Omega$ is μ-measurable if and only if there exist sequences of internal sets $\mathcal{A}_1 \subseteq \mathcal{A}_2 \subseteq \cdots \subseteq s^{-1}(\mathcal{S})$ and $\mathcal{B}_1 \subseteq \mathcal{B}_2 \subseteq \cdots \subseteq (\mathbb{T}\times\Omega)\setminus s^{-1}(\mathcal{S})$ such that $\pi[\mathcal{A}_m \cup \mathcal{B}_m] \geq 1 - \frac{1}{m}$. Such internal measures will be useful to us in Chapter 7 where we will let the time measure depend on ω; for now the reader may wish to think in terms of $\pi = \delta t \times P$ and μ-measurable = $(\lambda \times P)$-measurable.

(6.2.4) DEFINITION:

Let $\boldsymbol{\pi}$ be a bounded hyperfinite measure on $\mathbb{T} \times \Omega$. A function $X : [0,1] \times \Omega \longrightarrow \mathbb{M}$ is $\boldsymbol{\pi}$-almost progressively measurable if there is a progressively measurable process Y such that $Y(r,\omega) = X(r,\omega)$ a.e. $\mu(r,\omega) = \boldsymbol{\pi} \circ s^{-1}(r,\omega)$. We say that X is $\boldsymbol{\pi}$-almost previsibly measurable if there is a previsibly measurable process Z such that $X(r,\omega) = Z(r,\omega)$ a.e. $\mu(r,\omega) = \boldsymbol{\pi} \circ s^{-1}(r,\omega)$.

Suppose that X is $(\lambda \times \mathbf{P}) = (\delta t \otimes \delta P \circ s^{-1})$-measurable and X^r is $\mathcal{F}(r)$ [resp., $\mathcal{D}(r)$]-measurable for λ a.e. r. Then we may take $X'(r,\omega) = 0$ on the null set of r's where adaptation fails and have $X = X'$ a.e. $\lambda \times \mathbf{P}$. This means that we can weaken the hypothesis of the next result to $X^r \in \mathcal{F}(r)$ [resp., $\mathcal{D}(r)$] a.e. $\lambda(r)$ in the case when $\pi = \delta t \otimes \delta P$.

(6.2.5) ALMOST PROGRESSIVE AND PREVISIBLE MEASURABILITY THEOREM:

Let π *be a bounded hyperfinite measure on* $\mathbb{T} \times \Omega$ *and let* $\mu = \pi \circ s^{-1}$ *denote the completed* [Borel[0,1] × Loeb(Ω)]*-measure. If* $X : [0,1] \times \Omega \longrightarrow \mathbb{M}$ *is* μ*-measurable and each section* X^r *is* $\mathcal{F}(r)$*-measurable* [*resp.,* $\mathcal{D}(r)$*-measurable*]*, then* X *is* π*-almost progressively* [*resp.,* π*-almost previsibly*] *measurable, that is, there exists a* [Borel[0,1] × Loeb(Ω)]*-measurable* Y *that is* $\mathcal{F}$*-adapted* [*resp.,* $\mathcal{D}$*-adapted*] *and* $X(r,\omega) = Y(r,\omega)$ *a.e.* $\mu(r,\omega)$.

PROOF:

We shall give the proof only in the previsible case, the similar progressive case can be found in Keisler [1984, Lemma 7.5]. For any $\mathcal{S} \subseteq [0,1] \times \Omega$, let

$$(\mathcal{S}) = \{(r,\upsilon) : \upsilon \underset{r}{\sim} \omega \text{ for some } (r,\omega) \in \mathcal{S}\}.$$

Let $\mathcal{H} = \{\mathcal{S} \in \text{Borel}[0,1] \times \text{Loeb}(\Omega) : \mathcal{S} = (\mathcal{S})\}$. Note that $\mathcal{H}$ is a sigma algebra. Any $\mathcal{H}$-measurable function Y is previsible as the reader can easily check. We will show that X is measurable with respect to the completion $\mathscr{H}$ of $\mathcal{H}$ (with respect to μ), since then there will be an $\mathcal{H}$-measurable Y with $Y = X$ a.e. this proves our claim.

We know that sets of the form $\mathcal{S} = \{(r,\omega) : X(r,\omega) \in B\}$ are μ-measurable and satisfy $\mathcal{S} = (\mathcal{S})$ for any Borel $B \subseteq \mathbb{M}$. Let $\mathcal{J} = \{\mathcal{S} \in \text{Meas}(\mu) : \mathcal{S} = (\mathcal{S})\}$. We wish to show that $\mathcal{H} \subseteq \mathcal{J} \subseteq \mathscr{H}$ to finish our claim. Let $\mathcal{S} \in \mathcal{J}$. By (5.4.3),

$s^{-1}(\mathcal{S})$ is π-measurable on $\mathbb{T} \times \Omega$, so we may pick sequences of internal sets satisfying:

$$\mathcal{A}_1 \subseteq \mathcal{A}_2 \subseteq \cdots \mathcal{A}_m \subseteq s^{-1}(\mathcal{S})$$

$$\mathcal{B}_1 \subseteq \mathcal{B}_2 \subseteq \cdots \mathcal{B}_m \subseteq (\mathbb{T} \times \Omega)\setminus s^{-1}(\mathcal{S})$$

$$\pi[\mathcal{A}_m \cup \mathcal{B}_m] \geq 1 - \frac{1}{m}.$$

For each n in $\mathbb{N}$ define

$$\mathcal{A}_m^n = \{(t,\upsilon) : \upsilon^{[t-1/n]} = \omega^{[t-1/n]} \text{ for some } (t,\omega) \in \mathcal{A}_m\}$$

and

$$\mathcal{B}_m^n = \{(t,\upsilon) : \upsilon^{[t-1/n]} = \omega^{[t-1/n]} \text{ for some } (t,\omega) \in \mathcal{B}_m\}$$

where we understand $\upsilon^{[x]} = \upsilon^0$ if $x \leq 0$. Let $\mathcal{A} = \bigcup_m \bigcap_n \mathcal{A}_m^n$ and $\mathcal{B} = \bigcup_m \bigcap_n \mathcal{B}_m^n$. When m is fixed, $(t,\upsilon) \in \bigcap_n \mathcal{A}_m^n$, if and only if $(t,\omega) \in \mathcal{A}_m$ for some $\omega \underset{r}{\sim} \upsilon$ where $r = st(t)$. Since $\mathcal{S} = (\mathcal{S})$, this means

$$\mathcal{A} \subseteq s^{-1}(\mathcal{S}) \text{ and } \mathcal{B} \subseteq \mathbb{T} \times \Omega\setminus s^{-1}(\mathcal{S}),$$

so $\pi[\mathcal{A} \cup \mathcal{B}] = 1$ and $s(\mathcal{A}) \subseteq \mathcal{S} \subseteq s(\mathcal{B})$, in fact, $(s(\mathcal{A})) = s(\mathcal{A})$ and $(s(\mathcal{B})) = s(\mathcal{B})$, so Lemma (5.4.7) completes the proof by showing that $\mathcal{S}$ (from $\mathcal{J}$) is sandwiched between sets of $\mathcal{H}$. This is because $s(\mathcal{A}) = \bigcup_m \bigcap_n s(\mathcal{A}_m^n)$ and similarly for $\mathcal{B}$. The $\subseteq$ inclusion is easy to verify. Suppose that for some m, to each n corresponds a t_n such that $t_n \approx r$ and $(t_n,\omega) \in \mathcal{A}_m^n$, that is, $(r,\omega) \in \bigcup_m \bigcap_n s(\mathcal{A}_m^n)$. Then for each finite n there exists ω_n such that $\omega_n^{[t_n - 1/n]} = \omega^{[t_n - 1/n]}$ and $(t_n, \omega_n) \in \mathcal{A}_m$. We may extend the sequence (t_n, ω_n) internally using saturation (0.4) and obtain $t_p \approx r$, $(t_p, \omega_p) \in \mathcal{A}_m$ and $\omega^{[t_p - 1/p]} = \omega_p^{[t_p - 1/p]}$ for some infinite p (by the internal definition principle and Robinson's Sequential Lemma). For any finite n, $(t_p, \omega) \in \mathcal{A}_m^n$ because $\omega_p^{[t_p - 1/n]} = \omega^{[t_p - 1/n]}$, so $(r,\omega) = (st(t_p),\omega) \in s(\mathcal{A})$.

The reader should note the following consequence of the last result. If X is [Borel × Meas(P)]-measurable and adapted, then X is π-measurable and adapted *for every P-continuous bounded hyperfinite* π on $\mathbb{T} \times \Omega$. Hence for each π there exists a progressive [resp., previsible] Y_π with $X = Y_\pi$ a.e. $\pi(r,\omega)$.

(6.2.6) **PROPOSITION:**

If $X : [0,1] \times \Omega \longrightarrow \mathbb{M}$ *is* $\delta t \otimes \delta P$*-almost progressively measurable and a.s. has decent paths, then there is a progressively measurable process* Y *with a.s. decent paths such that* X *and* Y *are indistinguishable.*

PROOF:

Let Z be a progressive process such that $X(r,\omega) = Z(r,\omega)$ a.s. $(\lambda \times P)(r,\omega)$. There is a countable dense set $S \subseteq [0,1]$ and $\Lambda \subseteq \Omega$ with $P[\Lambda] = 1$ such that $X(s,\upsilon) = Z(s,\upsilon)$ on $S \times \Lambda$ and X_υ is a decent path for υ in Λ.

For each m in $\mathbb{N}$, choose a finite increasing sequence $S_m = \{s_m^0, s_m^1, \cdots, s_m^n\} \subseteq S$ with $0 \le s_m^0 < \frac{1}{m}$, $0 < s_m^{k+1} - s_m^k < \frac{1}{m}$, and $1 - s_m^n < \frac{1}{m}$. Let $Z_m(r,\omega)$ equal $Z(s_m^k,\omega)$ where

$s_m^k = \min_h\{s_m^h : s_m^h \geq r\}$, $Z(s_m^n,\omega)$ if $1 > r \geq s_m^n$ or $Z(1,\omega)$ if $r = 1$. Since $Z(s,\cdot)$ is progressively measurable, each Z_m is Borel × Loeb-measurable. Define

$$Y(r,\omega) = \limsup_{m\to\infty} Z_m(r,\omega).$$

If $\upsilon \in \Lambda$, since $X(s,\upsilon) = Z(s,\upsilon)$ and $\lim_{s\downarrow r} X(s,\upsilon) = X(r,\upsilon)$ $= \limsup_{m\to\infty} Z_m(r,\upsilon)$, we see that Y has the same paths as X on Λ.

For every m, $Z_m(r,\cdot)$ is $\mathcal{F}(r+\frac{1}{m})$-measurable, hence $Y(r,\cdot)$ is $\mathcal{F}(r)$-measurable by (6.1.5).

(6.2.7) **PROPOSITION:**

If $X : [0,1] \times \Omega \longrightarrow \mathbb{M}$ a.s. has left-continuous paths, if X is $\delta t \otimes \delta P$-almost progressively measurable and if $X(0)$ is $\mathcal{D}(0)$-measurable, then there is a previsible process Y with a.s. left-continuous paths such that X and Y are indistinguishable.

PROOF:

Let Z be progressive and satisfy $Z(r,\omega) = X(r,\omega)$ a.s. $(\lambda \times P)(r,\omega)$. There is a countable dense set $S \subseteq [0,1]$ and a $\Lambda \subseteq \Omega$ with $P[\Lambda] = 1$ such that $X(s,\upsilon) = Z(s,\upsilon)$ on $S \times \Lambda$ and X_υ is left-continuous for υ in Λ.

For each m in $\mathbb{N}$, choose a finite increasing sequence $S_m = \{s_m^0, s_m^1, \cdots, s_m^n\} \subseteq S$ with $0 \leq s_m^0 < \frac{1}{m}$, $0 < s_m^{k+1} - s_m^k < \frac{1}{m}$ and $1 - s_m^n < \frac{1}{m}$. Let $Z_m(r,\omega) = Z(s_m^k,\omega)$ where $s_m^k = \max_h\{s_m^h : s_m^h \leq r - \frac{1}{m}\}$, or s_m^0 if $r \leq s_m^0$. Each Z_m is (Borel × Loeb)-

measurable. Define $Y(0,\omega) = X(0,\omega)$ and if $r > 0$,

$$Y(r,\omega) = \limsup_{m\to\infty} Z_m(r,\omega),$$

so Y is (Borel × Loeb)-measurable.

If $\upsilon \in \Lambda$, since $X(s,\upsilon) = Z(s,\upsilon)$ for $s \in S$, when $r > 0$, $\lim_{s\uparrow r} X(s,\upsilon) = X(r,\upsilon) = \limsup_{m\to\infty} Z_m(r,\upsilon)$, so Y has the same paths as X on Λ.

Each $Z_m(r,\cdot)$ is $\mathcal{F}(r-\frac{1}{m})$-measurable and hence $\mathcal{D}(r)$-measurable, so $Y(r)$ is $\mathcal{D}(r)$-measurable when $r > 0$. By hypothesis, $X(0)$ is $\mathcal{D}(0)$-measurable and $Y(0) = X(0)$.

(6.3) Nonanticipating Processes

If an internal process only anticipates a sample sequence ω by an infinitesimal time into the future, then its projection will be progressively measurable. If X anticipates by a time ι, then $Y(t) = X(t-\iota)$ is a strictly nonanticipating process for $t \geq \iota$, but not all the way to zero. Separate treatment of $t = 0$ is incorporated in our next definition.

(6.3.1) DEFINITION:

An internal function X defined on $\mathbb{T} \times \Omega$ is nonanticipating if there is an infinitesimal $\iota \geq 0$ such that for each t in $\mathbb{T}$ with $t \geq \iota$, if $\omega^t = \upsilon^t$, then

$$X(t,\omega) = X(t,\upsilon).$$

Strictly speaking, we should say "X is nonanticipating after δ" if $t \geq \delta$ and $\omega^t = \upsilon^t$ imply that $X(t,\omega) = X(t,\upsilon)$. With δ fixed this is an internal property of X. "Nonanticipating" as we have defined it is external: 'for some *infinitesimal* ι, X is nonanticipating after ι'.

Recall the definition of Lebesgue lifting from (5.4.2). Here is the result you've been anticipating:

(6.3.2) NONANTICIPATING LIFTING THEOREM:

A stochastic process $X : [0,1] \times \Omega \longrightarrow \mathbb{M}$ is $\delta t \otimes \delta P$-almost progressively measurable if and only if X has a nonanticipating Lebesgue lifting.

This is an important result because it shows that progressive measurability with respect to the filtration we have defined corresponds to a simple property of internal processes. Nevertheless, we shall not give the proof which can be found in Keisler [1984, Theorem 7.6]. We shall describe the main lemma from Keisler's proof and then give the decent path versions of the result.

(6.3.3) **DEFINITION:**

If Λ *is an internal subset of* Ω *and* $t \in \mathbb{T}$, *define*

$$P[\Lambda|\omega^t] = \frac{P[\Lambda \cap [\omega^t]]}{P[[\omega^t]]} = \frac{{}^{\#}[\Lambda \cap [\omega^t]]}{{}^{\#}[[\omega^t]]}.$$

The uniform internal conditional probability $P[\ |\omega^t]$ *is only defined on internal sets. The hyperfinite extension of* $P[\ |\omega^t]$ *is denoted* $\mathbf{P}[\ |\omega^t]$.

In general, a **P**-measurable set might not be $\mathbf{P}[\ |\omega^t]$-measurable.

If Λ is **P**-measurable and Λ' is internal with $\mathbf{P}[\Lambda \triangledown \Lambda'] = 0$, then

$$P[\Lambda'|\omega^t] = \Sigma[P(\upsilon) : \upsilon \in \Lambda' \ \&\ \upsilon^t = \omega^t]/P[\omega^t]$$

$$= \Sigma[I_{\Lambda'}(\upsilon)P(\upsilon) : \upsilon^t = \omega^t]/P[\omega^t]$$

$$= E[I_{\Lambda'}|\omega^t]$$

where $I_{\Lambda'}$ is the indicator of Λ' and the internal conditional expectation was defined in (1.5.7) with $\rho(\omega) = [\omega^t]$ in this case. We know from (1.5.9-10) that

$$E[I_{\Lambda'}|\omega^t] \text{ lifts } E[I_\Lambda|\mathcal{M}(t)],$$

so $P[\Lambda'|\omega^t]$ almost equals $E[I_\Lambda|\mathcal{M}(t)]$. Unfortunately, the null sets matter so we must distinguish between these two things.

(6.3.4) **PROPOSITION:**

Let $t \in \mathbb{T}$. *If* $\Lambda \subseteq \Omega$ *is* P-*measurable*, $\Lambda \in$ Meas(P), *then for* P-*almost all* ω, Λ *is* $P[\ |\omega^t]$-*measurable*.

PROOF:

We express $\Omega = W^{\mathbb{T}}$ as a hyperfinite product,

$$\Omega^{\mathbb{T}} = \Omega^{\mathbb{T}[0,t]} \times \Omega^{\mathbb{T}(t,1]},$$

associating the pair $((\omega_0,\cdots,\omega_t);(\omega_{t+\delta t},\cdots,\omega_1))$ in Ω. The uniform measures on each factor, μ and υ, can be related to $P[\ |\omega^t]$ by taking sections in the product, for Λ thought of as in $W^{\mathbb{T}[0,t]} \times W^{\mathbb{T}(t,1]}$,

$$\Lambda^{\omega^t} = \{(\lambda_{t+\delta t},\cdots,\lambda_1) : \lambda^t = \omega^t\}.$$

Hence, $P[\Lambda|\omega^t] = \upsilon[\Lambda^{\omega^t}]$. Now, by Keisler's Fubini Theorem, μ-almost every ω^t produces a measurable section, so $P[\Lambda|\omega^t]$ is defined. If Ω_1 is an internal first factor of small

μ-measure where the sections are bad, then $\Omega_1 \times \Omega^{\mathbb{T}(t,1]}$ has the same P-measure and lies in Ω. This proves that $P[\Lambda|\omega^t]$ exists a.s. [P].

(6.3.5) DEFINITION:

For any $\Lambda \subseteq \Omega$, *the derived set* Λ' *is defined by*

$$\Lambda' = \{\omega \in \Lambda : (\forall t \in \mathbb{T})P[\Lambda|\omega^t] = 1\}.$$

Recall that if Λ is Loeb, then $P[\Lambda|\omega^t]$ is always defined. Keisler's main tool in the proof of nonanticipating lifting is the

(6.3.6) DERIVED SET LEMMA:

If $P[\Lambda] = 1$, *then* $P[\Lambda'] = 1$ *and* $\Lambda'' = \Lambda'$. *Moreover, if* Λ *is also internal, then there is an internal* $\Phi \subseteq \Lambda'$ *with* $P[\Phi] = 1$.

For proof, see Keisler [1984, Derived Set Lemma 6.6] and (6.5.16).

The S-Integrable Lebesgue Lifting Theorem (5.4.5) has a nonanticipating version, Theorem 7.8 of Keisler [1984]. We prove the decent path form in (6.3.7-8)(c).

(6.3.7) NONANTICIPATING DECENT PATH PROJECTION THEOREM:

Suppose the internal nonanticipating process $X : \mathbb{T} \times \Omega \longrightarrow {}^*\mathbb{R}$ *a.s. has a decent path sample. Then:*

(a) *The decent path projection* $\tilde{X}$ *is indistinguishable from a progressively measurable process.*

(b) *If each section* X^t *is S-integrable, then* $\tilde{X}$ *is indistinguishable from a uniformly integrable progressive process.*

(c) *If* $\|X\|^p$ *is* $\delta t \otimes \delta P$*-S-integrable for* $p \geq 1$, *then* $\tilde{X} \in L^p(\lambda \times P)$.

PROOF:

(a) Let X have a Δt-sample except for a Loeb null set $\Lambda_0 \subseteq \Omega$. For each $m \geq 1$ in $\mathbb{N}$ and each k, $1 \leq k \leq m$, there is a $t_m^k = n\Delta t$, for some n in ${}^*\mathbb{N}$, satisfying $X(t_m^k) \approx \tilde{X}(\frac{k}{m})$ except on the null set Λ_0. Define

$$Y_m(r,\omega) = \begin{cases} \overline{st}\, X(t_m^1,\omega) , & 0 \leq r \leq t_m^1 \\ \overline{st}\, X(t_m^k,\omega) , & \frac{k-1}{m} < r \leq \frac{k}{m} , \ 1 < k \leq m \\ \overline{st}\, X(1,\omega) , & r = 1 \end{cases}$$

and let $Y(r,\omega) = \liminf_m Y_m(r,\omega)$. Each Y_m is (Borel×Loeb)-measurable, so Y is also. If X_ω is a Δt-decent path, then $st_k X_\omega = Y_\omega$.

If $\omega \overset{r}{\approx} \upsilon$, then $\omega^t = \upsilon^t$ for some $t \gg r$, hence for all m such that $t-r > \frac{1}{m}$, $Y_m(r,\omega) = \overline{st}(X(t_m^k,\omega))$ where $t_m^k < t$, so $X(t_m^k,\omega) = X(t_m^k,\upsilon) \approx Y_m(r,\upsilon)$. This shows that $Y(r,\omega) = Y(r,\upsilon)$ so Y is progressive.

(b) Suppose that each X^t is S-integrable. The family $\{st\, X^t : t \in \mathbb{T}\}$ is uniformly integrable by an argument like Lemma (1.6.2). Hence each Y_m is a uniformly integrable process and by Fatou's lemma, Y is also (compare to (5.3.26)).

(c) By (5.4.6), X is a Lebesgue lifting of $\tilde{X}$, so by (5.4.5), $\tilde{X} \in L^p(\lambda\times P)$.

(6.3.8) **NONANTICIPATING DECENT PATH LIFTING THEOREM:**

Suppose a stochastic process $X : [0,1] \times \Omega \longrightarrow \mathbb{R}$ *is* $\mathcal{F}$*-adapted and a.s. has decent paths. Then:*

(a) X *has a nonanticipating decent path lifting, that is, there is an internal process* $Y : \mathbb{T} \times \Omega \longrightarrow {}^*\mathbb{R}$, *nonanticipating after* Δt *that a.s. has a* Δt*-decent path sample for some infinitesimal* Δt *in* $\mathbb{T}$ *such that the decent path projection* $\tilde{Y}$ *is indistinguishable from* X. *In particular,* X *is indistinguishable from a progressively measurable process.*

(b) *If* X *is uniformly integrable, the* Y *in part* (a) *may be chosen with each section,* Y^t, *S-integrable.*

(c) *If* $\|X\| \in L^p(\lambda\times P)$ *for real* $p \geq 1$, *then the* Y *in part* (a) *may be chosen so that* $\|Y\|^p$ *is* $\Delta t\otimes\delta P$*-S-integrable.*

PROOF:

(a) Let $W : \mathbb{T} \times \Omega \longrightarrow {}^*\mathbb{R}$ be a D-lifting of X, appealing to (5.3.23). For each $m \geq 1$ in $\mathbb{N}$ and $1 \leq k \leq m$ use (6.1.7) to choose Z^k_m an $\mathcal{E}(t^k_m)$-measurable random variable such that $Z^k_m \approx X(\frac{k}{m})$ a.s. and $t^k_m \approx \frac{k}{m}$, while $st[W(t^k_m)] = X(\frac{k}{m})$ a.s. (increasing t^k_m, if necessary). Define

$$Z_m(t,\omega) = \begin{cases} Z_m^1(\omega) , & 0 \le t < t_m^1 \\ Z_m^k(\omega) , & t_m^k \le t < t_m^{k+1} \end{cases}$$

(we may take $t_m^m = 1$ and $Z_m(1,\omega) = Z_m^m(\omega)$). The process Z_m is nonanticipating after t_m^1, constant on the intervals $\mathbb{T}[t_m^k, t_m^{k+1})$ with $\frac{1}{2m} < t_m^{k+1} - t_m^k < \frac{2}{m}$, while

$$P\left[\max_{1 \le k \le m} |W(t_m^k) - Z(t_m^k)| > \frac{1}{m}\right] < \frac{1}{m}.$$

The last statement is internal and holds for every finite m, hence there is an infinite m for which the statement holds. The process $Y(t,\omega) = Z_m(t,\omega)$ for such an m. It has decent paths because W does.

(b) If X is uniformly integrable, we may choose W and Z_m^k S-integrable and conclude moreover that

$$E\left[\max_{1 \le k \le m} |W(t_m^k) - Z_m(t_m^k)| > \frac{1}{m}\right] < \frac{1}{m}$$

in the internal statement above. This proves the theorem (b). Alternately, see the proof of (5.3.25).

(c) If Y is as in part (a), define truncations

$$Y_m(t,\omega) = \begin{cases} Y(t,\omega) & , \text{ for } |Y(t,\omega)| \le m \\ m\frac{Y(t,\omega)}{|Y(t,\omega)|} & , \text{ for } |Y(t,\omega)| \ge m \end{cases}$$

for m in ${}^*\mathbb{N}$ and similarly for X when m is finite. By (5.4.6) and (5.4.5) we know $|Y_m|^p$ is an S-Cauchy sequence, so there is an infinite m which makes $|Y_m|^p$ S-integrable [see the proof of (1.4.8)]. This shows (c).

(6.3.9) **REMARK:**

The first part of the nonanticipating lifting and projecting theorems can be proved (with little change) for $X : [0,1] \times \Omega \longrightarrow \mathbb{M}$ and $Y : \mathbb{T} \times \Omega \longrightarrow K$ where K is an S-dense internal subset of ${}^*\mathbb{M}$, a complete separable metric space entity. The integrable parts require linear structure on $\mathbb{M}$, but note that the integrable $\mathbb{M} = \mathbb{R}^d$ case follows from these results one coordinate at a time.

(6.3.10) **NONANTICIPATING DOMINATED LEBESGUE LIFTING THEOREM:**

Suppose $X : [0,1] \times \Omega \longrightarrow \mathbb{R}$ *is an* $\delta t \otimes \delta P$*-almost progressively measurable stochastic process. If* $X(r,\omega) \leq c(\omega)$ *a.s.* $[\lambda \times P]$ *for some function* $c : \Omega \longrightarrow \mathbb{R}$, *then there is a nonanticipating Lebesgue lifting* Y *of* X *satisfying*

$$\max_t Y(t,\omega) \lesssim c(\omega) \quad a.s. \quad P.$$

The complete proof of this result is in Keisler [1984, Theorem 7.9], but we shall start it off for contrast with the next exercise. Let

$$b(\omega) = \|X(\cdot,\omega)\|_{L^\infty(\lambda)}$$
$$= \inf[b \in \mathbb{R} : |X(r,\omega)| \leq b \text{ a.s. } \lambda(r)].$$

For each ω, $X(r,\omega) \leq b(\omega)$ a.s. λ and

$$\{\omega : b(\omega) \leq a\} = \{\omega : X(r,\omega) \leq a \ \text{ a.s.}\},$$

so b is P-measurable. (Use the ordinary Fubini Theorem.) Also, $b(\omega) \leq c(\omega)$ a.s. by the ordinary Fubini Theorem. So we can work with the measurable function $b(\omega)$ and a P-lifting $a(\omega)$. The rest of the proof can be found in Keisler [1984].

(6.3.11) **EXERCISE:**

State and prove the analog of (6.3.8) for decent path liftings. Assume X is $\delta t \otimes \delta P$-almost progressively measurable, a.s. has decent paths and $X(r,\omega) \leq c(\omega)$ a.s. What is the function corresponding to $b(\omega)$ in the discussion after (6.3.8)?

(6.3.12) **EXERCISE:**

(a) Show that a stochastic process $X : [0,1] \times \Omega \longrightarrow \mathbb{R}$ is $\mathcal{F}$-adapted and a.s. has nondecreasing paths in $D[0,1]$ if and only if there is a nonanticipating increasing internal $Y : \mathbb{T} \times \Omega \longrightarrow {}^*\mathbb{R}$ that a.s. has a D-sample and X is indistinguishable from $\tilde{Y}$.

(b) Prove part (a) in the special setting where both X and Y are restricted to have values in the two point space $\{0,1\}$.

(6.4) Measurable and Internal Stopping

Here is the point of Exercise (6.3.12). If Y is nonanticipating, increasing and takes only the values 0 and 1, then

$$\tau(\omega) = \min\{t \in \mathbb{T} : Y(t,\omega) = 1\}$$

has the property that if $\tau(\omega) = t$ and $\upsilon^t = \omega^t$, then $Y(t,\omega) = 1 = Y(t,\upsilon)$, so $\tau(\upsilon) \leq t$ while a symmetric argument interchanging υ and ω shows $\tau(\upsilon) = t$. This means τ satisfies the next definition with $\Delta t = \delta t$.

In numbers (5.3.8-11) we developed some special properties of particular stopping times associated with an approximate Poisson process. At that time we did not need to know that they were stopping times in the technical sense of the next definition but they are examples that we have used.

If Δt is a positive time $\Delta t \in \mathbb{T}$, we shall denote the set of multiples of Δt by

$$\mathbb{T}_\Delta = \{t \in \mathbb{T} : k\Delta t \textit{ for some } k \in {}^*\mathbb{N}\} \cup \{1\}.$$

(6.4.1) DEFINITION:

Let $\Delta t \in \mathbb{T}$. *A* Δt*-stopping time is an internal function* $\tau : \Omega \longrightarrow \mathbb{T}_\Delta$ *such that if* $\tau(\omega) = t$ *and* $\upsilon^t = \omega^t$, *then* $\tau(\upsilon) = t$.

If τ is a Δt-stopping time, then the internal process

$$Y(t,\omega) = \begin{cases} 0, & \text{if } \tau(\omega) > t \\ 1. & \text{if } \tau(\omega) < t \end{cases}$$

is increasing, takes values in $\{0,1\}$ and if $\upsilon^t = \omega^t$ then $Y(t,\omega) = Y(t,\upsilon)$, that is, Y is *nonanticipating*. To reiterate, the remarks before and after the definition give a correspondence between stopping times and increasing nonanticipating indicator processes.

If $Z(t)$ is simply an internal nonanticipating process for $t \geq \iota$, then functions like

$$\tau'(\omega) = \min\{t : Z(t,\omega) \geq a\}$$

could fail to be δt-stopping times. However, $\tau = \tau' \vee \iota$ is a δt-stopping time. This basic annoyance is caused by the fact that $\mathcal{F}(0) \supset \mathcal{D}(0)$. If $\Delta t > \delta t$, then $\sigma(\omega) = \min[t \in \mathbb{T}_{\Delta} : t > \tau(\omega)]$ is a Δt-stopping time and if $\Delta t \approx 0$, $st(\sigma) = st(\tau)$.

If $X : [0,1] \times \Omega \longrightarrow \{0,1\}$ is $\mathcal{F}$-adapted (not just a.s.) and increasing, then letting

$$\rho(\omega) = \inf\{r \in [0,1] : X(r,\omega) = 1\}$$

we see that $\{\omega : \rho(\omega) \leq r\} = \{\omega : X(r,\omega) = 1\} \in \mathcal{F}(r)$.

(6.4.2) DEFINITION:

A function $\rho : \Omega \longrightarrow [0,1]$ *is called an* $\mathcal{F}$-*stopping time if for each* r, $\{\omega : \rho(\omega) \leq r\} \in \mathcal{F}(r)$.

If we suppose that ρ is an $\mathcal{F}$-stopping time and define

$$X'(r,\omega) = \begin{cases} 0, & \text{if } \rho(\omega) > r \\ 1, & \text{if } \rho(\omega) \leq r \end{cases}$$

then X' is increasing and when $\omega \overset{r}{\sim} \upsilon$ either both $\rho(\omega) \lesssim r$ and $\rho(\upsilon) \leq r$ or $\rho(\omega) > r$ and $\rho(\upsilon) > r$, so $X'(r,\omega) = X'(r,\upsilon)$ or X $\mathcal{F}$-adapted. If $X'(s,\omega) = 1$ for all $s > r$, then $\rho(\omega) \leq s$ for all such s and $\rho(\omega) \leq r$, so X' is right continuous. Since X' has decent paths, (6.3.12)(a) says it is indistinguishable from an increasing progressively measurable process X''. Let

$$X(r,\omega) = \begin{cases} 0, & \text{if } X''(r,\omega) < \frac{1}{2} \\ 1, & \text{if } X''(r,\omega) \geq \frac{1}{2} \end{cases}.$$

The $\mathcal{F}$-stopping time $\tau(\omega) = \inf[r : X(r,\omega) \geq 1]$ satisfies $\tau = \rho$ a.s.

(6.4.3) THE STOPPING TIME LIFTING LEMMA:

If $\tau : \Omega \longrightarrow \mathbb{T}$ *is a* Δt*-stopping time, then* $\rho(\omega) = \mathrm{st}[\tau(\omega)]$ *is an* $\mathcal{F}$*-stopping time. For each* $\mathcal{F}$*-stopping time* $\rho : \Omega \longrightarrow [0,1]$*, and each infinitesimal* ∇t*, there is a* ∇t*-stopping time* τ *such that* $\mathrm{st}(\tau) = \rho$ *a.s.*

PROOF:

Let τ be a Δt-stopping time and $\rho = \mathrm{st}(\tau)$. If $\rho(\omega) \leq r$ then $\tau(\omega) = s$ for some $s \lesssim r$ If $\omega \overset{r}{\sim} \upsilon$ then $\omega^t = \upsilon^t$ for some $t \gg r$ and $\omega^s = \upsilon^s$ so $\tau(\upsilon) = s \lesssim r$ and $\rho(\upsilon) \leq r$, thus $\{\omega : \rho(\omega) \leq r\} \in \mathcal{F}(r)$.

If ρ is an $\mathcal{F}$-stopping time, define the process $X(r,\omega) = I_{\{\rho(\omega)\leq r\}}(r,\omega)$ as in the discussion above. Then X is progressive, has increasing paths in D and takes only the values 0 and 1. Let Y be an increasing nonanticipating (after Δt) lifting of X with an a.s. Δt-sample. The $\mathcal{F}$-stopping time $\sigma(\omega) = \inf\{r \in [0,1] : \tilde{Y}(r,\omega) = 1\}$ equals $st(\tau')$ where $\tau'(\omega) = \min[t \in \mathbb{T}_\Delta : Y(t,\omega) = 1]$. We make this a ∇t-stopping time by defining $\tau(\omega) = \min[t \in \mathbb{T}_\Delta : t \geq \tau'(\omega)]$. For almost all ω, $\tilde{Y}_\omega$ equals the path X_ω, so $st(\tau) = \rho$ a.s.

A good way to think of a stopping time is 'a pruning of the tree of samples'. The schematic variation on Figure (6.1.1) is as follows.

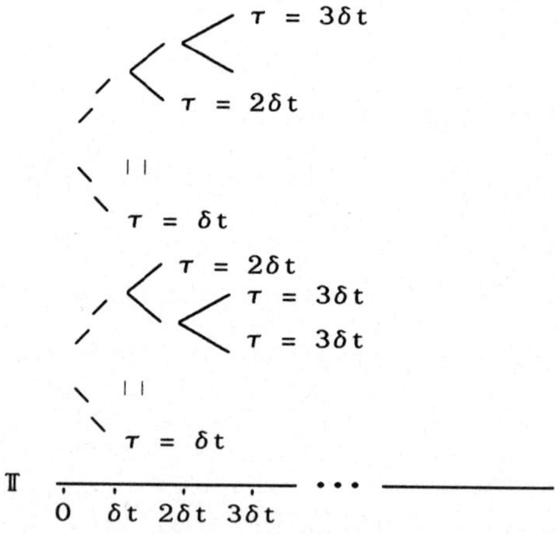

$\tau = 3\delta t$

$\tau = 2\delta t$

$\tau = \delta t$

$\tau = 2\delta t$

$\tau = 3\delta t$

$\tau = 3\delta t$

$\tau = \delta t$

$\mathbb{T}$ ——————— ... ————

0 δt $2\delta t$ $3\delta t$

Figure (6.4.4)

If Y is a decent path lifting of X, then $X(r) = S\text{-}\lim_{t\downarrow r} Y(t)$ a.s., so a simple argument shows that there is an $s \approx r$ such that whenever $t \approx s$ and $t \geq s$, $X(r) = \text{st } Y(t)$ a.s., in particular, $\text{st } Y(s) = X(\text{st}(s))$. Our next result from Hoover & Perkins [1983] extends this to random times.

(6.4.5) **THE PATH STOPPING LEMMA:**

(a) *Let* Y *be a decent path lifting of* X *and let* $\rho : \Omega \longrightarrow [0,1]$ *be an* $\mathcal{F}$*-stopping time. If* Y *has a* Δt*-decent path sample for some infinitesimal* Δt*, then there is a* Δt*-stopping time* τ *and a null set* $\Lambda \subseteq \Omega$ *such that whenever* $\omega \notin \Lambda$*, then* $\tau(\omega) \approx \rho(\omega)$ *and whenever* $\omega \notin \Lambda$ *and* $0 < \iota = k\Delta t \approx 0$*,* $\tau(\omega)+\iota \leq 1$*, then* $Y(\tau(\omega)+\iota,\omega) \approx X(\rho(\omega),\omega)$*, in particular,*

$$\text{st}[Y(\tau)] = X(\rho) \quad a.s.$$

(b) *Suppose in addition to the hypotheses of* (a) *that* X *is* $\mathcal{F}$*-adapted and*

$$\rho(\omega) = \inf[r : |X(r,\omega)| > a]$$

while Y *is also nonanticipating after* Δt*. Then there is a* $b \approx a$ *so that the* Δt*-stopping time of the conclusion of* (a) *has the form:*

$$\tau(\omega) = \min[t \in \mathbb{T}_\Delta : |Y(t,\omega)| > b].$$

PROOF:

Part (a).

Let Z lift the random variable $X(\rho(\omega),\omega)$. Let σ be a Δt-stopping time such that $st\ \sigma = \rho$ a.s. where Δt makes Y a.s. have a Δt-sample. For every p in $\mathbb{N}$ there is an $m_p \geq p$ in $\mathbb{N}$ such that the following set contains every finite $n \geq m_p$:

$$\left\{n \in {}^*\mathbb{N} : \frac{1}{n} \in \mathbb{T} \ \&\ P\left[\max_{\frac{1}{n}\leq t\leq\frac{1}{m_p}} |Y(\sigma+t)-Z| \geq \frac{1}{p}\right] \leq \frac{1}{p}\right\}.$$

Since the set is internal it contains an infinite n_p and the countable intersection $\bigcap_p {}^*\mathbb{N}[m_p.n_p]$ contains an infinite n by saturation. Let $\tau = \sigma + \frac{1}{n}$.

Let Λ be the union of the null sets where $Z(\omega) \not\approx X(\rho(\omega),\omega)$, Y doesn't have a Δt-sample, $\tau(\omega) \not\approx \rho(\omega)$ and $\lim st[\max_{\frac{1}{n}\leq t\leq\frac{1}{m_p}} |Y(\sigma+t)-Z|] \neq 0$. If $\omega \notin \Lambda$, $Y(\tau(\omega)+\iota,\omega) \approx Z(\omega) \approx X(\rho(\omega),\omega)$ and $\tau(\omega) \approx \rho(\omega)$, so this proves part (a).

Part (b).

Define Δt-stopping times

$$\tau_m(\omega) = \min[t \in \mathbb{T}_\Delta : |Y(t,\omega)| \geq a + \frac{1}{m}],$$

so $\tau_1 \geq \tau_2 \geq \cdots$. If Y has a Δt-decent path on ω and $\tilde{Y}(\cdot,\omega) = X(\cdot,\omega)$, then we will show that $\lim_{m\to\infty} st\ \tau_m = \rho$ for such random samples ω. To see this, suppose $\rho(\omega) << t$. Then there is an r, $\rho(\omega) \leq r << t$ such that $|X(r.\omega)| > a + \frac{1}{m}$.

This means that for all large enough m, $\rho(\omega) \le \tau_m < t$, so $\tau_m \downarrow \rho$ a.s. and moreover $\lim_m Y(\tau_m) = Y(\rho)$ a.s.

We need to 'internalize' these limits as follows. Let σ be a Δt-stopping time such that $st(\sigma) = \rho$ and $st[Y(\sigma)] = X(\rho)$ (both a.s., using part (a)). Then for every finite $p \in \mathbb{N}$ there is a finite $m_p > p$ such that for all finite $m > m_p$,

$$P[|\tau_m - \sigma| > \frac{1}{p} \text{ or } |Y(\tau_m) - Y(\sigma)| > \frac{1}{p}] < \frac{1}{p}.$$

Since the definition of $\{\tau_m\}$ is internal, there exists an infinite n_p satisfying the probability above. Use saturation to pick an infinite $n \in \bigcap_p [m_p, n_p]$. This τ_n is the claimed Δt-stopping time with $b = a + \frac{1}{n}$.

(6.4.6) **EXERCISE:**

Show that if ρ is only P-measurable we may still select an internal (nonstopping) τ so that the remaining conclusions in (6.3.14) are true, in particular, $st\, Y(\tau) = X(st(\tau))$.

(6.4.7) **EXAMPLE:**

This example shows one kind of difference between $st\, X(\tau)$ and $\tilde{X}(st(\tau))$. It is based on a remark in Hoover & Perkins [1983] and an example in Lindstrom [1980].

On the *finite set W (where $\Omega = W^{\mathbb{T}}$) choose an internal jump function $j : W \longrightarrow {}^*\mathbb{R}$ with a symmetric distribution which is a.s. finite but not S-integrable. For example, apply (4.1.5) to obtain a distribution:

$$P[0 \le j(w) \le x] = P[-x \le j(w) \le 0] \approx \int_2^x \frac{dy}{y^2}.$$

Notice that $P[|\tilde{j}| = \infty] = 0$, but $E[|\tilde{j}|] = \infty$. Define a process X by

$$X(t,\omega) = \begin{cases} 0 & , \quad t < \frac{1}{2} \\ \iota \dfrac{j(\omega_{1/2})}{1+|j(\omega_{1/2})|} & , \quad t = \frac{1}{2} \\ j(\omega_{1/2}) & , \quad t > \frac{1}{2} \end{cases}$$

where ι is a positive infinitesimal.

The infinitesimal jump at $t = \frac{1}{2}$ announces the finite jump at $t = \frac{1}{2} + \delta t$. The functions

$$\tau_m(\omega) = \begin{cases} \frac{1}{2}, & |j(\omega_{1/2})| \le m \\ 1, & \text{otherwise} \end{cases}$$

are δt-stopping times for each m and

$$P[\tau_m < 1] = P[|j| > m] \longrightarrow 0 \quad \text{as} \quad m \longrightarrow \infty.$$

Since $\tilde{X}(r) = j(\omega_{1/2})$ for $r \ge \frac{1}{2}$ and 0 for $r < \frac{1}{2}$, we see that $\tilde{X}(st(\tau_m)) = \tilde{X}(1)$ with the same distribution as $\tilde{j}$ while $P[st\ X(\tau_m) = 0] \to 1$ as $m \to \infty$, since $X(\frac{1}{2}) \approx 0$ and $\tau_m = \frac{1}{2}$, when $|j| \le m$.

(6.5) **Martingales**

The notion of *martingale on our evolution scheme is obtained by $*$-transform. This makes well known facts about finite martingales available for the study of hypermartingales.

(6.5.1) **DEFINITIONS:**

An internal $X : \mathbb{T} \times \Omega \longrightarrow {}^*\mathbb{R}^d$ *is called a* **martingale after* Δt *if* $s < t$ *in* $\mathbb{T}_\Delta^+$ *and* ω *in* Ω *imply*

$$E[X(t) \mid \omega^s] = X(s,\omega).$$

A *nonanticipatting internal* $X : \mathbb{T} \times \Omega \longrightarrow {}^*\mathbb{R}$ *is called a* **submartingale (resp.,* **supermartingale) after* Δt *if* $s < t$ *in* $\mathbb{T}_\Delta^+$ *and* ω *in* Ω *imply*

$$E[X(t) \mid \omega^s] \geq X(s,\omega) \qquad \textit{(resp. } \leq).$$

A *martingale is automatically nonanticipating after Δt because $X(t,\omega) = E[X(1) \mid \omega^t] = E[X(1) \mid \upsilon^t] = X(t,\upsilon)$ when $\omega^t = \upsilon^t$. Also, we could simply say X is a *martingale after Δt if

$$E[X(1) \mid \omega^t] = X(t,\omega)$$

for t in $\mathbb{T}_\Delta^+$ and ω in Ω, since a simple *finite calculation yields the defining property above from this.

If $M : \mathbb{T} \times \Omega \longrightarrow {}^*\mathbb{R}^d$ is a *martingale and $\varphi : {}^*\mathbb{R}^d \longrightarrow {}^*\mathbb{R}$ is *convex, then $X = \varphi(M)$ is a *submartingale: $\varphi(\Sigma c_j x_j) \leq \Sigma c_j \varphi(x_j)$, so

$$E[\varphi(M(t)) \mid \omega^s] \geq \varphi\left\{ \sum_{\upsilon^s=\omega^s} M(t,\upsilon) \frac{\delta P(\upsilon)}{P[\omega^s]} \right\}$$

$$= \varphi(M(s,\omega)).$$

If X is a *submartingale and φ is increasing and *convex, then $Y = \varphi(X)$ is also a *submartingale.

An important kind of *martingale is given by a sum of *independent mean zero terms,

$$X(t,\omega) = \Sigma[f(s,\omega_s) : 0 < s \leq t, \text{ step } \delta t]$$

where $f : \mathbb{T} \times W \longrightarrow {}^*\mathbb{R}$ satisfies $\Sigma[f(s,w) : w \in W] = 0$ for each s in $\mathbb{T}$, see (4.3.1). Anderson's random walk

$$B(t,\omega) = \Sigma[\sqrt{\delta t}\ \beta(\omega_s) : 0 < s \leq t, \text{ step } \delta t]$$

is such an example where $f(s,w)$ does not depend on s. It is a central example leading to Brownian motion. Example (4.3.3) only differs from a *martingale by a 'drift' term, that is,

$$J(t,\omega) - pt$$

is a *martingale related to the Poisson process (5.3.17). Notice that the full generality of (4.3.4) does not lead to a *martingale if those increments are anticipating. (We condition on ω^t, not $X(t_1),\cdots,X(t_n)$.)

Our next two results are *-transforms of well known theorems.

(6.5.2) THE *MARTINGALE MAXIMAL INEQUALITY:

If $X : \mathbb{T} \times \Omega \longrightarrow {}^*\mathbb{R}$ *is a* **submartingale after* Δt, *then for any* $x > 0$

(a) $P[\max_{\Delta t \leq t \leq 1} X(t) > x] \leq \frac{1}{x} E|X(1)|$

and

(b) $P[\min_{\Delta t \leq t \leq 1} X(t) \leq x] \leq \frac{1}{x}(E[|X(1)|] - E[X(1)])$.

If $M : \mathbb{T} \times \Omega \longrightarrow {}^*\mathbb{R}^d$ *is a* **martingale after* Δt *and* $x > 0$, *then*

(c) $P[\max_{\Delta t \leq t \leq 1} |M(t)| > x] \leq \frac{1}{x} E[|M(1)|]$.

PROOF:

See Breiman [1968, 5.13] or Doob [1953] or Helms & Loeb [1980]. Note $|M|$ is a *submartingale by remarks above.

(6.5.3) THE UPCROSSING LEMMA:

Let $X : \mathbb{T} \times \Omega \longrightarrow {}^*\mathbb{R}$ *be a* **submartingale after* Δt. *Let* $\beta_a^b(\omega)$ *equal the number of times that the sequence* $X(\Delta t, \omega), \ X(\Delta t+\delta t, \omega), \cdots, X(1, \omega)$ *crosses the interval* $[a,b]$ *from left to right. Then*

$$E[\beta_a^b] \leq \frac{E[(X(1)-a)^+]}{b-a} \leq \frac{E[|X(1)|]+|a|}{b-a}.$$

PROOF:

See Breiman [1968, 5.17] or Doob [1953].

Our next result, (6.5.6), is a preliminary for the definition of Δt-martingales. It shows that paths of *martingales are not too ill-behaved, but first we give an example that shows that it is the best we can hope for.

(6.5.4) EXAMPLE:

Let $\beta : W \longrightarrow \{-1,1\}$ be $+1$ on one half of W and -1 on the other half, as in (4.3.2). Define

$$N(t,\omega) = \beta\left[\omega_{\frac{1}{2}}\right] I_{[\frac{1}{2},1]}(t) + \beta\left[\omega_{\frac{1}{2}+\delta t}\right] I_{[\frac{1}{2}+\delta t,1]}(t).$$

N is a nasty martingale with the paths as shown in Figure (6.5.5)—it splits a jump in infinitesimal time.

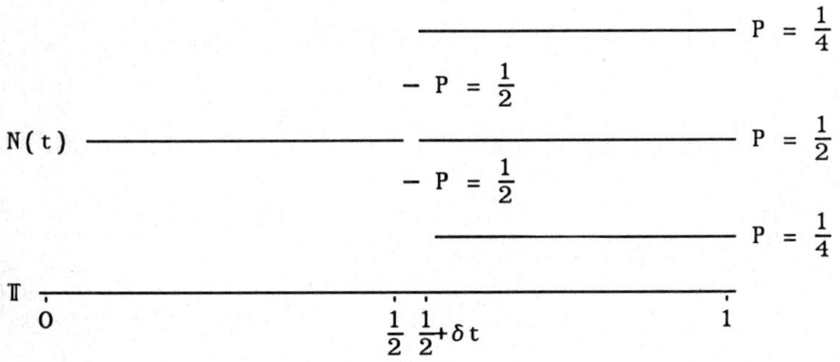

$P = \frac{1}{4}$

$P = \frac{1}{2}$

$N(t)$ $P = \frac{1}{2}$

$P = \frac{1}{2}$

$P = \frac{1}{4}$

$\mathbb{T}$ 0 $\frac{1}{2}$ $\frac{1}{2}+\delta t$ 1

Figure (6.5.5)

(6.5.6) PROPOSITION:

Suppose $M : \mathbb{T} \times \Omega \longrightarrow {}^*\mathbb{R}^d$ *is a* **martingale after* $\Delta t \approx 0$ *and* $E[|M(1)|]$ *is limited,* $M(1) \in FL^1(\Omega)$. *Then there is an infinitesimal* ∇t *in* $\mathbb{T}$ *such that* M *a.s. has a* ∇t*-decent path sample.*

PROOF:

By (6.5.2), $\max_{0 \le t \le 1} |M(t)|$ is finite a.s., so for the rest of the proof we avoid the null set where any $M(t)$ is infinite.

We use (6.5.3) to show that both

$$\underset{t\uparrow r}{\text{S-lim}}\, M(t) \quad \text{and} \quad \underset{t\downarrow r}{\text{S-lim}}\, M(t)$$

exist and are finite for every r a.s. It suffices to show that the limits exist for each coordinate M_j, $1 \le j \le d$, of M.

Let $X(t)$ be one coordinate of $M(t)$. If either $\underset{t\uparrow r}{\text{S-lim}}\, X(t,\omega)$ or $\underset{t\downarrow r}{\text{S-lim}}\, X(t,\omega)$ fail to exist for some r in $[0,1]$ then there are sequences $t_m^1(\omega)$ and $t_m^2(\omega)$ that tend monotonically to r and $\underset{m}{\text{S-lim}}\, X(t_m^1(\omega),\omega) = \text{S-lim inf}\, X(t,\omega) < \text{S-lim sup}\, X(t,\omega) = \underset{m}{\text{S-lim}}\, X(t_m^2(\omega),\omega)$. As a result there must be a rational interval $[a,b]$ that $X(t,\omega)$ crosses an infinite number of times, or $\beta_a^b(\omega)$ is infinite (in ${}^*\mathbb{N}$) for β_a^b as in (6.5.3). Since $E[\beta_a^b]$ is finite, by (6.5.3), $P[\cap\{\beta_a^b > m\}] = 0$ and $\cup[\{\beta_a^b \approx \infty\} : a < b \text{ in } \mathbb{Q}]$ is a P-null set.

As a result of the last two paragraphs, except for a null set all the coordinate limits exist and are finite, so by (5.3.25) M has a ∇t-decent path sample.

(6.5.7) **PROPOSITION:**

Let X be a positive *submartingale after $\Delta t \approx 0$ and let τ be a Δt-stopping time with $\tau \le u \le 1$. If $X(u)$ is S-integrable, $X(u) \in SL^1(\Omega)$, then $X(\tau) \in SL^1(\Omega)$. In particular, if M is a *martingale after Δt and $M(1) \in SL^1(\Omega)$, then $M(\tau) \in SL^1(\Omega)$.

PROOF:

Let $E_\tau[X(u)] = E[X(u) \mid \omega^{\tau(\omega)}]$. By the submartingale definition, $E_\tau[X(u)](\omega) \geq X(\tau(\omega),\omega) \geq 0$. Let h be positive infinite and $k = \sqrt{h}$. The following calculation shows that $X(\tau)$ is S-integrable.

$$\sum_{[X(\tau)\geq h]} X(\tau)\delta P \leq \sum_{[E_\tau[X(u)]\geq h]} E_\tau[X(u)]\delta P$$

$$\leq \sum_{[E_\tau[X(u)]\geq h]} X(u)\delta P$$

$$\leq \sum_{[h\leq E_\tau[X(u)]\&X(u)\leq k]} X(u)\delta P + \sum_{[X(u)>k]} X(u)\delta P$$

$$\lesssim kP[E_\tau[X(u)]\geq h] \leq \frac{k}{h} E[X(u)] \lesssim 0.$$

For the second inequality, break the sum into a sum over classes $[\omega^{\tau(\omega)}]$. For the second to last use the *martingale maximal inequality, (6.5.2) $E_\tau[X(u)](\omega) \leq \max_{0\leq t\leq u} X(t,\omega)$.

The last part follows from the first by taking the *submartingale $X = |M| \geq 0$.

(6.5.8) REMARK:

If M is an L^2-finite *martingale, $M(1) \in FL^2(\Omega)$, or $E[|M|^2] \in \mathcal{O}$, then M is S-integrable. Suppose I_Λ is the indicator function of $\Lambda \subseteq \Omega$, then

$$E[I_\Lambda M(1)] \leq P[\Lambda] \sqrt{E[|M|^2]}$$

by the *Cauchy inequality, so if $P[\Lambda]$ is infinitesimal, then the left hand side is also. The condition S-AC, which we have just shown, implies S-integrability for P.

When $\Delta t \geq \delta t$ belongs to $\mathbb{T}$, let

$$\mathbb{T}_\Delta = \{t \in \mathbb{T} : (\exists k \in {}^*\mathbb{N})[t = k \cdot \Delta t]\} \cup \{1\}$$

$$= \{0, \Delta t, 2 \cdot \Delta t, 3 \cdot \Delta t, \cdots\cdot, 1\}.$$

(6.5.9) **DEFINITION:**

We shall say that an internal process $M : \mathbb{T} \times \Omega \longrightarrow {}^*\mathbb{R}^d$ *is a* Δt*-martingale for some infinitesimal* $\Delta t \geq \delta t$ *in* $\mathbb{T}$ *if* M *a.s. has a* Δt*-decent path sample and if whenever* $0 < s < t$ *belong to* $\mathbb{T}_\Delta$, *then*

$$E[M(t) \mid \omega^s] = M(s, \omega)$$

and $M(t)$ *is S-integrable.*

Our results above show that if $M(1)$ is S-integrable and M is a *martingale after $\forall t \approx 0$, then M is a Δt-martingale for some infinitesimal Δt in $\mathbb{T}$. The Δt-martingales correspond under projection to the uniformly integrable martingales with respect to the progressive filtration. The last phrase is such a mouthful, that we introduce a term for it.

(6.5.10) **DEFINITION:**

A uniformly integrable progressive process $M : [0,1] \times \Omega \longrightarrow \mathbb{R}^d$ *is called a hypermartingale if* M *a.s. has decent paths and for each* $r < s$ *in* $[0,1]$,

$$E[M(s) \mid \mathcal{F}(r)] = M(r) \quad \text{a.s. } P.$$

(6.5.11) **THE Δt-MARTINGALE PROJECTION THEOREM:**

If $M : \mathbb{T} \times \Omega \longrightarrow {}^*\mathbb{R}^d$ *is a* Δt*-martingale, then its decent path projection,* $\tilde{M}$, *is a hypermartingale.*

PROOF:

By (6.3.7)(b) we see that $\tilde{M}$ is indistinguishable from a uniformly integrable progressive process. By (1.5.10), $E[M(1) \mid \omega^t]$ lifts $E[\tilde{M}(1) \mid \mathcal{M}(t)]$, and $\underset{t \downarrow r}{\text{S-lim}}\, E[M(1) \mid \omega^t] = \underset{t \downarrow r}{\text{S-lim}}\, M(t) = \tilde{M}(r)$. We could finish by using (6.1.5)(e) and the reverse martingale convergence theorem from Doob [1953, p. 328], but here is a direct proof.

If $\Lambda' \in \mathcal{F}(r)$, we wish to show that $E[\tilde{M}(1)I_{\Lambda'}] = E[\tilde{M}(r)I_{\Lambda'}]$. Let Λ be an internal event determined at t_1, with $t_1 \approx r$ and $P[\Lambda \nabla \Lambda'] = 0$. Let $t_2 \approx r$ be such that $\tilde{M}(r) = \text{st}\, M(t_2)$ a.s. Take $t = \max(t_1, t_2)$.

$$\begin{aligned} E[M(t)I_\Lambda] &= \sum_{[\omega^t] \subseteq \Lambda} M(t,\omega)\delta P[\omega^t] \\ &= \sum_{[\omega^t] \subseteq \Lambda} \left\{ \sum_{\upsilon \in [\omega^t]} M(1,\upsilon)\frac{\delta P(\upsilon)}{P[\omega^t]} \right\} P[\omega^t] \\ &= E[M(1)I_\Lambda]. \end{aligned}$$

where the first sum is over a selection of the equivalence classes contained in Λ, which is well-defined since $M(t)$ is nonanticipating and $\Lambda \in \mathcal{M}(t)$. $E[\tilde{M}(r)I_{\Lambda'}] \approx E[M(t)I_{\Lambda}] \approx E[M(1)I_{\Lambda}] \approx E[\tilde{M}(1)I_{\Lambda'}]$.

(6.5.12) **DEFINITION:**

If $M : [0,1] \times \Omega \longrightarrow \mathbb{R}^d$ *is a hypermartingale, then we say* $N : \mathbb{T} \times \Omega \longrightarrow {}^*\mathbb{R}^d$ *is a martingale lifting of* M *provided there is a positive infinitesimal* $\Delta t \in \mathbb{T}$ *so that* N *is a* Δt*-martingale whose decent path projection,* $\tilde{N}$, *is indistinguishable from* M.

(6.5.13) **THE HYPERMARTINGALE LIFTING THEOREM:**

Every hypermartingale M *has a martingale lifting* N. *If* $M^p(1)$ *is integrable, for* $p \geq 1$, *we may chose* N *so that* $N^p(t)$ *is S-integrable.*

PROOF:

Let $N(1)$ be an S-integrable lifting of $M(1)$ (resp. an SL^p-lifting) and define $N(t) = E[N(1) \mid \omega^t]$. We know that N a.s. has a Δt-decent path sample for some infintesimal Δt in $\mathbb{T}$ and that $E[N(1) \mid \omega^t]$ lifts $E[M(1) \mid \mathcal{M}(t)]$, when $t = k\Delta t$ (1.5.10). (Integrability of $N^p(t)$ follows from (6.5.7) when $N^p(1)$ is S-integrable.)

One way to prove that N is a Δt-martingale lifting of M is to use the reverse martingale convergence theorem of Doob [1953, p. 328]. Since $\mathcal{F}(r) = \cap[\mathcal{M}(t) : t \gg r]$, $N(t) \longrightarrow \tilde{N}(r)$, and $M(r) = E[M(1) \mid \mathcal{F}(r)]$ a.s. we get $\tilde{N}(r) = N(r)$ a.s. for all rational r. This implies indistinguishability since both have decent paths.

Another way to prove that N is a Δt-martingale lifting of M is to show that the left limits, $\tilde{N}(1^-) = M(1^-)$ a.s., because then $E[\tilde{N}(1^-) \mid \mathcal{F}(r)] = E[M(1^-) \mid \mathcal{F}(r)]$ a.s. (Note that the Δt-sample could fall short of $t = 1$ by Δt.) Since $\underset{t\uparrow r}{\text{S-lim}}\, N(t) = \tilde{N}(1^-)$ a.s., every sequence $t_m \uparrow 1$ gives $E[M(1) \mid \mathcal{M}(t_m)] \longrightarrow \tilde{N}(1^-)$ a.s. Also, both $M(1^-)$ and $\tilde{N}(1^-)$ are $\mathcal{D}(1)$-measurable. Exercise (6.1.11) shows that $\overline{\Sigma\mathcal{M}(t_m)} \supseteq \mathcal{D}(1)$, so ordinary martingale convergence, Doob [1953, p. 319] or see the proof of Helms & Loeb [1980], shows $M(1^-) = \tilde{N}(1^-)$ a.s. In fact, we only need the uniformly integrable martingale convergence theorem. Helms & Loeb's [1980] proof of the special case (as well as their proof of the full theorem) is based on a result like (6.5.14) below.

A third way to complete the argument is as follows. We know $\tilde{N}(r) \longrightarrow \tilde{N}(1^-)$ a.s. as $r\uparrow 1$, since decent paths have left limits; in fact, for $t = 1$ on the D-sample, $\text{st}\, N(t) = \tilde{N}(1^-)$ a.s. We claim that $E[M(1) \mid \mathcal{D}(1)] = \tilde{N}(1^-)$. Proof of this claim establishes the result, because if $r < 1$ is rational,

$$\begin{aligned} M(r) &= E[M(1) \mid \mathcal{F}(r)] \\ &= E[E[M(1) \mid \mathcal{D}(1)] \mid \mathcal{F}(r)] \\ &= E[\tilde{N}(1^-) \mid \mathcal{F}(r)] \\ &= E[E[\tilde{N}(1) \mid \mathcal{D}(1)] \mid \mathcal{F}(r)] \\ &= E[\tilde{N}(1) \mid \mathcal{F}(r)] = \tilde{N}(r), \end{aligned}$$

and the paths of M and N are decent. The claim above is

established as follows. If $\Lambda' \in \mathcal{D}(1)$ then for each $t \approx 1$ there is an internal $\Lambda \in \mathcal{M}(t)$ such that $P[\Lambda \triangledown \Lambda'] = 0$ by (6.1.12). We know $N(1)$ is an integrable lifting of $M(1)$, so

$$\begin{aligned} E[M(1)I_{\Lambda'}] &\approx E[N(1)I_{\Lambda}] \\ &= E[E[N(1) \mid \omega^t]I_{\Lambda}] \\ &= E[N(t)I_{\Lambda}] \\ &\approx E[\tilde{N}(1^-)I_{\Lambda}] = E[\tilde{N}(1^-)I_{\Lambda'}]. \end{aligned}$$

The next result (mentioned above) is included because it seems potentially useful.

(6.5.14) **ALMOST EVERYWHERE CONVERGENCE LEMMA:**

Let $\tilde{X}_m$ *be a countable sequence of measurable random variables. Without loss of generality, we may assume that* $\{\tilde{X}_m : m \in {}^{\sigma}\mathbb{N}\}$ *is the projection of the finitely subscripted elements of an internal sequence* $\{X_k : k \in {}^{*}\mathbb{N}\}$. *The following are equivalent:*

(a) $\tilde{X}_m \longrightarrow 0$ *a.s.*

(b) *For some infinite* n *and all infinite* $k \leq n$,

$$M_k^n = \max_{k \leq j \leq n} |X_j| \approx 0 \qquad \text{n.s.}$$

PROOF:

There is no loss of generality in the internal sequence because each $\tilde{X}_m$ could be lifted to X_m and then the sequence could be extended using countable comprehension.

Let $\Omega_m^{\epsilon} = \{\omega : |X_m(\omega)| > \epsilon\}$. If $\tilde{X}_m \longrightarrow 0$ a.s. and $\epsilon >> 0$, then

$$P[\bigcap_h \bigcup_{m>h} \Omega_m^\epsilon] = 0,$$

where the intersection and union are over ${}^\sigma\mathbb{N}$. Thus $\text{S-}\lim_h P[\bigcup_{m>h} \Omega_m^\epsilon] = 0$ or, for each p in ${}^\sigma\mathbb{N}$, there exists $m_p(\epsilon) \geq p$ such that $P[\bigcup_{m>m_p} \Omega_m^\epsilon] < \frac{1}{h}$, so for each finite n, $P[\bigcup_{m_p}^{n} \Omega_m^\epsilon] < \frac{1}{h}$. Since the last statement is internal it must also hold for some infinite $n_p(\epsilon)$. Let $n \in \bigcap\{[m_p(\epsilon), n_p(\epsilon)] : p \in {}^\sigma\mathbb{N},\ \epsilon \in {}^\sigma\mathbb{Q}^+\} \cap {}^*\mathbb{N}$, which is nonempty by saturation (0.4). For any $\epsilon >> 0$ and infinite k, $P[\bigcup_{j=k}^{n} \Omega_j^\epsilon] \approx 0$. Therefore the internal set $\{\epsilon \in {}^*\mathbb{R}^+ : P[\max_{k\leq j\leq n} |X_j| > \epsilon] < \epsilon\}$ contains all finite ϵ and (b) holds for a particular infinitesimal ϵ from the set.

If (b) holds, then whenever $\epsilon >> 0$ and $\vartheta >> 0$ the internal set $\{k \in {}^*\mathbb{N} : P[\max_{k\leq j\leq n} |X_j| > \epsilon] < \vartheta\}$ contains a finite $k(\vartheta)$. Therefore $P[\bigcup_{m>k(\vartheta)}^{n} \Omega_m^\epsilon] < \vartheta$. so $P[\bigcap_k \bigcup_{m>k} \Omega_m^\epsilon] = 0$ and finally,

$$P[\bigcup_{\epsilon\in\mathbb{Q}^+} \bigcap_k \bigcup_{m>k} \Omega_m^\epsilon] = 0.$$

The last null set is precisely the set where $X_j \not\to 0$, so (b) implies (a).

(6.5.15) EXERCISE:

Let $\Lambda \subseteq \Omega$ be an internal set and define $M(t,\omega) = P[\Lambda \mid \omega^t]$. Show that M is a *martingale.

(6.5.16) **EXERCISE:**

Prove the Derived Set Lemma (6.3.6) with the following hints of Perkins. If $P[\Lambda] = 1$, for each finite m, we may pick an internal $\Omega_m \supseteq \Lambda^c$ such that $P[\Omega_m] < (\frac{1}{m})^3$. Use (6.5.15) and (6.5.2) to show that

$$P\left[\max_t P[\Omega_m \mid \omega^t] > \frac{1}{m}\right] \leq (\frac{1}{m})^2.$$

The Borel-Cantelli Lemma (cf. Breiman [1968]) says that if $\Sigma P[\Lambda_m] < \infty$, then $P[\bigcap_m \bigcup_{k\geq m} \Lambda_k] = 0$. (In other words the probability of Λ_m "infinitely often" is zero.)

We let $\Lambda_m = \{\omega : \max_t P[\Omega_m \mid \omega^t] > \frac{1}{m}\}$ and apply the Borel-Cantelli lemma to obtain a Loeb null set $\Omega_0 \supseteq (\bigcap_m \bigcup_{k\geq m} \Lambda_k)$. Also, let $\Omega_\infty = \bigcap \Omega_m$, so $\Omega_\infty \supseteq \Lambda^c$ and being a Loeb set $P[\Omega_\infty \mid \omega^t]$ is always defined. Show that $P[\Omega_\infty \mid \omega^t] = 0$ provided $\omega \notin \Omega_0$ (because $\max_t P[\Omega_m \mid \omega^t] \leq \frac{1}{m}$ for sufficiently large m). Use this to conclude that $\Lambda' \supseteq \Lambda \setminus \Omega_0$.

(6.5.17) **PROPOSITION:**

Let M be a Δt-martingale and let τ be a Δt-stopping time. The process $N(t,\omega) = M(t \wedge \tau(\omega),\omega)$ is also a Δt-martingale.

PROOF:

Exercise with these hints: Compute $E[N(t+\delta t) \mid \omega^t]$ in two cases, $\tau(\omega) \leq t$ and $\tau(\omega) > t$. If M has an S-decent path for the sample ω, what about N along ω? Is $N(1)$ S-integrable?

Doob's inequality is a well-known result which we could obtain for *martingales by transfer. However, a variation on the standard proof yields extra S-integrability information which we develop now.

(6.5.18) DEFINITION:

Let $X : \mathbb{T} \times \Omega \longrightarrow {}^*\mathbb{R}$ *be internal and let* $0 < \Delta t \in \mathbb{T}$. *Define the* Δt*-maximal function of* X *by*

$$X^{\Delta}(t,\omega) = \max[|X(s,\omega)| : 0 < s \leq t,\ s \in \mathbb{T}_{\Delta}], \quad \textit{for} \quad t \in \mathbb{T}_{\Delta}$$

where

$$\mathbb{T}_{\Delta} = \{t \in \mathbb{T} \mid (\exists k \in {}^*\mathbb{N})[t = k\Delta t]\} \cup \{1\}.$$

Here is a useful basic fact about maximal functions.

(6.5.19) LEMMA:

For every positive **submartingale* X,

$$aP[X^{\Delta}(t) > a] \leq E[X(t)I_{\{X^{\Delta}(t)>a\}}].$$

PROOF:

Whenever τ is a δt-stopping time, the reader should verify that

$$X(\tau) \leq E[X(t) \mid \tau] \quad \text{when} \quad \tau(\omega) \leq t$$

Let $\tau(\omega) = \min[t \in \mathbb{T}_{\Delta} : X(t) > a] \wedge 1$, so

$$\begin{aligned} aP[\tau<t] &\le \Sigma[X(\tau(\omega),\omega)\delta P(\omega) \;:\; \tau(\omega) < t] \\ &\le \Sigma[E[X(t,\omega) \mid \tau(\omega)]\delta P(\omega) \;:\; \tau(\omega) < t] \\ &\le \Sigma[X(t,\omega)\delta P(\omega) \;:\; \tau(\omega) < t]. \end{aligned}$$

Finally, $X^{\Delta}(t) > a$ if either $\tau < t$ or $X^{\Delta}(t) = X(t) > a$, so

$$aP[X^{\Delta}(t) > a] \le \sum_{\omega}[X(t,\omega)\delta P(\omega) \;:\; X^{\Delta}(t,\omega) > a].$$

This proves the lemma.

The standard form of Doob's inequality is obtained from the lemma as follows. Let M be a *martingale, so $|M|$ is a positive *submartingale. Write the result of the previous lemma as

$$aE[I_{\{M^{\Delta}(t)>a\}}] \le E[|M(t)|I_{\{M^{\Delta}(t)>a\}}]$$

and integrate both sides with respect to the measure $pa^{p-2}da$ on $[0,\infty)$, obtaining

$$\begin{aligned} E[(M^{\Delta}(t))^p] &\le \int_0^{\infty} pa^{p-2}\, E[|M(t)|I_{\{M^{\Delta}(t)>a\}}]da \\ &\le \frac{p}{p-1}\, E[|M(t)|(M^{\Delta}(t))^{p-1}] \end{aligned}$$

by Fubini's Theorem. Hölder's inequality gives

$$E[|M(t)|(M^{\Delta}(t))^{p-1}] \le E[|M(t)|^p]^{\frac{1}{p}}\, E[(M^{\Delta}(t))^p]^{\frac{p-1}{p}}.$$

This proves the first part of the next result.

(6.5.20) **DOOB'S INEQUALITY:**

Let M *be a* **martingale after* ∇t, $1 < p < \infty$ *and* $\nabla t \leq \Delta t \in \mathbb{T}$, $\Delta t \approx 0$. *Then for* $t \in \mathbb{T}_\Delta$,

$$(E[(M^\Delta(t))^p])^{\frac{1}{p}} \leq \frac{1-p}{p}(E[|M(t)|^p])^{\frac{1}{p}}.$$

If $|M(t)|^p$ *is S-integrable,* $M(t) \in SL^p(\Omega)$, *then* $|M^\Delta(t)|^p$ *is also S-integrable.*

PROOF:

Let $b > 0$ be infinite and compare

$$\int_b^\infty pa^{p-1}\ E[I_{\{M^\Delta(t)>a\}}]\ da \leq \int_b^\infty pa^{p-1}\ E[|M(t)|I_{\{M^\Delta(t)>a\}}]\ da.$$

Use Fubini's Theorem and Hölder's inequality as above to obtain

$$E[(M^\Delta(t))^p I_{\{M^\Delta(t)>b\}}] \leq \frac{p-1}{p}\ E[|M(t)|^p I_{\{M^\Delta(t)>b\}}]^{\frac{1}{p}} E[(M^\Delta(t))^p]^{\frac{p-1}{p}}.$$

S-integrability of $|M(t)|^p$ means that

$$E[|M(t)|^p I_{\{M^\Delta(t)>b\}}] \approx 0,$$

because of Lemma (6.5.19) applied to $|M| = X$ with the infinite constant $b = a$.

(6.6) Predictable Processes

The internal and S(tandardizable)-notions of "predictable" are as follows.

(6.6.1) DEFINITION:

For each $\iota > 0$ *and* $t \in \mathbb{T}$, *let*

$$[t-\iota] = 0 \vee \max[s \in \mathbb{T} : s \leq t-\iota].$$

An internal function $G : \mathbb{T} \times \Omega \longrightarrow {}^*\mathbb{M}$ *is called* ι*-predictable if for all* $t \in \mathbb{T}$, *whenever two samples* ω, υ *satisfy*

$$\omega^{[t-\iota]} = \upsilon^{[t-\iota]}, \quad \textit{then} \quad G(t,\omega) = G(t,\upsilon).$$

An internal set $\mathcal{U}$ *is* ι*-predictable if the indicator process* $I_{\mathcal{U}}(t,\omega)$ *is, or in other words, if for all* $t \in \mathbb{T}$, *the sections* $\mathcal{U}^t$ *are determined at* $[t-\iota]$. *Let* ι-Pred *denote the internal algebra of* ι*-predictable sets.*

An internal set is called finitely predictable if it is ϵ*-predictable for some* $\epsilon \gg 0$. *Let* F-Pred *denote the external algebra consisting of all finitely predictable internal sets.*

Finally, we say a subset $\mathcal{V} \subseteq \mathbb{T} \times \Omega$ *is predictable if it belongs to the smallest sigma algebra containing the finitely predictable sets. We denote the family of predictable sets by* Pred.

The 0-predictable sets are simply the sets which are nonanticipating for all t, not just after some infinitesimal time. When $0 < \iota \approx 0$, we have the inclusions

$$\text{F-Pred} \subset \iota\text{-Pred} \subset 0\text{-Pred}.$$

(6.6.2) **NOTATION:**

Let $\mathcal{K}\cdot\mathcal{D}$ *denote the collection of all subsets of* $[0,1] \times \Omega$ *of the form* $K \times \Lambda$ *where* K *is a compact subset of* $[0,1]$ *and* Λ *is an internal set determined before the instant* $r = \min K$, $\Lambda \in \mathcal{D}(r)$.

Every set in $\mathcal{K}\cdot\mathcal{D}$ is Borel × Loeb measurable and $\mathcal{D}$-adapted or previsible. The family $\mathcal{K}\cdot\mathcal{D}$ is a semicompact paving in the sense of (2.2.2).

If $\mathcal{F}$ is a family of sets, we denote the smallest sigma algebra containing $\mathcal{F}$ by $\Sigma(\mathcal{F})$. Here is the connection between finitely predictable sets and $\mathcal{K}\cdot\mathcal{D}$.

(6.6.3) **LEMMA:**

If $\mathcal{U}$ *is a finitely predictable internal set,* $\mathcal{U} \in$ F-Pred, *then its (half) standard part is previsible, in fact,* $s(\mathcal{U}) \in \Sigma(\mathcal{K}\cdot\mathcal{D})$.

PROOF:

The proof is a refinement of the proof of (5.4.1). Suppose that $\mathcal{U}$ is ϵ-predictable. Let $\{(a_m, b_m)\}$ be an enumeration of all standard open rational endpoint intervals of length less than ϵ. Define a sequence of internal sets by

$$\mathcal{U}_m = \{\omega \in \Omega : (t,\omega) \in \mathcal{U} \text{ for some } t \in {}^*(a_m,b_m)\}.$$

Since $b_m - a_m < \epsilon$ and $\mathcal{U}$ is ϵ-predictable, $\mathcal{U}_m$ is determined before a_m. Each of the sets

$$\mathcal{U}^m = ([0,a_m] \times \Omega) \cup ([a_m,1] \times \mathcal{U}_m) \cup ([b_m,1] \times \mathcal{U}_m^c)$$

is a finite union of $\mathcal{H}\cdot\mathcal{D}$-sets by the remarks above. We claim

$$s(\mathcal{U}) = \cap\, \mathcal{U}^m.$$

Suppose that $(r,\omega) \in s(\mathcal{U})$ and m is fixed. If $\omega \in \mathcal{U}_m$, then either $(r,\omega) \in [0,a_m] \times \Omega$ or $(r,\omega) \in [a_m,1] \times \mathcal{U}$. If $\omega \notin \mathcal{U}_m$, then $r \notin (a_m,b_m)$ because some $t \approx r$ satisfies $st(t,\omega) = (r,\omega)$ and $(t,\omega) \in \mathcal{U}$ which would make ω an element of $\mathcal{U}_m$. Hence, $(r,\omega) \in [0,a_m] \times \Omega$ or $(r,\omega) \in [b_m,1] \times \mathcal{U}_m^c$.

To show the opposite inclusion of our claim, suppose that $(r,\omega) \in \mathcal{U}^m$. Choose a subsequence of intervals (a_m,b_m) decreasing to r. Since $(r,\omega) \in \mathcal{U}^m$ and $r \in (a_m,b_m)$ for the subsequence, we must have $(r,\omega) \in (a_m,b_m) \times \mathcal{U}_m$, so there is a $t_m \in {}^*(a_m,b_m)$ such that $(t_m,\omega) \in \mathcal{U}$. Now choose a sub-subsequence so that $|t_m - r|$ is decreasing. Extend this $\{t_m\}$ to an internal sequence and choose an infinite n so that for all $m \le n$, $(t_m,\omega) \in \mathcal{U}$ and $|t_m - r| < |t_{m-1} - r|$. Then $(t_n,\omega) \in \mathcal{U}$ and $st(t_n) = r$. This proves the claim and hence the lemma.

(6.6.4) **PROPOSITION:**

$\mathscr{W} \in \Sigma(\mathscr{K}\bullet\mathscr{D})$ *if and only if* $s^{-1}(\mathscr{W}) \in$ Pred.

PROOF:

A proof very similar to (5.4.8) works here. Consider the algebra of $\Sigma(\mathscr{K}\bullet\mathscr{D})$ sets consisting of finite disjoint unions of sets of the form $I \times \Lambda$ where I is an interval (either open or closed) and Λ is internal and determined before the left instant of I. Since

$$s^{-1}(I\times\Lambda) = \begin{cases} \bigcup_m \mathbb{T}(a+\frac{1}{m},\ b-\frac{1}{m}) \times \Lambda, & I = (a,b) \\ \bigcap_m \mathbb{T}(a-\frac{1}{m},\ b+\frac{1}{m}) \times \Lambda, & I = [a,b] \end{cases}$$

and Λ is determined finitely before a in $\mathbb{T}$, these sets are in Pred, the sigma algebra generated by finitely predictable sets. The Monotone Class Lemma completes the proof of the forward implication, because the monotone class of this algebra has $s^{-1}(\mathscr{W}) \in$ Pred.

The proof of the converse is nearly the same as (5.4.8) except now we may derive $s^{-1}(\mathscr{W})$ and $s^{-1}(\mathscr{W}^c)$ from the Souslin operation (2.2.1) on finitely predictable sets $\mathscr{I}_{\sigma|n}$ and $\mathscr{I}_{\sigma|n}$, so

$$\mathscr{W} = s(s^{-1}(\mathscr{W})) = s(\bigcup_\sigma \bigcap_m \mathscr{I}_{\sigma|m}) = \bigcup_\sigma \bigcap_m s(\bigcap_{k=1}^{m} \mathscr{I}_{\sigma|k})$$

and

$$\mathscr{W}^c = s(s^{-1}(\mathscr{W}^c)) = s(\bigcup_\sigma \bigcap_m \mathscr{I}_{\sigma|m}) = \bigcup_\sigma \bigcap_m s(\bigcap_{k=1}^{m} \mathscr{I}_{\sigma|k}).$$

Thus, by the Separation Theorem (2.2.3) and Lemma (6.6.3), $\mathcal{W} \in \Sigma(\mathcal{H}\cdot\mathcal{D})$.

Recall the Souslin operation from (2.2.1). We say that a set is *$\mathcal{H}\cdot\mathcal{D}$-Souslin* if it can be derived from $\mathcal{H}\cdot\mathcal{D}$ sets by the Souslin operation.

(6.6.5) **LEMMA:**

If $\mathcal{U} \subseteq \mathbb{T} \times \Omega$ is internal, then the $\mathcal{D}$-adaptation of $s(\mathcal{U})$ is $\mathcal{H}\cdot\mathcal{D}$-Souslin; in fact,

$$\mathcal{D}(\mathcal{U}) = \{(r,\omega) : \exists(t,\upsilon) \in \mathcal{U} \text{ with } t \approx r \text{ and } \upsilon \underset{r}{\sim} \omega\} \in \Sigma(\mathcal{H}\cdot\mathcal{D}).$$

PROOF:

The set $\mathcal{D}(\mathcal{U})$ is simply the smallest $\mathcal{D}$-adapted set containing $s(\mathcal{U})$.

Define a decreasing sequence of internal sets by

$$\mathcal{U}_m = \{(t,\omega) : \exists\upsilon \in \Omega \text{ with } (t,\upsilon) \in \mathcal{U} \text{ and } \upsilon^{[t-1/m]} = \omega^{[t-1/m]}\}.$$

Each $\mathcal{U}_m$ is finitely predictable, so by Lemma (6.6.3) each $s(\mathcal{U}_m) \in \Sigma(\mathcal{H}\cdot\mathcal{D})$.

A simple extension of the proof of Lemma (2.2.4) shows that whenever $\mathcal{U}_m$ is a decreasing sequence of internal sets $s(\cap\mathcal{U}_m) = \cap s(\mathcal{U}_m)$, thus it is sufficient to prove that $s(\cap\mathcal{U}_m) = \mathcal{D}(\mathcal{U})$ for the $\mathcal{U}_m$'s defined above.

Each $\mathcal{U}_m \supseteq \mathcal{U}$ and $\Sigma(\mathcal{H}\cdot\mathcal{D})$-sets are $\mathcal{D}$-adapted, so $\cap s(\mathcal{U}_m) \supseteq \mathcal{D}(\mathcal{U})$. Conversely, suppose $(r,\omega) \in s(\cap\mathcal{U}_m)$. Then there is a $t \approx r$ such that for every m, $(t,\omega) \in \mathcal{U}_m$ or there is a υ_m so that $(t,\upsilon_m) \in \mathcal{U}$ and $\upsilon_m^{[t-1/m]} = \omega^{[t-1/m]}$. Extend the sequence $\{\upsilon_m\}$ to an internal sequence and select an infinite n such that $(t,\upsilon_n) \in \mathcal{D}(\mathcal{U})$. This proves the lemma.

Recall from section (2.2) that we call a set a *Henson set* if it can be derived from internal sets with the Souslin operation.

(6.6.6) **LEMMA:**

If $\mathcal{V}$ *is a Henson set and* $s(\mathcal{V})$ *is* $\mathcal{D}$*-adapted, then* $s(\mathcal{V})$ *is* $\mathcal{K}\cdot\mathcal{D}$*-Souslin.*

PROOF:

Let $\mathcal{V}$ be generated by Souslin's operation on internal sets,

$$\mathcal{V} = \bigcup_{\sigma} \bigcap_{n} \mathcal{I}_{\sigma|n}$$

where we may assume each intersection is decreasing (if necessary, replace $\mathcal{I}_{\sigma|n}$ by $\bigcap_{m=1}^{n} \mathcal{I}_{\sigma|m}$). Our lemma reduces to the claim

$$s(\mathcal{V}) \supseteq \bigcup_{\sigma} \bigcap_{n} \mathcal{D}(\mathcal{I}_{\sigma|n})$$

where $\mathcal{D}(\mathcal{I}_{\sigma|n})$ denotes the $\mathcal{D}$-adaptation of $s(\mathcal{I}_{\sigma|n})$ as in the previous lemma which says $\mathcal{D}(\mathcal{I}_{\sigma|n})$ is $\mathcal{K}\cdot\mathcal{D}$-analytic. In other words, the Souslin operation on $\mathcal{K}\cdot\mathcal{D}$-Souslin sets gives $\mathcal{K}\cdot\mathcal{D}$-Souslin.

Suppose that $(r,\omega) \in \bigcap_{n} \mathcal{D}(\mathcal{I}_{\sigma|n})$ so that for every n there exist $(t_n,\upsilon_n) \in \mathcal{I}_{\sigma|n}$ with $t_n \approx r$ and $\upsilon_n \underset{r}{\sim} \omega$. We may extend $\{(t_n, \upsilon_n), \mathcal{I}_{\sigma|n}\}$ to an internal sequence satisfying

$|t_n - r| < \frac{1}{n}$, $\omega^{[t_n - 1/n]} = \upsilon_n^{[t-1/n]}$, $(t_n, \upsilon_n) \in \mathcal{I}_{\sigma|n}$, and whenever $m < n$, $\mathcal{I}_{\sigma|m} \supseteq \mathcal{I}_{\sigma|n}$. Since the extension is internal there is an infinite n satisfying these conditions. Such a $(t_n, \upsilon_n) \in \mathcal{I}_{\sigma|m}$ for all finite m and $t_n \approx r$, so $(r, \upsilon_n) \in s(\mathcal{V})$. Finally, $(r, \omega) \in s(\mathcal{V})$ because $\upsilon_n \underset{r}{\sim} \omega$ and we have assumed that $s(\mathcal{V})$ is $\mathfrak{D}$-adapted.

(6.6.7) **THEOREM:**

A *set* $\mathcal{W} \subseteq [0,1] \times \Omega$ *is previsible, that is,* [Borel × Loeb]*-measurable and* $\mathfrak{D}$*-adapted, if and only if* $\mathcal{W} \in \Sigma(\mathcal{K} \cdot \mathfrak{D})$ *if and only if* $s^{-1}(\mathcal{W}) \in$ Pred, *its inverse standard part is predictable.*

PROOF:

Assume that $\mathcal{W}$ is previsible. By (5.4.8), both $s^{-1}(\mathcal{W})$ and $s^{-1}(\mathcal{W}^c)$ are Loeb($\mathbb{T}\times\Omega$)-measurable and hence analytic over the internal sets. By Lemma (6.6.6), $s(s^{-1}(\mathcal{W})) = \mathcal{W}$ and $s(s^{-1}(\mathcal{W}^c)) = \mathcal{W}^c$ are both $\mathcal{K}\cdot\mathfrak{D}$-Souslin sets. Therefore the Separation Theorem (2.2.3) says $\mathcal{W} \in \Sigma(\mathcal{K}\cdot\mathfrak{D})$.

If $\mathcal{W} \in \Sigma(\mathcal{K}\cdot\mathfrak{D})$, it is previsible because every $\mathcal{K}\cdot\mathfrak{D}$ set is.

Equivalence of the last two conditions was proved in Lemma (6.6.4).

Every previsible process is Borel × Loeb-measurable, hence is $\mu = (\boldsymbol{\pi} \circ s^{-1})$-measurable and therefore $\boldsymbol{\pi}$-almost previsible.

If Z is a μ-measurable process and there is a Borel set $B \subseteq [0,1]$ such that $Z(r)$ is $\mathfrak{D}(r)$-measurable unless $r \in B$ and such that $\mu[B\times\Omega] = 0$, then the process $X(r) = Z(r)$ for

$r \notin B$, $X(r) = 0$ for $r \in B$ is μ-measurable and $\mathfrak{D}$-adapted. In this case (6.2.6) says that X is μ-almost previsible and the result above implies that X and therefore Z has an ι-predictable π-lifting. If $\pi = \upsilon$ is the path measure of a differential process from Chapter 7 and if b is a point of a.s. continuity of that process (e.g., $b = 0$), then $B = \{b\}$ satisfies the conditions above.

(6.6.8) THE PREDICTABLE LIFTING THEOREM:

Let π be any bounded hyperfinite measure on $\mathbb{T} \times \Omega$ and let ι be an infinitesimal, $\iota \geq 0$. If $X : [0,1] \times \Omega \longrightarrow \mathbb{M}$ is π-almost previsible, then there is an internal ι-predictable process Y such that

$$\pi\{(t,\omega) : \text{st } Y(t,\omega) \neq X(s,t(t),\omega)\} = 0,$$

that is, X has an ι-predictable π-lifting. If $\mathbb{M}$ is a normed space and X is bounded, Y may be chosen with the same bound.

PROOF:

Since X is π-almost previsible, there is a previsible process W such that

$$\pi\{(t,\omega) : W(st(t),\omega) \neq X(st(t),\omega)\} = 0.$$

(We may choose W bounded if X is, for example; $W \wedge b$ is previsible if W is.) By Theorem (6.6.7), $W(st(t),\omega)$ is Pred-measurable. The sigma algebras generated respectively by finitely predictable, ι-predictable, 0-predictable and internal sets satisfy the inclusions:

$$\text{Pred} \subseteq \Sigma(\iota\text{-Pred}) \subseteq \Sigma(0\text{-Pred}) \subseteq \text{Loeb}(\mathbb{T}\times\Omega), \quad 0 \approx \iota \geq 0.$$

Therefore $W(st(t),\omega)$ is $\Sigma(\iota\text{-Pred})$-measurable so by (1.1.8) and (1.3.12), $W \circ s$ has an ι-predictable lifting Y. (Also, Y may be chosen with the same bound as W if W is bounded.) This completes the proof.

Consider Borel × Loeb-sets of the forms

$$\{0\} \times \Lambda, \quad \Lambda \in \mathcal{D}(0) \cap \text{Loeb}(\Omega)$$

or

$$(r_1, r_2] \times \Lambda, \quad \Lambda \in \mathcal{F}(r_1) \cap \text{Loeb}(\Omega).$$

We shall call these sets *basic previsible sets*. Because of the open left end on the intervals, it is easy to see that they are $\mathcal{D}$-adapted. On the other hand, if Λ is internal and in $\mathcal{D}(r_1)$, then $\Lambda \in \mathcal{F}(r_1 - \frac{1}{m})$ for sufficiently large finite m, so that

$$[r_1, r_2] \times \Lambda = \bigcap_m (r_1 - \tfrac{1}{m}, r_2] \times \Lambda.$$

This implies that each $\Sigma(\mathcal{H}\cdot\mathcal{D})$-set is in the sigma algebra generated by the basic previsible sets. By Theorem (6.6.7), basic previsible sets generate all previsible sets. This is not very important, but now we relax the Loeb measurability requirement and more.

(6.6.9) **DEFINITIONS:**

We call sets of the following forms

$\{0\} \times \Lambda$, *for* $\Lambda \in \mathcal{M}(0)$ *(the P-completion of* $\mathfrak{D}(0)$*),*

or

$(r_1, r_2] \times \Lambda$, *for* Λ *in the P-completion of* $\mathcal{F}(r_1)$*,*

basic almost-previsible sets. Sets in the sigma algebra generated by the basic almost-previsible sets are called almost-previsible sets.

A bounded hyperfinite measure π *on* $\mathbb{T} \times \Omega$ *is called P-continuous if whenever* $P[\Lambda] = 0$, *then*

$$\pi[\mathbb{T} \times \Lambda] = 0.$$

The total path variation measures of Chapter 7 are all P-continuous.

Proposition (6.1.6) tells us that basic almost-previsible sets are π-almost previsible for every P-continuous measure π. For example, if $\Lambda \in \mathcal{F}(r_1)$, then there is an internal set Ω_1 determined at a time $t_1 \approx r_1$ such that $P[\Lambda \triangledown \Omega_1] = 0$. The set

$$(r_1, r_2] \times \Omega_1 \text{ is a countable union of } \mathcal{X}\cdot\mathfrak{D}\text{-sets}$$

and

$$\pi \circ s^{-1}[(r_1, r_2] \times (\Lambda \triangledown \Omega_1)] = 0.$$

We may extend this observation as follows.

(6.6.10) **PROPOSITION:**

If $\mathcal{W}$ *is an almost previsible set, that is, belongs to the sigma algebra generated by the basic almost previsible sets and if* π *is a* P-continuous *bounded hyperfinite measure on* $\mathbb{T} \times \Omega$, *then* $\mathcal{W}$ *is* π-*almost previsible.*

PROOF:

Disjoint finite unions of basic almost previsible sets form an algebra, for example,

$$\{(r_1,r_2] \times \Lambda\}^c = [0,r_1] \times \Omega \cap (r_1,r_2] \times \Lambda^c \cap (r_2,1] \times \Omega.$$

Thus we may apply the Monotone Class Lemma (3.3.4) beginning with this algebra. The remarks before the statement show that the algebra has this property. If $\mathcal{W}_m$ is a monotone sequence of almost previsible sets such that for every π there exist previsible sets $\mathcal{V}_m$ such that $\pi \circ s^{-1}[\mathcal{W}_m \nabla \mathcal{V}_m] = 0$, then $\cup\mathcal{W}_m$ (resp., $\cap\mathcal{W}_m$) is π-almost previsible because $\pi[\cup\mathcal{W}_m \nabla \mathcal{V}_m] = 0$ and if $\mathcal{W} = \cup\mathcal{W}_m$ (resp., $\cap\mathcal{W}_m$) and $\mathcal{V} = \cup\mathcal{V}_m$ (resp., $\cap\mathcal{V}_m$), then

$$\pi[\mathcal{W} \nabla \mathcal{V}] \le \pi[\cup\mathcal{W}_m \nabla \mathcal{V}_m] = 0.$$

Therefore the collection of almost previsible sets which are π-almost previsible for each P-continuous π is a monotone class containing the algebra generated by basic almost-previsible sets. This proves the result.

The almost-previsible sets are very nearly the sets which are "predictable with respect to the usual augmentation of $\mathcal{F}$" in the sense of Dellacherie & Meyer [1978]. They would also permit basic sets of the form

$$\{0\} \times \Lambda, \quad \text{for } \Lambda \text{ in the completion of } \mathcal{F}(0).$$

We say that the measure π is *continuous at zero* if $\pi \circ s^{-1}[\{0\} \times \Omega] = 0$. The total path variation measures from Chapter 7 are continuous at zero.

(6.6.11) **PROPOSITION:**

If $K : [0,1] \times \Omega \longrightarrow \mathbb{M}$ *is previsible with respect to the usual augmentation of* $\mathcal{F}$ *and* π *is* P-*continuous and continuous at zero, then there exists a previsible* $G : [0,1] \times \Omega \longrightarrow \mathbb{M}$ *such that*

$$\pi \circ s^{-1}\{(r,\omega) \mid K(r,\omega) \neq G(r,\omega)\} = 0.$$

PROOF:

Let $a \in \mathbb{M}$ and define $H(r,\omega) = a$, if $r = 0$, and $H(r,\omega) = K(r,\omega)$ if $r > 0$. Then H is almost previsible and

$$\pi \circ s^{-1}[K(r,\omega) \neq H(r,\omega)] = 0.$$

We may apply (6.6.10) and the usual simple function approximations to obtain a previsible G which is π-almost equal to H.

(6.6.12) **DEFINITION:**

A basic almost-previsible process is a bounded process of the form

$$H(r,\omega) = h_0(\omega)I_{\{0\}}(r) + \sum_{j=1}^{m} h_j(\omega)I_{(r_j,r_{j+1}]}(r),$$

where $0 = r_1 < r_2 < \cdots < r_m < r_{m+1} = 1$, h_0 *is measurable with respect to the completion of* $\mathfrak{D}(0)$, *and* h_j *is measurable with respect to the completion of* $\mathcal{F}(r_j)$ *for* $j = 1,\cdots,m$ *(and all* h_j*'s are bounded).*

An almost previsible process is a process that is measurable with resepct to the almost previsible sigma algebra generated by the basic almost-previsible sets.

Notice that we have extended our use of the term "basic." The indicator function of a basic almost-previsible set is a basic almost-previsible process, but so is any linear combination of disjoint indicator functions.

(6.6.13) **LEMMA:**

Let π *be any* P-continuous *bounded hyperfinite measure on* $\mathbb{T} \times \Omega$ *and let* ι *satisfy* $0 \approx \iota \geq 0$. A *(bounded) basic almost-previsible process* H *has an* ι*-predictible* π*-lifting of the form*

$$G(t,\omega) = g_0(\omega)I_{[0,t_1]}(t) + \sum_{j=1}^{m} g_j(\omega)I_{(t_j,t_{j+1}]}(t),$$

where $t_j \approx r_j$ *for* $j = 1,\cdots,m,$ $t_{m+1} = 1,$ *and the internal functions* g_j *have the same bound as* H.

PROOF:

By Lemma (6.1.7) there exist bounded internal $\mathcal{M}(s_j)$-measurable functions, $g_j(\omega)$, determined at times $0 = s_0$, $s_j \approx r_j$, for $j = 1,\cdots,m$, satisfying

$$P[g_j(\omega) \not\approx h_j(\omega)] = 0.$$

We use the fact that each of these sets $st^{-1}(r_j) \times \Omega$ is π-measurable to show how to select t_j.

Let $p = \pi[st^{-1}(a) \times \Omega]$ for any $a \in [0,1]$. There exist times $s \leq a \leq t$, $s \approx t$ such that $\pi[[s,t] \times \Omega] \approx p$. To see this, for each k in $\mathbb{N}$ use inner-measurability to find an internal $\mathcal{U}_k \subseteq st^{-1}(a) \times \Omega$ with $\pi[\mathcal{U}_k] > p - \frac{1}{k}$. Let $s_k = \min[t \in \mathbb{T} : (t,\omega) \in \mathcal{U}_k$ for some $\omega]$ and $t_k = \max[t \in \mathbb{T} : (t,\omega) \in \mathcal{U}_k$ for some $\omega]$. (If $\mathcal{U}_k$ is empty, any $t_k \approx s_k \approx a$ will do.) The set $[s_k,t_k] \times \Omega$ is a larger inner approximation to $st^{-1}(a) \times \Omega$, because $s_k \approx t_k \approx a$ since $\mathcal{U}_k \subseteq st^{-1}(a) \times \Omega$. This means that

$$\pi[[s_k,t_k] \times \Omega] > p - \frac{1}{k}.$$

Extend the sequence $\{s_k,t_k\}$ to an internal sequence and use

Robinson's Sequential Lemma and the Internal Definition Principle to select an infinite n so that $s_n \approx t_n \approx a$ and $\pi[[s_k,t_k] \times \Omega] > p - \frac{1}{k}$ for all $k \leq n$. Let $s = \min[s_k : k \leq n]$ and $t = \max[t_k : k \leq n]$. Then

$$\pi[[s,t] \times \Omega] \approx p.$$

Now, having chosen g_j's determined at s_j with $s_0 = 0$ and $s_j \approx r_j$, let $t_0 = 0$. Choose $t_1 > s_1+\iota$, but $t_1 \approx s_1$, so g_1 is determined at $[t_1-\iota]$. Also increase t_1 infinitesimally if necessary, so that $\pi[(st^{-1}(0) \cap [t_1,1]) \times \Omega] = 0$. The latter is possible by the remarks of the previous paragraph. Choose each t_j for $j = 1,\cdots,m$ in this manner, that is, $t_j > s_j+\iota$, $t_j \approx s_j$ and $\pi[(st^{-1}(r_j) \cap [t_j,1]) \times \Omega] = 0$. Let $t_{m+1} = 1$. Then for each $j = 0,1,\cdots,m$,

$$\pi[\{(t_j,t_{j+1}] \bigtriangledown st^{-1}(r_j,r_{j+1}]\} \times \Omega] = 0$$

(replacing $(t_j,t_{j+1}]$ by $[0,t_1]$ and replacing $(r_j,r_{j+1}]$ by $\{0\}$ if $j = 0$). This means that for these t_j's and g_j with G as above,

$$\pi[st\ G(t,w) \neq H(st(t),w)] = 0$$

by the P-continuity of π.

The next result is also helpful in stochastic integration.

(6.6.14) **LEMMA:**

Suppose that $\mathbb{V}$ *is a vector space of bounded almost-previsible processes with values in a separable normed space. If* $\mathbb{V}$ *contains the basic almost-previsible processes and is closed under bounded pointwise convergence, then* $\mathbb{V}$ *contains all bounded almost-previsible processes with values in that space.*

PROOF:

We will show that for any almost previsible set $\mathcal{W}$ and any vector b in the range space, the function $bI_{\mathcal{W}}(r,w) \in \mathbb{V}$. This proves the lemma because all bounded measurable functions are bounded pointwise limits of sums of these "simple" functions.

Let b be an arbitrary but fixed range vector and consider the collection of sets

$$\Sigma(b) = \{H^{-1}(b) : H \in \mathbb{V} \ \& \ H \text{ takes at most the values } 0 \ \& \ b\}.$$

Every basic almost-previsible set is in $\Sigma(b)$ because the function $bI_{\Lambda}(w)I_{(q,s]}(r)$ belongs to $\mathbb{V}$. Finite disjoint unions of basic almost-previsible sets belong to $\Sigma(b)$ because $\mathbb{V}$ is a vector space. Finally, $\Sigma(b)$ is a monotone class because if $H_m^{-1}(b)$ is either increasing or decreasing, $\lim H_m$ takes only the values 0 and b and belongs to $\mathbb{V}$. The Monotone Class Lemma (3.3.4) shows that $\Sigma(b)$ is the whole sigma algebra of almost-previsible sets for each vector b. This proves the lemma as remarked above, since all bounded functions are limits of "simple" functions (partition the range).

(6.6.15) REMARKS ON EXTENSION TO $[0,\infty)$:

Only minor technical changes are required to extend the results of this section to $[0,\infty) \times \Omega$ in the framework of (5.5.4). One change which we mention explicitly is this. In order for finite disjoint unions of basic almost-previsible sets to form an algebra we must also include sets of the form

$$(r,\infty) \times \Lambda, \quad \text{for } \Lambda \text{ in the P-completion of } \mathcal{F}(r).$$

Also, a *basic almost-previsible process on* $[0,\infty)$ is one of the form

$$H(r,\omega) = h_0(\omega)I_{\{0\}}(r) + \sum_{j=1}^{m-1} h_j(\omega)I_{(r_j,r_{j+1}]}(r) + h_m(\omega)I_{(r_m,\infty)}(r)$$

where $0 = r_1 < r_2 < \cdots < r_m$, h_0 is measurable with respect to the completion of $\mathcal{D}(0)$ and h_j is measurable with respect to the completion of $\mathcal{F}(r_j)$ for $j = 1,\cdots,m$.

One would also expect the measures π to satisfy

$$\pi\{O \times \Omega\} \approx \pi\{\mathbb{T} \times \Omega\},$$

as the path variation measures of Chapter 7 will; however, this is not required for the results we have stated. (They become meaningless, not false.)

Extension of some of the progressive notions to $[0,\infty)$ is more technical and is outlined in detail in the next section.

(6.7) Beyond $[0,1]$ with Localization

In section (5.5) we indicated how the infinitesimal analysis of paths of processes on $[0,1]$ extends to paths on $[0,\infty)$. We will refer to the same internal time scheme (5.5.4) in this section. The definitions and results of this chapter carry over to this setting with little formal change other than replacing the condition $r \in [0,1]$ with $r \in [0,\infty)$ and sometimes adding 'when t is finite' or ignoring the special role $r = 1$ plays in $[0,1]$. Some of the results can be proved the same way or by a change of scale, while most require one more countable sequence in a saturation argument. This would have cluttered our proofs.

This section is only intended to highlight the new ingredients needed for this extension. It does not give many details.

Our primary interest here is the extension of our treatment of martingales. The uniform integrability assumption we made on $[0,1]$ is too strong for $[0,\infty)$. We could localize our martingale with a deterministic sequence of times, but use of random times is important and requires the additional technical details that we wish to outline. Randomly localized martingales are the main topic of this section. The proof of the lifting theorem (6.7.5) is quite tehcnical, but we omit it anyway. The extra stopping time in the coarser sample theorem (6.7.6) is a special feature of hyperfinite local martingales.

(6.7.1) NONANTICIPATING DECENT PATH LIFTING THEOREM FOR $[0,\infty)$:

Suppose a stochastic process $X : [0,\infty) \times \Omega \longrightarrow \mathbb{R}$ *is* $\delta t \otimes \delta P$*-almost progressively measurable and a.s. has decent paths. Then:*

(a) X *has a nonanticipating decent path lifting, that is, there is an infinitesimal* Δt *in* $\mathbb{T}$ *and an internal process* $Y : \mathbb{T} \times \Omega \longrightarrow {}^*\mathbb{R}$ *that is nonanticipating after* Δt *and a.s. has a* Δt*-decent path sample whose decent path projection,* $\tilde{Y}$, *is indistinguishable from* X.

(b) *If* $X \mid [0,r] \times \Omega$ *is uniformly integrable for each* $r \in [0,\infty)$, *then the* Y *of part* (a) *may be chosen so that for each finite* t, *the section* Y^t *is S-integrable.*

The proof follows the steps in the proof of (6.3.8). First one extends the results used in that proof, (6.2.7), (5.3.23) and (6.1.7). These extensions are made by rescaling $[0,m]$ to $[0,1]$ and taking a limit.

PROOF SKETCH:

Referring to the steps of the proof of (6.3.8), first we know by the extension of (6.2.7) that we may assume X has decent paths and is progressive. Next, by (5.5.12) take W a $D[0,\infty)$-lifting of X (make W S-integrable at each finite t in case (b)) with sample δt. For each $m \geq 1$ in $\mathbb{N}$ and $0 \leq k \leq m^2$, use the extension of (6.1.7) to choose a random variable $Z_m^k \approx X(\frac{k}{m})$ a.s. where Z_m^k is determined at $t_m^k \approx \frac{k}{m}$

for $0 \leq k \leq m^2$ and $Z_m^k \approx W(t_m^k)$ a.s. (make Z_m^k S-integrable in case (b)). Define $Z_m(t,\omega)$ as before except take $Z_m(t,\omega) = Z_m^{m^2}$ for $t \geq m$. This gives us a "Z_m nonanticipating after t_m^1 constant on $\mathbb{T}[t_m^k, t_m^{k+1})$, with $\frac{1}{2m} < t_m^{k+1} - t_m^k < \frac{2}{m}$ and

$$P[\max_{0 \leq k \leq m^2} |W(t_m^k) - Z(t_m^k)| > \frac{1}{m}] < \frac{1}{m} \quad \text{(resp., } E \text{ in case (b))."}$$

We may extend the Z_m to an internal family of processes by saturation. Choose an infinite m satisfying the whole internal probabilistic formula in quotes above. The process $Y = Z_m$ is our nonanticipating decent path lifting. The inequalities on the t_m^k keep Z decent because W is. Finally we may take $\Delta t = t_m^1$ so that Z has a Δt-sample nonanticipating after Δt. This sketches the proof of (6.7.1).

FILTRATIONS:

The relations of Definition (6.1.4) and the filtrations $\mathcal{F}$ and $\mathcal{D}$ have the same formal definition on our larger time scheme (5.5.4). Similarly, the definition (6.4.2) of an $\mathcal{F}$-stopping time carries over with the only change that r is any instant of $[0,\infty)$ rather than just $[0,1]$. When $\Delta t \in \mathbb{T}$ is infinitesimal and $\mathbb{T}_\Delta = \{t \in \mathbb{T} : t = k\Delta t,\ k \in {}^*\mathbb{N}\}$, a Δt-stopping time is an internal function $\tau : \Omega \longrightarrow \mathbb{T}_\Delta$ such that whenever $\tau(\omega) = t$ and $\upsilon^t = \omega^t$, then $\tau(\upsilon) = t$, formally just as in (6.4.1). The Stopping Lemma (6.4.3) carries over to our larger time scheme as well since (6.7.1) is just what is needed to extend the proof of (6.4.3) to this setting. (This is why we chose (6.7.1) to illustrate the simple localization technique.)

The Definitions (6.2.1) of adapted, progressive, etc. extend via the same formulas to $[0,\infty)$.

(6.7.2) **DEFINITION:**

A decent path process $M : [0,\infty) \times \Omega \longrightarrow \mathbb{R}^d$ *is called a (d-dimensional) local hypermartingale provided that* M *is progressively measurable and there is a reducing sequence* $\{\rho_m\}$ *of* $\mathcal{F}$*-stopping times for* M. *A reducing sequence for* M *means* $\rho_m \leq \rho_{m+1}$, $\lim_m \rho = \infty$ *a.s. and for each* m, $M_m(r) = M(r \wedge \rho_m)$ *is a uniformly integrable* $\mathcal{F}$*-martingale,* $E[M(s \wedge \rho_m) \mid \mathcal{F}(r)] = M(r \wedge \rho_m)$ *a.s.* P *for* $r < s$.

Again, the definition of $\mathcal{F}$-martingale is just the formal extension of (6.5.10) to the case $r < s$ in $[0,\infty)$ rather than only $[0,1]$. Uniform integrability was defined in (5.3.25) and studied from the internal point of view as early as section (1.6).

(6.7.3) **DEFINITION:**

An internal process $M : \mathbb{T} \times \Omega \longrightarrow {}^*\mathbb{R}^d$ *is called a (d-dimensional)* Δt*-local martingale for some infinitesimal* $\Delta t \in \mathbb{T}$ *provided that* M *is a* **martingale, that is, provided*

$$E[M(t) \mid \omega^s] = M(s,\omega), \quad \textit{for} \quad \Delta t \leq s \leq t,$$

M *has a* Δt*-decent path sample a.s. and* M *has a* Δt-

reducing sequence $\{\tau_m\}$. *We say that a (countable, external) sequence* $\{\tau_m\}$ *of internal* Δt*-stopping times is a* Δt*-reducing sequence for* M *provided:*

(a) $\tau_{m-1} \leq \tau_m \lesssim m$ *and a.s.* $\lim_{m\to\infty} st(\tau_m) = \infty$.

(b) *The* Δt*-maximal function*

$$M^{\Delta}(t) = max[\,|M(s)| : \Delta t \leq s \leq t,\ s \in \mathbb{T}_{\Delta}],\quad t \in \mathbb{T}_{\Delta}$$

satisfies:

$$M^{\Delta}(\tau_m - \Delta t) \lesssim m \quad \textit{for all } \omega \textit{ with } \tau_m(\omega) > \Delta t,$$

(paths are bounded by m *before* τ_m*)*

and

$$M^{\Delta}(\tau_m) \text{ is S-integrable}$$

(so $M(t \wedge \tau_m)$ *is also S-integrable).*

(c) *The decent path projection* $\tilde{M}$ *satisfies*

$$\tilde{M}(st\ \tau_m) = st[M(\tau_m)] \quad a.s.$$

The lifting theorem (6.7.5) contains the additional facts about turning mere uniform integrability of the standard reducing sequence into the technically useful bounds on a Δt-reducing sequence without loss of generality.

Example (6.4.7) shows why we add the last requirement that 'the standard part of the localization equals the localization of the standard part'. This is an essential extra technicality because X of (6.4.7) is a *martingale with a decent path sample while $\tilde{X}$ is not a local martingale.

The purpose of this section is to show that Δt-local martingales and local hypermartingales are corresponding internal and measurable notions.

(6.7.4) LOCAL MARTINGALE PROJECTION THEOREM:

Let M *be a* Δt*-local martingale. The decent path projection* $\tilde{M}$ *is a local hypermartingale.*

PROOF:

The standard parts, $\rho_m = \mathrm{st}\ \tau_m$, form an increasing sequence of $\mathcal{F}$-stopping times satisfying $\rho_m \leq m$, a.s. $\lim \rho_m = \infty$ and $\tilde{M}(\rho_m) = \mathrm{st}\ M(\tau_m)$ a.s. Since each τ_m is bounded by $m+1$ we can rescale and apply (6.5.11), that is, for each m in $\mathbb{N}$, let $M_m'(t) = M(t(m+1) \wedge \tau_m)$. The decent path projection, $\lim_{t \downarrow r} M_m'(t) = \tilde{M}'(r)$ is a (uniformly integrable) hypermartingale on $[0,1]$. Hence $\tilde{M}(r \wedge \rho_m) = \lim_{t \downarrow r} M(t \wedge \tau_m)$ is a uniformly integrable martingale on $[0,m+1]$. Since $\tilde{M}(r \wedge \rho_m) = \tilde{M}(\rho_m)$, a.s., for all $r \geq m$, $\tilde{M}$ is a local hypermartingale.

(6.7.5) LOCAL MARTINGALE LIFTING THEOREM:

Let M *be a local hypermartingale. Then there is an infinitesimal* $\Delta t \in \mathbb{T}$, *and a* Δt*-local martingale* N *whose decent path projection,* $\tilde{N}$, *is indistinguishable from* M. *Such an* N *is called a local martingale lifting of* M.

PROOF:

This proof is quite technical and hence omitted, see Hoover & Perkins [1983].

In order to keep a path sample of one process comparable to another, we frequently want to take a 'coarser sample'. When we do this with local martingales we must also modify the reducing sequence. In order to maintain S-integrability we may need to stop M at a time not in the coarser time sample, for example, in case M jumps an infinite amount δt after one of the times τ_m and before $\tau_m+\Delta t$. We need to do this on a set of infinitesimal probability (a.s. is not enough).

We call the internal N from the next theorem the ∇t-sample of M and later abuse the notation slightly by writing $\nabla M(t)$ for $N(t+\nabla t)-N(t)$, $t \in \mathbb{T}_\nabla$.

(6.7.6) **COARSER SAMPLING LEMMA FOR LOCAL MARTINGALES:**

Suppose that M *is a* Δt*-local martingale and* $\nabla t \in \mathbb{T}_\Delta^+$ *is infinitesimal. Then there is a* ∇t*-stopping time* τ *such that* $N(t) = M(t \wedge \tau)$ *is a* ∇t*-local martingale, while* $\tilde{\tau} = \infty$ *a.s.*

PROOF:

Let τ_m be a Δt-reducing sequence for M. By Lemma (6.4.5) (details omitted), for each finite n, an increasing finite sequence of ∇t-stopping satisfies: the maximal function $M^\nabla(\tau_m^n-\nabla t) \lesssim m$, $\tau_m^n \lesssim m$ and $\tilde{\tau}_m^n = \tilde{\tau}_m$ a.s., $\tilde{M}(\tau_m^n \wedge \tau_n) = \text{st}\, M(\tau_m^n \wedge \tau_n) = \tilde{M}(\tilde{\tau}_m) = \text{st}\, M(\tau_m)$ a.s. Since $M(\tau_m)$ is S-integrable,

$$E[\max\{|M(\tau_m^n \wedge \tau_n) - M(\tau_m)| : m \le n\}] < \frac{1}{n}$$

and

$$P[\max\{|\tau_m^n-\tau_n| : m \le n\} > \frac{1}{n}] < \frac{1}{n}.$$

We may extend $\{\tau_n, \{\tau_m^n : m \leq n\}\}$ to an internal sequence and select an infinite n satisfying the expectation and probability inequalities above, so that the Δt-stopping times satisfy $\tau_k \leq \tau_{k+1}$ for $k \leq n$, and each sequence of ∇t-stopping times $\{\tau_m^n : m \leq n\}$, is increasing. Let $\tau = \tau_n$. Since $\tau \geq \tau_k$ for finite k and $\tilde{\tau}_k \longrightarrow \infty$, $\tilde{\tau} = \infty$ a.s. The sequence $\{\tau_m^n : m \in \mathbb{N}\}$ reduces $M(t \wedge \tau)$ by the formulas above since $M(\tau_m)$ is S-integrable.

CHAPTER 7: STOCHASTIC INTEGRATION

In this chapter we study pathwise integration with respect to a process that is a sum of a martingale and a process of bounded variation. The infinitesimal analysis of the latter is similar to section (2.3) and analogous to the classical analysis of Lebesgue-Stieltjes path-integrals. The new feature of this approach is that infinitesimal Stieltjes sums also work in the general case.

(7.1) Pathwise Stieltjes Sums

In section (2.3) we showed how to represent every Borel measure on $[0,1]$ by choosing an internal measure on $\mathbb{T}$. In section (4.1) we saw how an internal measure can arise first and how to make a standard Borel measure from it. We also saw that the classical Stieltjes measure $d\tilde{F}$ equals $dF \circ st^{-1}$, where $\tilde{F} = \underset{s \downarrow r}{S\text{-}\lim} F(s)$ and dF is the hyperfinite projection measure. We saw a hint of some problems with jumps of F in Chapter 4 (where F was increasing and finite). In section (5.3) we resolved similar problems for more general processes by taking Δt-decent path samples where Δt was a coarser infinitesimal time increment. We will use the same basic approach in this chapter. In this section the first step is to show how to sample a process simultaneously with its pathwise variation. We begin by fixing some basic notation that we shall use for the rest of the chapter.

(7.1.1) NOTATION:

The infinitesimal time axis, $\mathbb{T}$, *the sample space,* Ω, *and the uniform probability* P *on* Ω *are the same as in Chapters 5 and 6 except that now we let* δt *denote any infinitesimal element of* $\mathbb{T}$. *(For example, if* $\frac{1}{n!}$ *is the smallest positive element of* $\mathbb{T}$, *we might have* $\delta t = \left(\frac{1}{n}\right)^2$ *and* $\Delta t = \frac{1}{n}$, *a larger increment.) For any* δt, Δt *in* $\mathbb{T}$, *let*

$$\mathbb{T}_\delta = \{t \in \mathbb{T} : t = k\delta t,\ k \in {}^*\mathbb{N}\} \cup \{1\}$$

and

$$\mathbb{T}_\Delta = \{t \in \mathbb{T} : t = k\Delta t,\ k \in {}^*\mathbb{N}\} \cup \{1\}.$$

If $g : \mathbb{T} \longrightarrow {}^*\mathbb{R}^d$ *is internal* (d *finite) let*

$$\delta g(t) = g(t+\delta t) - g(t), \quad \text{for} \quad t \in \mathbb{T}_\delta, \quad t \leq 1-\delta t$$

and

$$\Delta g(t) = g(t+\Delta t) - g(t), \quad \text{for} \quad t \in \mathbb{T}_\Delta, \quad t \leq 1-\Delta t$$

denote the forward differences of g *corresponding to* δt *and* Δt. *Also let*

$$\delta\mathrm{Var}\ g(t) = \sum_{\delta t}^{t} |\delta g|, \quad \text{for} \quad t \in \mathbb{T}_\delta$$

$$= \sum[\,|\delta g(s)| : \delta t \leq s < t,\ s \in \mathbb{T}_\delta]$$

and

$$\Delta\mathrm{Var}\ g(t) = \sum[\,|\Delta g(u)| : \Delta t \leq u < t,\ u \in \mathbb{T}_\Delta]$$

denote the variations of g *in steps of* δt *or* Δt *up to time* t, *where* $|\cdot|$ *denotes the* d*-dimensional* *euclidean norm on* ${}^*\mathbb{R}^d$. *Finally, if* $f : \mathbb{T} \longrightarrow {}^*\mathbb{R}$ *[or* $f : \mathbb{T} \longrightarrow {}^*\mathrm{Lin}(\mathbb{R}^d, \mathbb{R}^m)$*] is an internal function, let*

$$\sum_s^t f\delta g = \sum_s^t f(u)\delta g(u)\ , \quad \textit{for} \quad s,t \in \mathbb{T}_\delta^+$$
$$= \sum[f(u)\delta g(u) : s \le u < t,\ u \in \mathbb{T}_\delta]$$

and

$$\sum_s^t f\Delta g = \sum_{\substack{u=s \\ \text{step } \Delta t}}^t f(u)\Delta g(u)\ , \quad \textit{for} \quad s,t \in \mathbb{T}_\Delta^+$$

[When the values of $f(u)$ *are linear maps,* $f(u)\delta g(u)$ *means the map evaluated at the vector* $\delta g(u)$.*]*

Our convention to start the sums defining the internal variations at δt or Δt is to make it compatible with our definition of δt-decent path samples especially in the case of progressive processes whose liftings are only nonanticipating after δt.

Our (artificial D-space convenience-) convention of having a right-most instant $r = 1$ causes us an extra headache when $\max[\mathbb{T}_\delta \setminus \{1\}] = \tau < 1$. (We could ignore this problem on $[0,\infty)$.) In this case we take $\delta g(\tau) = [g(1) - g(\tau)]$ and interpret $\delta t = [1-\tau]$ if necessary and also let

$$\delta \mathrm{Var}\ g(1) = |\delta g(\tau)| + \sum_{\delta t}^{\tau} |\delta g| \quad \text{and} \quad \sum_s^1 f\delta g = f(\tau)\delta g(\tau) + \sum_s^{\tau} f\delta g$$

with a similar convention for $\mathbb{T}_\Delta$.

The last convention will allow us to place a final jump at

$r = 1$ on our internal paths and account for the corresponding measure. We simply lift $X(1)$ separately.

Suppose that we begin with an internal function $g : \mathbb{T} \longrightarrow {}^*\mathbb{R}$ whose δt-variation, $\Sigma^1|\delta g(s)|$, is limited. We would like to say that $\Sigma f(s)\delta g(s)$ represents the integral of $st(f)$ against $d(st(g))$, but that is too simple. Suppose $g(k\delta t) = (-1)^k\delta t$, so $st\ g = 0$ but $\Sigma^t|\delta g| = 2t$. The standard part of the function is zero, while the variation isn't, so $\Sigma(\ )\delta g$ does not properly represent $\int(\ )dh$ when $h = st\ g$. On the other hand, if we let $\Delta t = 2\delta t$, then $\Delta g = 0$ and the variation is O.K. Coarser sampling always works; perhaps it even works too well. If $f(k\delta t) = (-1)^k - 1$, then the δt-variation $\Sigma|\delta f|$ is infinite and the standard part is not 'Borelable', but if $\Delta t = 2\delta t$, then $f(k\Delta t) = 0$ and $\Delta f = 0$ so the standard part is zero. The following results show how sampling along coarser infinitesimal time axes works for Stieltjes integration.

(7.1.2) PROPOSITION:

If $X : \mathbb{T} \times \Omega \longrightarrow {}^*\mathbb{R}^d$ has a δt-decent path sample and if its decent path projection $\tilde{X} : [0,1] \times \Omega \longrightarrow \mathbb{R}^d$ almost surely has finite classical variation, $\text{var}\ \tilde{X}(\cdot,\omega) < \infty$, a.s., then there is an infinitesimal $\Delta t \geq \delta t$ in $\mathbb{T}_\delta$ such that the $(d+1)$-dimensional process $(X, \Delta\text{Var}\ X)$ has a Δt-decent path sample and the decent path projection of $\Delta\text{Var}\ X$ is indistinguishable from $\text{var}\ \tilde{X}$.

Recall that the classical variation of a path $\tilde{X}(\cdot,\omega)$ from 0 to r is defined to be the sup of all sums

$$\sum_{j=1}^{m} |\tilde{X}(r_j,\omega)-\tilde{X}(r_{j-1},\omega)|$$

over the set of all finite partitions of $[0,r]$. Finiteness of this sup is equivalent to saying that each component, $\tilde{X}_k$, of the vector $\tilde{X}$ is the difference of two increasing functions.

We can supplement (7.1.2) with the hypothesis in the next result.

(7.1.3) PROPOSITION:

If $X : \mathbb{T} \times \Omega \longrightarrow {}^*\mathbb{R}^d$ *almost surely has limited* δt*-variation,* $\delta\mathrm{Var}\ X(1) \in \mathcal{O}$ *a.s., for some infinitesimal* δt *in* $\mathbb{T}$*, then there is an infinitesimal* $\Delta t \geq \delta t$ *in* $\mathbb{T}_\delta$ *such that* $(X,\ \Delta\mathrm{Var}\ X)$ *has a* Δt*-decent path sample and the projection of that sample of* $(X,\ \Delta\mathrm{Var}\ X)$ *is indistinguishable from* $(\tilde{X},\ \mathrm{var}\ \tilde{X})$.

PROOF:

First we shall prove that the hypothesis of (7.1.3) implies that X has a Δt-sample and $\tilde{X}$ has bounded variation. Then we shall prove (7.1.2).

Suppose $\Lambda \subseteq \Omega$ is measurable, $P[\Lambda] = 1$, and whenever $\omega \in \Lambda$, $\delta\mathrm{Var}\ X_\omega(1)$ is finite. For each $\omega \in \Omega$ define internal measures on $\mathbb{T}_\delta^+$ by

$$\upsilon_\omega(t) = \delta X(t,\omega)$$

$$\upsilon_\omega^+(t) = (\delta X_1^+(t,\omega),\cdots,\delta X_d^+(t,\omega))$$

$$\upsilon_\omega^-(t) = (\delta X_1^-(t,\omega),\cdots,\delta X_d^-(t,\omega))$$

where $a^+ = \max(a,0)$ and $a^- = -[\min(a,0)]$. We know that whenever S is an internal subset of $\mathbb{T}_\delta^+$ and $\sigma = +, -$ or blank, then

$$|\upsilon_\omega^\sigma[S]| \leq \sum|\delta X(s,\omega)| : s \in S].$$

Therefore the machinery of Chapter 2 can be used to see that whenever $\omega \in \Lambda$, the formulas $\mu_\omega^\sigma = \boldsymbol{\upsilon}_\omega^\sigma \circ st^{-1}$ ($\sigma = +,-$) define d-tuples of Borel measures on $[0,1]$. Let $\mu_\omega = \mu_\omega^+ - \mu_\omega^-$. For $r \in (0,1)$ and any countable sequence t_m strictly finitely decreasing to r,

$$\mu_\omega^\sigma[0,r] = \operatorname*{S-lim}_{m\to\infty} \upsilon_\omega^\sigma[\mathbb{T}_\delta[0,t_m]], \quad \sigma = +,-,\text{blank},$$

and for any sequence strictly finitely increasing to r,

$$\mu_\omega^\sigma[0,r) = \operatorname*{S-lim}_{m\to\infty} \upsilon_\omega^\sigma[\mathbb{T}_\delta[0,t_m]], \quad \sigma = +,-,\text{blank}.$$

This shows us that $\operatorname*{S-lim}_{t\downarrow r} X(t,\omega) = \mu_\omega[0,r]$, that the S-limit as t increases to r equals $\mu_\omega[0,r)$ and that each component of $\mu_\omega[0,r]$ is the difference of two right continuous increasing functions. Existence of increasing and decreasing S-limits on Λ implies that X has a Λt-sample whose projection, $\tilde{X}$, is indistinguishable from a decent path process, Lemma (5.3.25). This shows that the hypothesis of (7.1.3) implies that of (7.1.2), but our sampling convention at Δt means that $\tilde{X}(0) = \text{st } X(\Delta t)$ may not equal $\text{st } X(\delta t)$ if $\mu[0] \neq 0$. Notice that close jumps of ν^+ and ν^- can also cancel so that we may need to choose Δt even larger. For example, let ν^+ be unit mass at $\frac{1}{2} + \delta t$ and ν^- be unit mass at $\frac{1}{2} - \delta t$, for each ω. Then $\mu = \mu^+ - \mu^- = 0$. The proof of (7.1.2) given next completes this part of the argument.

PROOF OF (7.1.2):

Suppose $\Delta t \geq \delta t$ is infinitesimal and ω is such that $\tilde{X}(\cdot,\omega) = \text{st}_k X(\cdot,\omega)$ and $\text{var } \tilde{X}(1,\omega) < \infty$. Then for every $q, r \in [0,1]$ and $\epsilon \gg 0$, there exist $q = r_0 < r_1 < \cdots < r_m = r$ such that

$$\text{var } \tilde{X}(r,\omega) - \text{var } \tilde{X}(q,\omega) - \epsilon \leq \sum_{j=1}^{m} |\tilde{X}(r_j,\omega) - \tilde{X}(r_{j-1},\omega)|.$$

There are also times $s, t_j, t \in \mathbb{T}_\Delta$ such that $X(s,\omega) \approx \tilde{X}(q,\omega)$, $X(t_j,\omega) \approx \tilde{X}(r_j,\omega)$ and $X(t,\omega) \approx \tilde{X}(r,\omega)$, so

$$\text{var } \tilde{X}(r,\omega) - \text{var } \tilde{X}(q,\omega) - \epsilon \lesssim \sum |X(t_j,\omega) - X(t_{j-1},\omega)| \leq \sum_s^t |\Delta X|.$$

Hence for each infinitesimal $\Delta t \geq \delta t$,

$$\text{var } \tilde{X}(r) - \text{var } \tilde{X}(q) \lesssim \sum_s^t |\Delta X| \quad \text{a.s.}$$

Next we find one infinitesimal time sample satisfying the opposite inequality. Let $V(t)$ be a δt-lifting of $\operatorname{var} \tilde{X}$. We know $S\text{-}\lim(X,V) = (\tilde{X}, \operatorname{var} \tilde{X})$ a.s., so for every finite natural number m, there exists $\Delta t \approx \frac{1}{m}$ in $\mathbb{T}_\delta$ such that for $0 < j \leq m$, $X(j\Delta t \wedge 1) \approx \tilde{X}(\frac{j}{m})$ and $V(j\Delta t \wedge 1) \approx \operatorname{var} \tilde{X}(\frac{j}{m})$ a.s. For this Δt, whenever $s,t \in \mathbb{T}_\Delta^+$,

$$\begin{aligned}\sum_s^t |\Delta X| &\approx \sum_{j=h}^{k-1} |\tilde{X}(\tfrac{j+1}{m}) - \tilde{X}(\tfrac{j}{m})| \quad \text{a.s.} \\ &\leq \operatorname{var} \tilde{X}(\tfrac{k}{m}) - \operatorname{var} \tilde{X}(\tfrac{h}{m}) \\ &\lesssim V(t) - V(s) \quad \text{a.s.}\end{aligned}$$

Thus the internal set of Δt's $\geq \delta t$ in $\mathbb{T}_\delta$ such that

$$P[\max(\left|V(t)-V(s)-\sum_s^t |\Delta X|\right| : s,t \in \mathbb{T}_\Delta^+) > \Delta t] < \Delta t$$

contains an infinitesimal. Such an infinitesimal Δt makes $(X(t), \Sigma_{\Delta t}^t |\Delta X|)$ a Δt-lifting of $(\tilde{X}, \operatorname{var} \tilde{X})$.

(7.1.4) DEFINITIONS:

If $U : \mathbb{T} \times \Omega \longrightarrow {}^*\mathbb{R}^d$ *is an internal process such that* $(U, \delta\mathrm{Var}\, U)$ *has a* δt*-decent path sample with projection indistinguishable from* $(\tilde{U}, \operatorname{var} \tilde{U})$, *then we say* U *has S-bounded* δt*-variation or* δt*-bounded variation.*

If $W : [0,1] \times \Omega \longrightarrow \mathbb{R}^d$ *a.s. has bounded classical variation and* U *has S-bounded* δt*-variation with the* *projection* $\tilde{U}$ *indistinguishable from* W, *then we say that* U *is a* δt*-bounded variation lifting of* W.

When U *has* S-*bounded* δt*-variation, we define internal pathwise measures on* $\mathbb{T}_\delta$ *by the weight functions*

$$\delta\kappa_\omega(t) = |\delta U_\omega|(t) = |\delta U(t,\omega)|$$

and

$$\delta\mu_\omega(t) = \frac{|\delta U(t,\omega)|}{1 \vee \delta \mathrm{Var}\ U(1,\omega)}, \qquad (\vee = \max)$$

as well as a measure on $\mathbb{T}_\delta \times \Omega$,

$$\delta\upsilon(t,\omega) = \delta\mu_\omega(t)\delta P(\omega).$$

Extend these measures to either all of $\mathbb{T}$ or all of $\mathbb{T} \times \Omega$ by taking $\delta\mu_\omega(t) = 0$ if $t \notin \mathbb{T}_\delta$.

The measures we have just introduced play a role in showing the connection between internal summation and classical pathwise integration. The hyperfinite measures $\kappa_\omega \circ \mathrm{st}^{-1}$ are the total variation measures of the paths of $\tilde{U}$, while μ_ω is normalized so that both $\mu_\omega \lesssim 1$ and $\upsilon \lesssim 1$. This makes the function $f(\omega) = \mu_\omega[\mathbb{T}]$ S-integrable with respect to P (of course, weaker conditions would suffice for this).

If $\mathcal{U} \subseteq \mathbb{T} \times \Omega$, $\mathcal{U}_\omega$ denotes the section,

$$\mathcal{U}_\omega = \{t \in \mathbb{T} \mid (t,\omega) \in \mathcal{U}\}.$$

(7.1.5) THE ITERATED INTEGRATION LEMMA FOR PATH MEASURES:

Let $\delta\mu : \mathbb{T} \times \Omega \longrightarrow {}^*[0,1]$ *be an internal function. For each* ω *the weight function* $\delta\mu_\omega(t)$ *defines a measure on* $\mathbb{T}$. *Suppose that the function* $f(\omega) = \mu_\omega[\mathbb{T}]$ *is S-integrable with respect to* P. *Let* ν *be given by the weight function* $\delta\nu(t,\omega) = \delta\mu_\omega(t)\delta P(\omega)$. *The hyperfinite extension measures satisfy:*

(a) *If* $\mathcal{U} \in \text{Loeb}(\mathbb{T} \times \Omega)$, *then the function* $\boldsymbol{\mu}_\omega(\mathcal{U}_\omega)$ *is* Loeb(Ω)*-measurable and*

$$E[\boldsymbol{\mu}_\omega(\mathcal{U}_\omega)] = \boldsymbol{\nu}[\mathcal{U}].$$

(b) *If* $\mathcal{W}$ *is* $\boldsymbol{\nu}$*-measurable, then for a.a.* ω, $\mathcal{W}_\omega$ *is* $\boldsymbol{\mu}_\omega$*-measurable,* $\boldsymbol{\mu}_\omega(\mathcal{W}_\omega)$ *is* P*-measurable and*

$$E[\boldsymbol{\mu}_\omega(\mathcal{W}_\omega)] = \boldsymbol{\nu}[\mathcal{W}].$$

(c) *If* $X : \mathbb{T} \times \Omega \longrightarrow [-\infty,\infty]$ *is* $\boldsymbol{\nu}$*-integrable, then for almost all* ω, X_ω *is* $\boldsymbol{\mu}_\omega$*-integrable and*

$$E\left[\int X_\omega(t)\,d\boldsymbol{\mu}_\omega(t)\right] = \int X(t,\omega)\,d\boldsymbol{\nu}(t,\omega).$$

PROOF:

(a) If $\mathcal{U}$ is internal, then $\mathcal{U}_\omega$ is internal for each ω and $\mu_\omega[\mathcal{U}_\omega] \le \mu_\omega[\mathbb{T}]$. Moreover, $\nu[\mathcal{U}] = E[\mu_\omega(\mathcal{U}_\omega)]$, so $\boldsymbol{\nu}[\mathcal{U}] = E[\boldsymbol{\mu}_\omega(\mathcal{U}_\omega)]$. The rest of (a) follows easily from the Monotone Class Lemma (3.3.4), because $\lim E[\boldsymbol{\mu}_\omega(\mathcal{U}^m_\omega)] = E[\lim \boldsymbol{\mu}_\omega(\mathcal{U}^m_\omega)]$ by the Dominated Convergence Theorem and the hypothesis that $\mu_\omega[\mathbb{T}]$ is P-S-integrable.

(b) If $\mathcal{W}$ is ν-measurable, then by (1.2.13) there is an internal $\mathcal{U}$ such that $\nu[\mathcal{W} \triangledown \mathcal{U}] = 0$. Therefore $\mathcal{W} \triangledown \mathcal{U}$ is contained in a Loeb null set $\mathcal{N}$. By part (a) $\mu_\omega[\mathcal{N}_\omega] = 0$ a.s. and since μ_ω is complete, $\mathcal{W}_\omega \triangledown \mathcal{U}_\omega$ has measure zero a.s. Using (1.2.13) for these ω, we see that $\mathcal{W}_\omega$ is μ_ω-measurable a.s. and $\mu_\omega[\mathcal{W}_\omega] = \mu_\omega[\mathcal{U}_\omega]$ a.s. Since P is complete, $\mu_\omega[\mathcal{W}_\omega]$ is P-measurable if we take any value for the null set of ω's where it may fail to be defined (for example, we may take the outer measure $\bar{\mu}_\omega[\mathcal{W}_\omega]$). Finally, $\nu[\mathcal{W}] = \nu[\mathcal{U}] = E[\mu_\omega(\mathcal{U}_\omega)]$ $= E[\mu_\omega(\mathcal{W}_\omega)]$.

(c) If $X \geq 0$ is ν-integrable, let $\{X^k\}$ be a monotone sequence of simple functions with $X^k \uparrow X$. By monotone convergence we know $\int X^k d\nu \uparrow \int X d\nu$. By part (b) we know $\int X^k d\nu = E[\int X^k_\omega d\mu_\omega]$. Again by monotone convergence $\lim E[\int X^k_\omega d\mu_\omega] = E[\lim \int X^k_\omega d\mu_\omega] = E[\int X d\mu_\omega]$. Thus part (c) holds for positive integrable functions. Finally, we may write $X = X^+ - X^-$ and apply the positive part to X^+ and X^- in order to finish part (c).

(7.1.6) THE STIELTJES DIFFERENTIAL LIFTING LEMMA:

Let $W : [0,1] \times \Omega \longrightarrow \mathbb{R}^d$ *a.s. have decent paths of bounded variation. Then* W *has a* δt*-bounded variation lifting* U. *If* U *is such a lifting, then for a.a.* ω,

(a) *the Borel measures* $\pi_\omega = \delta U_\omega \circ st^{-1}$ *equal the Lebesgue-Stieltjes measures generated by* dW_ω;

(b) *the total variation measure,* $|dW_\omega|$ *equals* $\kappa_\omega \circ st^{-1}$;

(c) $\pi_\omega(0) = 0$.

PROOF:

Let U be a Δt-decent path lifting of W. Apply (7.1.2) to obtain a $\delta t \geq \Delta t$ so that U has S-bounded δt-variation. Let Λ be the null set where $st_k[U, \delta Var\ U] \neq [W, var\ W]$.

We know that if $\omega \notin \Lambda$, then

$$\pi_\omega[0,r] = \underset{t \downarrow r}{S\text{-}\lim}\ U(t)-U(\delta t) = W_\omega(r)-W_\omega(0) = dW_\omega[0,r]$$

and

$$|\pi_\omega|[0,r] = \underset{t \downarrow r}{S\text{-}\lim}\ \delta Var\ U(t) = var\ W(r)$$

so (a) and (b) hold.

Since $\lim_{r \downarrow 0} W(r) = W(0)$ a.s., $\pi_\omega(0) = 0$. This proves the lemma.

Next we deal with the measures from (7.1.4) that we are most interested in for stochastic integration.

(7.1.7) DEFINITION:

Let $H : [0,1] \times \Omega \longrightarrow \mathbb{R}$ *be a function. An internal* $G : \mathbb{T} \times \Omega \longrightarrow {}^*\mathbb{R}$ *such that for a.a.* ω, *the hyperfinite measure*

$$\kappa_\omega\{t : st\ G(t,\omega) \neq H(st(t),\omega)\} = 0$$

is called a δU*-path lifting of* H.

In order to compute martingale integrals by internal summation in (7.3), we will need to require additional properties on G and hence also on H.

(7.1.8) **THE δU-PATH LIFTING LEMMA:**

Let $U : \mathbb{T} \times \Omega \longrightarrow {}^*\mathbb{R}^d$ *have S-bounded* δt*-variation. If* $H : [0,1] \times \Omega \longrightarrow \mathbb{R}$ *is* [Borel[0,1] × Meas(P)]-*measurable, then* H *has a* δU*-path lifting* G. *Moreover, if* H *is bounded by* b, *we may choose* G *so it is bounded by* b.

PROOF:

Let K be a (Borel × Loeb)-measurable function indistinguishable from H (see (5.4.10)). By (5.4.9), $K(st(t),\omega)$ is (Loeb × Loeb)-measurable and hence ν-measurable. Let G be a ν-lifting of $K(st(t),\omega)$ (resp. a bounded ν-lifting, see (1.3.9)). By the Iterated Integration Lemma (7.1.5),

$$0 = E[\mu_\omega\{st\ G(t,\omega) \neq K(st(t),\omega)\}]$$

so for a.a. ω,

$$\mu_\omega\{st\ G(t,\omega) \neq K(st(t),\omega)\} = 0$$

Except for a null set $\Lambda \subseteq \Omega$, $\delta Var\ U(1,\omega)$ is limited so that κ_ω is a simple multiple of μ_ω on the Loeb sets of $\mathbb{T}$. Hence for a.a. ω,

$$\kappa_\omega\{t : st\ G(t,\omega) \neq K(st(t),\omega)\} = 0.$$

Finally, $K(st(t),\omega) = H(st(t),\omega)$ for all t, a.s. ω, so our lemma is proved.

(7.1.9) THEOREM:

Let $H : [0,1] \times \Omega \longrightarrow \mathbb{R}$ *be* $[\text{Borel}[0,1] \times \text{Meas}(P)]$-*measurable and bounded by* b. *Let* $W : [0,1] \times \Omega \longrightarrow \mathbb{R}^d$ *be a process with a.a. decent paths of bounded variation. Let* $U : \mathbb{T} \times \Omega \longrightarrow {}^*\mathbb{R}^d$ *be a* δt-*bounded variation lifting of* W *and let* $G : \mathbb{T} \times \Omega \longrightarrow {}^*\mathbb{R}$ *be a* δU-*path lifting of* H *also bounded by* b. *Then the internal process* $S(t,\omega) = \sum_{\delta t}^{t} G(s,\omega)\delta U(s,\omega)$ *is a* δt-*bounded variation lifting of the pathwise classical integral* $I(r,\omega) = \int_0^r H \cdot dW$.

PROOF:

First we show that S has a δt-decent path sample. Let $b \ll \infty$ denote a bound for G and H. If $t, t+\Delta t \in \mathbb{T}_\delta$,

$$|S(t+\Delta t) - S(t)| = \left|\sum_{s=t}^{t+\Delta t} G(s)\delta U(s)\right| \leq b \sum_{s=t}^{t+\Delta t} |\delta U(s)|,$$

so S only has finite jumps where $\delta \text{Var}\, U(t)$ does and S is right-S-continuous where $\delta \text{Var}\, U(t)$ is.

By (7.1.6), for a.a. ω, $\delta U_\omega \circ st^{-1} = \pi_\omega$ is the Borel measure generated by dW_ω. By change of variables,

$$\underset{t\downarrow r}{\text{S-lim}} \sum^{t} G(s,\omega)\delta U(s,\omega) = \int_{st^{-1}[0,r]} H(st(s),\omega)d(\delta U_\omega)(s)$$

$$= \int_{[0,r]} H(q,\omega)d\pi_\omega(q) = \int_{[0,r]} H(q,\omega)dW_\omega(q).$$

We apply a similar argument to $\Sigma G \cdot \delta U$. This proves the theorem.

(7.1.10) **DEFINITION:**

We call the function G *of the next theorem a* δU*-summable path lifting of the process* H.

(7.1.11) **THE STIELTJES SUMMABLE LIFTING THEOREM:**

Let U *be a* δt*-bounded variation lifting of* W. *If* H *is* $[\text{Borel}[0,1] \times \text{Meas}(P)]$*-measurable and if*

$$\int_0^1 |H_\omega| \cdot |dW_\omega| < \infty \quad a.s. \quad \omega,$$

then H *has a* δU*-lifting* G *such that* G_ω *is* κ_ω*-S-integrable for a.a.* ω. *In this case*

$$S(t,\omega) = \sum_{s=\delta t}^{t-\delta t} G(s,\omega)\delta U(s,\omega)$$

is a δt*-bounded variation lifting of*

$$I(r,\omega) = \int_0^r H(q,\omega)dW(q,\omega).$$

PROOF:

Let G' be any δU-lifting of H. We will show that all sufficiently small infinite truncations of G' are pathwise S-integrable. Of course, infinite truncations of a lifting are also liftings.

Define

$$H^m = \begin{cases} m, & H \geq m \\ H, & |H| < m \\ -m, & H < -m \end{cases} \quad \text{and} \quad G^m = \begin{cases} m, & G \geq m \\ G', & |G| < m. \\ -m, & G < -m \end{cases}$$

for finite m in the case of H and all m for G'. For a.a.

ω, $\{H^m\}$ is a Cauchy sequence in the $L^1(\text{var } W)$-norm, so that for every ϵ in $\mathbb{Q}^+$, there is a finite $m(\epsilon) \geq \frac{1}{\epsilon}$ such that if $k \geq m(\epsilon)$,

$$P[\int |H^m_\omega - H^k_\omega||dW_\omega| > \frac{\epsilon}{2}] < \frac{\epsilon}{2}.$$

Thus, the internal set

$$\{n \in {}^*\mathbb{N} : m(\epsilon) \leq k \leq n \Rightarrow P[\Sigma|G^n_\omega - G^k_\omega||\delta U_\omega| > \epsilon] < \epsilon\}$$

contains an infinite $n = n(\epsilon)$. By saturation the countable intersection $\cap {}^*\mathbb{N}[m(\epsilon), n(\epsilon)]$ contains an infinite n. We claim that $G = G^n$ is our summable lifting for H. This follows from the definition of n because $P[\Sigma|G-G^k||\delta U| > \epsilon] < \epsilon$ for all standard ϵ.

By the bounded lifting theorem (7.1.9) above, for each finite m we know $\tilde{S}_m(t,\omega) = I_m(r,\omega)$ where

$$S_m(t,\omega) = \sum_{\delta t}^{t} G^m(s,\omega)\delta U(s,\omega) \quad \text{and} \quad I_m(r,\omega) = \int_0^r H^m(q,\omega)dW(q,\omega).$$

We also know $I_m \longrightarrow \int HdW$ in probability in $D[0,1]$ and $st_k S_m \longrightarrow st_k \Sigma G\delta U$ in probability, hence $st_k \sum G\delta U = \int HdW$ a.s.

By the bounded case (7.1.8), for each finite m we know that the decent path projection of

$$\sum^t |G^m(s,\omega)||\delta U(s,\omega)| \quad \text{and} \quad \int^r |H^m(q,\omega)||dW(q,\omega)|$$

are indistinguishable. Since $|G^n-G^k| \geq ||G^n|-|G^k||$, the same convergence estimates show that the decent path projection of

$$\sum^t |G||\delta U| \quad \text{and} \quad \int^r |H||dW|$$

are indistinguishable. Hence $\sum^t G\delta U$ is a δt-bounded variation lifting of $\int^r HdW$.

(7.1.12) **EXAMPLE:**

Consider the process $J(t,\omega)$ of (5.3.8), (4.3.3), and (0.3.6) whose decent path projection $\tilde{J}$ is a classical Poisson process. We wish to calculate

$$\int_0^r \tilde{J}d\tilde{J} = \frac{\tilde{J}(r)[\tilde{J}(r)+1]}{2}$$

as an example. Since J is finite and increasing by jumps of one on a δt-sample, it has S-bounded δt-variation. Since $\delta J(t,\omega)$ is a function of $\omega_{t+\delta t}$ alone (=1 with probability $p\delta t$) and since only the times when it jumps count for δJ-liftings, we see that $J(t+\delta t,\omega)$ *is a* δJ-*lifting of* $\tilde{J}(r,\omega)$. (This depends on our right continuous path convention too.) The important fact that we are trying to illustrate is that the lifting $J(t+\delta t,\omega)$ *must anticipate* the next toss the "coin," $\omega_{t+\delta t}$. Now we compute

$$\sum_{s=\delta t}^{t-\delta t} J(s+\delta t,\omega)\delta J(s,\omega) = 1\cdot 1 + 2\cdot 1 + \cdots + J(\tau_n,\omega)\cdot 1$$
$$= \frac{J(t,\omega)[J(t,\omega)+1]}{2}$$

where τ_n is the time of the last jump of $J(\cdot,\omega)$ at or before time t.

Notice that if we want to lift the Brownian motion $\tilde{B}$ obtained from projecting Anderson's infinitesimal random walk, any infinitesimal time advance or delay can be tolerated in a δJ-lifting because the paths of B are S-continuous. Hence, $B(t,\omega)$ is a nonanticipating δJ-lifting of $\tilde{B}$. Path liftings need to be done more carefully when the differential process is a martingale with paths of infinite variation as we shall see below.

Here is a result that adds the stochastic evolution structure to our *finite representation of Stieltjes path integrals.

(7.1.13) THE NONANTICIPATING STIELTJES LIFTING THEOREM:

Let $W : [0,1] \times \Omega \longrightarrow \mathbb{R}^d$ *be a progressive process with a.a. decent paths of bounded variation,* $\operatorname{var} W(1) < \infty$ *a.s. Let* $H : [0,1] \times \Omega \longrightarrow \mathbb{R}$ *be a previsible process with*

$$\int_0^1 |H_\omega|\cdot|dW_\omega| < \infty \quad a.s.$$

There is an infinitesimal $\delta t > 0$ *such that* W *has a* δt*-bounded variation lifting* U *which is nonanticipating after* δt *and* H *has a* 0*-predictable* δU*-summable path* *lifting* G. *These liftings make*

$$\sum^{t} G\cdot\delta U \quad a \quad \delta t\text{-bounded variation lifting of} \quad \int^{r} H\cdot dW$$

which is nonanticipating after δt.

PROOF:

Apply the Nonanticipating Decent Path Lifting Theorem (6.3.8) to W obtaining an internal process U which is nonanticipating after Δt and has a Δt-decent path sample for some infinitesimal Δt. Next, apply (7.1.2) to U to find an infinitesimal $\delta t \in \mathbb{T}_{\Delta}$ so that U has S-bounded δt-variation. The coarser sample still has decent paths, so U is our lifting of W.

If H is bounded we apply (6.6.8) with the measure υ of (7.1.4) associated with U to obtain a similarly bounded O-predictable G. In the integrable case, apply the truncation argument of the proof of (7.1.11) to a sequence of processes which are called O-predictable. This proves the theorem.

(7.1.14) **SUMMARY:**

There are two main ideas in this section. The first is that coarse enough time samples of a process whose standard part has finite classical variation have a variation that can be interchanged with the standard part (in D[0,1].) The second idea is that Iterated Integration allows us to connect a.s. path approximation to internal sums. The following two exercises test your understanding of the second idea on internal summands which need not be liftings of any standard process. Showing this internal stability, separate from lifting, makes the development of more general integrals easier in the following sections. The internal sums also have "standard" applications.

(7.1.15) **EXERCISE:**

Let U *have* δt*-bounded variation. Suppose that* $G_1(t,\omega)$ *and* $G_2(t,\omega)$ *are internal, bounded by* $b \in \mathcal{O}$, *and for* a.a. ω

$$\kappa_{\omega}\{t \mid st\ G_1(t,\omega) \neq st\ G_2(t,\omega)\} = 0.$$

Then for a.a. ω

$$\max\left[\left|\sum_{s=\delta t}^{t-\delta t}(G_1(s,\omega)-G_2(s,\omega))\delta U(s,\omega)\right| : \delta t \leq t \leq 1\right] \approx 0.$$

In other words, both summands give nearly the same Stieltjes sums.

(7.1.16) **EXERCISE:**

Let U *have* δt*-bounded variation. Suppose that* G *is internal and has a limited bound. Then* $S(t,\omega) = \sum_{\delta t}^{t} G\cdot\delta U$ a.s. *only jumps where* U *does.*

Hence S(t) has a decent path sample. See the proof of (7.3.8) if you have trouble formulating the jump condition. It is easy to see that S(t) need not have δt-bounded variation.

(7.2) Quadratic Variation of Martingales

One of the main ingredients in martingale integration is the quadratic variation process associated with a martingale. It is used in estimates similar to the classical domination of a signed measure by its total variation in the previous section, but there are also surprises. Consider these curious heuristic formulas for one-dimensional Brownian motion:

$$(db)^2 = dt \quad \& \quad dbdt = 0,$$

so

$$\begin{aligned} d(f(b)) &= f'(b)db + \frac{1}{2} f''(b)(db)^2 + \cdots \\ &= f'(b)db + \frac{1}{2} f''(b)dt, \end{aligned}$$

for example,

$$d(b^2) = 2bdb + dt.$$

No doubt our reader will see some tempting analogies for Andersonn's infinitesimal random walk, for example,

$$\begin{aligned} \delta(B^2) &= (\delta B+B)^2 - B^2 = (\delta B)^2 + 2B\delta B \\ &= (\pm\sqrt{\delta t})^2 + 2B\delta B \\ &= 2B\delta B + \delta t. \end{aligned}$$

Such calculations are made precise by the (generalized) Itô transformation formula given below. In this section we take up the generalized study of the $(db)^2$ term. Brownian motion is an important test case.

The following is an extension of (7.1.1) where δM and ΔM, etc. are defined (for the internal functions $t \to M(t,\omega)$).

(7.2.1) NOTATION:

Let δt *and* Δt *be time increments. If* M *and* N *are internal processes with values in* $^*\mathbb{R}^d$ *and if* (x,y) *denotes the euclidean inner product on* $^*\mathbb{R}^d$, *we define the joint quadratic variation processes for the respective time increments by:*

$$[\delta M,\delta N](t,\omega) = (M(\delta t,\omega),N(\delta t,\omega)) + \sum[(\delta M(s,\omega),\delta N(s,\omega)) : 0 < s < t,\ s \in \mathbb{T}_\delta]$$

$$\text{for } \delta t \leq t \in \mathbb{T}_\delta,$$

and

$$[\Delta M,\Delta N](t,\omega) = (M(\Delta t,\omega),N(\Delta t,\omega)) + \sum[(\Delta M(s,\omega),\Delta N(s,\omega)) : 0 < s < t,\ s \in \mathbb{T}_\Delta]$$

$$\text{for } \Delta t \leq t \in \mathbb{T}_\Delta.$$

We also define maximal functions for the respective increments by:

$$M^\delta(t,\omega) = \max[|M(s,\omega)| : 0 < s \leq t,\ s \in \mathbb{T}_\delta],$$

$$\text{for } \delta t \leq t \in \mathbb{T}_\delta,$$

and

$$M^\Delta(t,\omega) = \max[|M(s,\omega)| : 0 < s \leq t,\ s \in \mathbb{T}_\Delta],$$

$$\text{for } \Delta t \leq t \in \mathbb{T}_\Delta.$$

All the little details in our definition of quadratic variation are important. We include $M(\delta t)$ in the quadratic variation, while no such term was needed in first variation. This term corresponds to the standard term $\tilde{M}(0)$ because of our convention of starting δt-decent path samples at δt.

If $M = B(t,\omega)$ is Anderson's infinitesimal random walk $(d = 1)$ and δt is as in (5.2.3), then

$$[\delta B,\delta B](t,\omega) \approx t = \sum[(\pm\sqrt{\delta t})^2 : 0 < s \leq t, \text{ step } \delta t].$$

It is well known that the paths of classical Brownian motions such as $\tilde{B}(r,\omega)$ are nowhere differentiable (see one of the books by Breiman, Doob or Loeve from the references). In fact, various classical formulations of the idea that finite increments of Brownian motion tend toward $\pm\sqrt{\Delta t}$ could be given. We would like to turn the question around and ask: What is $[\Delta B,\Delta B](t,\omega)$ when Δt is much larger than δt, but still infinitesimal? This is answered by Lemma (7.2.10).

We begin with some simple, but illustrative calculations.

(7.2.2) **EXERCISE** (Cauchy's inequality for quadratic variation): *For internal* d*-dimensional processes* M *and* N,

$$|[\delta M,\delta N](t,\omega)| \leq \sqrt{[\delta M,\delta M](t,\omega)}\ \sqrt{[\delta N,\delta N](t,\omega)}.$$

HINT: Apply the *transform of Cauchy's inequality to vectors with components $\delta M_i(s)$, $\delta N_i(s)$ for $1 \leq i \leq d$ and $t > s \in \mathbb{T}_\delta^+$.

The next result frequently allows us to focus our attention on single martingales, yet conclude results about pairs.

(7.2.3) **EXERCISE** (Polarization identities)

For internal d-dimensional processes M *and* N,

$$[\delta M,\delta N](t,\omega)$$
$$= \frac{1}{2}\{[\delta(M+N),\delta(M+N)](t,\omega)-[\delta M,\delta M](t,\omega)-[\delta N,\delta N](t,\omega)\}$$
$$= \frac{1}{4}\{[\delta(M+N),\delta(M+N)](t,\omega)-[\delta(M-N),\delta(M-N)](t,\omega)\}.$$

HINT: Sum the corresponding identity of ${}^*\mathbb{R}^d$.

(7.2.4) **EXERCISE:**

Let M *be a one-dimensional* δt*-martingale and let* τ *be a* δt*-stopping time. Show that* $E[M^2(\tau(\omega),\omega)] = E\{[\delta M,\delta M](\tau(\omega),\omega)\}$.

General martingales require some coarser time sampling just as in the last section. The nasty martingale N(t) of (6.5.4) makes $N^\delta \neq N^\Delta$ and $[\delta N,\delta N] \neq [\Delta N,\Delta N]$. A main result about quadratic variation says that if (M,N) is a δt-martingale, then δt-sampling also works for the quadratic variation, that is, $[\delta M,\delta N]$ has a δt-decent path sample for the same δt as M and N; moreover, $[\delta M,\delta N] \approx [\Delta M,\Delta N]$ a.s. for any coarser infinitesimal Δt in $\mathbb{T}_\delta$. The path property is proved in (7.4.9) using estimates for stochastic integrals. Stability for bigger increments is Theorem (7.2.10).

We had a hard time deciding what level of generality to

present in hyperfinite stochastic integration. We shall present the L^2 case on $[0,1]$ in section (7.3). This reduces the technicalities that would be required for a treatment of the full local theory . We do offer notes on the extension to local martingales in sections (7.4) and (7.6) (which our reader may ignore). Lindstrom [1980] treats local L^2 martingales, but our outline is directed toward Hoover & Perkins [1983] more general theory. Our reader must consult their paper for more details of the local case. This section is not very technical in the local case, so we also give the local results. The reader may ignore the statements such as "t is limited in $\mathbb{T}$" if she is only interested in $[0,1]$. Some stopping times are needed anyway, so this should cause no trouble.

Definition (6.7.3) is set up so that a "δt-local martingale" automatically has a locally-S-integrable maximal function. The lifting theorem (6.7.5) shows that there is no loss in generality with this definition, or, to put it another way, the maximal functions of standard local martingales are always locally integrable (but localizing with a different sequence in general). Our next result gives integrability of the quadratic variation in both the local and "global" cases.

(7.2.5) **THEOREM:**

Let M *be a d-dimensional* **martingale after* δt *and let* τ *be a* δt*-stopping time. Then for each finite* $p \geq 1$, $(M^{\delta}(\tau))^p$ *is S-integrable if and only if* $([\delta M,\delta M](\tau))^{p/2}$ *is S-integrable. In particular,* $M^2(1)$ *is S-integrable if and only if* $[\delta M,\delta M](1)$ *is*

S-integrable. If M *is a* δt*-local martingale with the* δt*-reducing sequence* $\{\tau_m\}$, *then* $[\delta M, \delta M]^{1/2}(\tau_m)$ *is S-integrable for each* m.

PROOF:

The last remark follows from the first part with $p = 1$ by simply applying (6.7.3)(b).

Both implications of the first part are proved using (1.4.17) and the Burkholder-Davis-Gundy inequalities (7.2.6) given next. Assuming that $(M^{\delta}(\tau))^p$ (resp. $[\delta M, \delta M](\tau)^{p/2}$) is S-integrable, there is a convex increasing internal function Φ satisfying the conditions of (1.4.17) with $f(\omega) = (M^{\delta}(\tau(\omega)))^p$ (resp. $[\delta M, \delta M](\tau)^{p/2}$). The internal function $\Psi(x) = \Phi(x^p)$, $x \in {}^*[0,\infty)$, is convex, increasing, has $\Psi(0) = 0$ and satisfies $\Psi(2x) \le k\Psi(x)$, for all $x \in {}^*[0,\infty)$, where $k = 4^{p+1}$. The inequality (7.2.6) completes the proof of the first assertion (because one implication in (1.4.17) does not require part (b) as noted in its proof).

The S-integrable Doob inequality (6.5.20) shows that $M^2(1)$ is S-integrable if and only if $[M^{\delta}(1)]^2$ is S-integrable and the first part of the result connects this with quadratic variation using $\tau \equiv 1$. This completes the proof.

(7.2.6) THE BURKHOLDER-DAVIS-GUNDY INEQUALITIES:

For every standard real $k > 0$ *and* $d \in \mathbb{N}$, *there exist standard real constants* $c, C > 0$ *such that for every d-dimensional* **martingale* M *and every* δt *and* $t \in \mathbb{T}_{\delta}$ *and for every internal convex increasing function* $\Psi : {}^*[0,\infty) \longrightarrow {}^*[0,\infty)$ *satisfying*

$$\Psi(0) = 0 \quad \textit{and} \quad \Psi(2x) \le k\Psi(x) \quad \textit{for all} \quad x \in {}^*[0,\infty)$$

the following inequalities hold:

$$cE[\Psi(\{\delta M,\delta M](t)\}^{\frac{1}{2}})] \le E[\Psi(M^{\delta}(t))] \le CE[\Psi(\{[\delta M,\delta M](t)\}^{\frac{1}{2}})].$$

PROOF:

This result follows by taking the *transform of (the finite case of the d-dimensional extension of) Burkholder-Davis-Gundy's [1972] Theorem 1.1. While this is a cornerstone of our theory, we shall not give a proof since it is a "well-known standard result."

(7.2.7) **PATHWISE PROJECTION OF $[\delta M,\delta M]$:**

For any internal process M, the quadratic variation process $[\delta M,\delta M](t)$ is increasing for all ω since it is a sum of squares. The (local) S-integrability of $[\delta M,\delta M]$ proved in (7.2.5) means that except for ω in a single null set Λ, whenever t is finite $[\delta M,\delta M](t,\omega)$ is also finite. Hence for each $r \in [0,\infty)$ when $\omega \notin \Lambda$, the left and right limits along $\mathbb{T}_\delta$,

$$\operatorname*{S-lim}_{t\uparrow r}[\delta M,\delta M](t) = \inf_{t\approx r}\{st[\delta M,\delta M](t)\}, \quad t \in \mathbb{T}_\delta,$$

and

$$\operatorname*{S-lim}_{t\downarrow r}[\delta M,\delta M](t) = \sup_{t\approx r}\{st[\delta M,\delta M](t)\}, \quad t \in \mathbb{T}_\delta$$

both exist in $\mathbb{R}$. It follows, (5.3.25), that $[\delta M,\delta M]$ has a

Δt-decent path sample for some infinitesimal Δt in $\mathbb{T}_\delta$, but $[\delta M, \delta M]$ actually has a δt-decent path sample for the same δt as the process M. The proof that $[\delta M, \delta M]$ has a δt-decent path sample, Theorem (7.4.9), uses the machinery we develop for stochastic integration. We believe that there should be simple direct proofs of this basic fact, but do not know any.

We abuse notation and define a pathwise projected process using the extended standard part:

$$[\tilde{M}, \tilde{M}](r, \omega) = \sup_{t \approx r} [\overline{st}[\delta M, \delta M](t, \omega)\} \quad \text{for} \quad r \in [0, \infty).$$

By the preceding remarks $[\tilde{M}, \tilde{M}](r, \omega)$ is indistinguishable from a process with paths in $D[0, \infty)$. The abuse of notation is justified by (7.2.11)

The following is a key technical lemma that tells us some information about paths of quadratic variation processes.

(7.2.8) **LEMMA:**

Let M *be a* d-dimensional δt-*local martingale and suppose that* σ *is a* δt-*stopping time whose standard part,* $\tilde{\sigma}(\omega) = \overline{st}\ \sigma(\omega)$, *satisfies* $\tilde{\sigma} < \infty$ *and*

$$\text{S-}\lim_{t \downarrow \sigma} M(t) = st[M(\sigma)] \quad a.s.$$

Then

$$[\tilde{M}, \tilde{M}](\tilde{\sigma}) = st\{[\delta M, \delta M](\sigma)\} \quad a.s.$$

PROOF:

We will show that for any infinitesimal Δt in $\mathbb{T}_\delta$,

$$[\delta M, \delta M](\sigma+\Delta t) \approx [\delta M, \delta M](\sigma) \quad \text{a.s.}$$

The dependence of the exceptional null set on Δt (in the "a.s.") does not matter. The external almost sure statement means that the internal probability

$$P\{[\delta M, \delta M](\sigma+\Delta t) - [\delta M, \delta M](\sigma) > \epsilon\} < \epsilon$$

holds for all Δt in a finite interval of $\mathbb{T}_\delta$. Hence $st[\delta M, \delta M](\sigma+\Delta t)$ tends to $st[\delta M, \delta M](\sigma)$ in probability as Δt decreases finitely to zero. Thus it has an a.s. convergent subsequence, but since quadratic variation is increasing, the whole limit converges a.s. and $[\tilde{M}, \tilde{M}](\tilde{\sigma}) = st[\delta M, \delta M](\sigma)$ a.s.

We shall prove the lemma in the case where $M^2(1)$ is S-integrable and $\sigma \leq 1$. In this case, the martingale

$$N(t) = \begin{cases} 0 & , \quad t \leq \sigma \\ M(t)-M(\sigma) & , \quad \sigma \leq t \leq \sigma+\Delta t \\ M(\sigma+\Delta t)-M(\sigma) & , \quad t \geq \sigma+\Delta t \end{cases}$$

is also SL^2, so by Doob's inequality (6.5.20) has an SL^2 maximal function. That maximal function

$$N^\delta(1) = \max_t[\,|M(t)-M(\sigma)| \;:\; \sigma \leq t \leq \sigma+\Delta t]$$

is infinitesimal a.s. by the hypothesis that $\underset{t\downarrow\sigma}{S\text{-}\lim}\, M(t) = st[M(\sigma)]$. S-integrability means it has infinitesimal square expected value. Finally, applying the BDG inequality (7.2.6) to N yields

$$cE\{[\delta M,\delta M](\sigma+\Delta t)-[\delta M,\delta M](\sigma)\} \le E\{\max_t |M(t)-M(\sigma)|^2\} \approx 0.$$

This proves the lemma in the global $SL^2[0,1]$ case. (The reader can easily prove the local case by introducing a reducing sequence. Moreover, the bounded integrability of the reducing sequence is all that is needed, not the fact that M has a δt-decent path sample. This is helpful in Exercise (7.4.4).)

(7.2.9) **COROLLARY:**

If M is a d-dimensional δt-local martingale, then $[\delta M,\delta M]$ is S-continuous at $t = 0$, a.s.

PROOF:

The stopping time $\sigma(\omega) = \delta t$ satisfies $\tilde{M}(0) = st\ M(\delta t)$ a.s. so the lemma yields S-continuity of $[\delta M,\delta M]$ at zero.

(7.2.10) **THE QUADRATIC VARIATION LEMMA:**

Let M be a d-dimensional δt-martingale. If $\{t_j : j \in {}^*\mathbb{N},\ 0 \le j \le n\}$ is any S-dense internal subset of $\mathbb{T}_\delta^+$ with $t_0 < t_1 < \cdots < t_n$, then

$$\sup\{st([\delta M,\delta M](t_k)-|M(t_0)|^2-\sum_{j=1}^{k}|M(t_j)-M(t_{j-1})|^2) : t_k \in \mathcal{O}\}$$
$$= 0 \quad a.s.$$

PROOF:

The components of a martingale are also martingales and

$$[\delta M,\delta M](t) = \sum^{d} [\delta M_i,\delta M_i](t).$$

If we prove the lemma for one-dimensional martingales, it follows for d-dimensional ones by summing components. Hence we shall assume that M is a one-dimensional δt-local martingale for the rest of the proof.

Since M has a δt-decent path sample, it is right S-continuous at zero along $\mathbb{T}_\delta^+$, so $|M(\delta t)|^2 \approx |M(t_0)|^2 \approx |M(t_1)|^2$ a.s. Corollary (7.2.9) shows that $[\delta M,\delta M](t_0) \approx |M(\delta t)|^2$ a.s. Hence, we may as well assume that $t_0 = \delta t$ so that we only have to compare the difference between summing large and small increments beginning at the same time. Here is a useful formula for comparing large and small increments, starting with the first large one:

$$M(t_1) - M(t_0) = \sum[\delta M(s) : t_0 \leq s < t_1, \text{ step } \delta t],$$

so

$$|M(t_1)-M(t_0)|^2 = \sum|\delta M(s)|^2 + 2\sum_{s>r}\sum \delta M(r)\delta M(s)$$

$$= \sum|\delta M(s)|^2 + 2\sum_s\Big(\sum_{t_0}^{s-\delta t} \delta M(r)\Big)\delta M(s).$$

Hence,

$$|M(t_0)|^2 + |M(t_1)-M(t_0)|^2$$

$$= [\delta M,\delta M](t_1) + 2\sum_{s=t_0}^{t_1-\delta t}(M(s)-M(t_0))\delta M(s).$$

In general,

$$(|M(t_0)|^2 + \sum_{j=1}^{k} |M(t_j)-M(t_{j-1})|^2) - [\delta M,\delta M](t_k) = N(t_k)$$

where (using our convention on $\mathbb{T}_\delta$-sums; $t_{j-1} \leq s < t_j$, $s \in \mathbb{T}_\delta$)

$$N(t_k) = 2 \sum_{j=1}^{k} (\sum_{s=t_{j-1}}^{t_j} (M(s)-M(t_{j-1}))\delta M(s))$$

or, letting $[s] = \max[t_j : t_j \leq s, 0 \leq j \leq n]$,

$$N(t_k) = 2 \sum_{\delta t}^{t_k} (M(s)-M([s]))\delta M(s).$$

The same sum formula may be used to define $N(t)$ for any upper summand t in $\mathbb{T}_\delta$ yielding a *martingale along $\mathbb{T}_\delta^+$. By direct calculation the quadratic variation

$$[\delta N,\delta N](t) = 4 \sum^{t} |M(s)-M([s])|^2 |\delta M(s)|^2.$$

To conclude the proof we need a reducing sequence even when M^2 is S-integrable on $[0,1]$. In this case, apply the Path Stopping Lemma (6.4.5) to obtain stopping times τ_m such that $M^\delta(\tau_m-\delta t) \lesssim m$, $\tilde{M}(st\ \tau_m) = st(M(\tau_m)]$ and $\tau_m \uparrow 1$. In general, if

$\{\tau_m\}$ is a δt-reducing sequence for M with $M^\delta(\tau_m) \lesssim m$, then

$$[\delta N, \delta N]^{\frac{1}{2}}(\tau_m) < 5m[\delta M, \delta M]^{\frac{1}{2}}(\tau_m).$$

By (7.2.5) and (1.4.14), $[\delta N, \delta N]^{\frac{1}{2}}(\tau_m)$ is S-integrable. Burkholder-Davis-Gundy's (7.2.6) inequality and (1.4.13) tell us that

$$\begin{aligned} E(\max[\,|N(t)|\,:\,t \le \tau_m]) &\le CE([\delta N, \delta N]^{\frac{1}{2}}(\tau_m)) \\ &\lesssim CE(st[\delta N, \delta N]^{\frac{1}{2}}(\tau_m)), \end{aligned}$$

for a standard positive constant C. We will show that

$$E(st[\delta N, \delta N]^{\frac{1}{2}}(\tau_m)) = 0$$

and therefore $|N(t)| \approx 0$ for finite t a.s. proving our lemma.

For each n in $\mathbb{N}$

$$\begin{aligned} &E[st(\{\sum^{\tau_m} |M(s)-M([s])|^2 |\delta M(s)|^2\}^{\frac{1}{2}})] \\ &\le \frac{2}{n} E(st[\delta M, \delta M]^{\frac{1}{2}}(\tau_m)) + 3mE(st\{\sum[\,|\delta M(s)|^2 \,:\, s \in A_m^n(\omega)]\}^{\frac{1}{2}}), \end{aligned}$$

where

$$A_m^n(\omega) = \{s \in \mathbb{T}_\delta^+ \,:\, s < \tau_m(\omega) \;\&\; |M(s)-M([s])| > \frac{2}{n}\}.$$

estimating $|M(s)-M([s])| \lesssim 2m$ on A_m^n where it is large $(s \lesssim \tau_m)$. It is sufficient to prove that

$$E(st\{\Sigma[|\delta M(s)|^2 : s \in A_m^n(\omega)\}^{\frac{1}{2}}) = 0$$

for every $m,n \in \mathbb{N}$. Define δt-stopping times as follows: $\rho_0 = \sigma_0 = \delta t$ and

$$\rho_i = i^{\underline{th}} \text{ time} |M(t)-M(t-\delta t)| > \frac{1}{n}$$

and

$$\sigma_i = \min[t_j : t_j \geq \rho_i, \quad 0 \leq j \leq n].$$

If M has a δt-decent path for the sample ω and $s \in A_m^n(\omega)$, then $|M(s)-M([s])| > \frac{2}{n}$ and $s \lesssim m$. Therefore $M(\cdot,\omega)$ varies a finite amount infinitely near s, but since it has a decent path along $\mathbb{T}_\delta$ it must have jumped by an amount $\gtrsim \frac{2}{n} > \frac{1}{n}$ at one single time in $\mathbb{T}_\delta$ between $[s]$ and s. On the other hand, $M(\cdot,\omega)$ could only have jumped by more than $\frac{1}{n}$ a finite number of times before $s \lesssim m$ by (the $[0,\infty)$-version of) (5.3.4)(c). Thus for almost all ω, $A_m^n(\omega)$ is contained in the countable (external) union

$$A_m^n(\omega) \subseteq \bigcup_{i\in\mathbb{N}} [\rho_i \wedge \tau_m(\omega), \sigma_i \wedge \tau_m(\omega)).$$

Hence, by S-integrability and (7.2.6) applied as in the proof of (7.2.8),

$$E(\text{st}\sqrt{\Sigma[\,|\delta M(s)|^2 : s \in A_m^n(\omega)]})$$

$$\leq \sum_{i \in \mathbb{N}} E\{\text{st}\sqrt{[\delta M,\delta M](\sigma_i \wedge \tau_m) - [\delta M,\delta M](\rho_i \wedge \tau_m)}\}$$

$$\leq \sum_{i \in \mathbb{N}} \text{st } E\{\sqrt{[\delta M,\delta M](\sigma_i \wedge \tau_m) - [\delta M,\delta M](\rho_i \wedge \tau_m)}\}$$

$$\leq \frac{1}{c} \sum_{i \in \mathbb{N}} \text{st } E(\max[\,|M(t) - M(\rho_i \wedge \tau_m)| : \rho_i \wedge \tau_m \leq t \leq \sigma_i \wedge \tau_m,\ t \in \mathbb{T}_\delta])$$

$$= 0.$$

We get zero because $M^\delta(\tau_m)$ is S-integrable while

$$\max[\,|M(t) - M(\rho_i \wedge \tau_m)| : \rho_i \wedge \tau_m \leq t \leq \sigma_i \wedge \tau_m] \approx 0$$

whenever M has δt-decent paths for the sample ω, which happens a.s. This proves that $\max[\,|N(t)| : t \leq \tau_m] \approx 0$ a.s. Since $\tau_m \uparrow \infty$ a.s. as $m \longrightarrow \infty$, this concludes the proof of the lemma.

The primary consequence of this lemma is the fact that the quadratic variation of a standard local hypermartingale is independent of the lifting and the infinitesimal increment in particular (that is, once the increment is coarse enough to make the paths of M decent).

Fix an $r \in [0,\infty)$ and if $0 = r_0 < r_1 < \cdots < r_k = r$ is an increasing sequence, define

$$S(\tilde{M},\{r_j\}) = |\tilde{M}(0)|^2 + \sum_{j=1}^{k} |\tilde{M}(r_j)-\tilde{M}(r_{j-1})|^2.$$

(7.2.11) **COROLLARY:**

Let $\tilde{M}$ *be a local hypermartingale and let* M *be a* δt*-local martingale lifting of* $\tilde{M}$. *For each* $r \in [0,\infty)$, $S(\tilde{M},\{r_j\})$ *converges to* $[\tilde{M},\tilde{M}](r)$ *in probability as the mesh of the sequence,* $\max|r_j-r_{j-1}|$, *tends to zero. The standard quadratic variation of* $\tilde{M}$ *does not depend on the choice of lifting and we denote the unique (up to indistinguishability) decent path standard process by* $[\tilde{M},\tilde{M}](r)$.

PROOF:

Choose any $t \in \mathbb{T}$ such that $\tilde{M}(r) = \text{st}\, M(t)$ a.s. and $[\tilde{M},\tilde{M}](r) = \text{st}[\delta M,\delta M](t)$ a.s. Let m in $\mathbb{N}$ be finite. By Lemma (7.2.10), whenever the mesh of a *finite sequence $\delta t = t_0 < t_1 < \cdots < t_k = t$ is infinitesimal,

$$P\{([\delta M,\delta M](t)-|M(\delta t)|^2 - \sum_{j=1}^{k} |M(t_j)-M(t_{j-1})|^2) > \tfrac{1}{m}\} < \tfrac{1}{m}.$$

This is an internal statement so there is an $\epsilon_m >> 0$ such that whenever $\max|t_j-t_{j-1}| < \epsilon_m$, the probability above holds.

Let $0 = r_0 < r_1 < \cdots < r_k = r$ be a standard sequence in $[0,\infty)$ with $\max|r_j-r_{j-1}| << \epsilon_m$. Let $t_0 = \delta t$, $t_k = t$ and choose t_j, $0 < j < k$ so

$$\tilde{M}(r_j) = \text{st}\, M(t_j) \quad \text{a.s.}$$

The sequence $\{t_j\}$ satisfies the internal probability above, so standard parts yield

$$P\{[\tilde{M},\tilde{M}](r) - S(\tilde{M},\{r_j\}) > \tfrac{1}{m}\} < \tfrac{1}{m},$$

proving the corollary.

Two δt-martingales M and N could have finite jumps at distinct infinitely close times. This would mean that the sum $M+N$ is not a δt-martingale because the paths no longer have separated jumps. Of course $M+N$ is a *martingale, so some coarser infinitesimal time sample increment Δt in $\mathbb{T}_\delta$ would make M,N and $M+N$ all Δt-martingales. If we start with standard local hypermartingales $\tilde{M}$ and $\tilde{N}$ we may apply the martingale lifting theorem to the 2d-dimensional martingale $(\tilde{M},\tilde{N})$. If (M,N) is a 2d-dimensional δt-local martingale lifting of $(\tilde{M},\tilde{N})$, then M and N must jump together (or not at all), so $M+N$ is a δt-local martingale. In this case we may define $[\tilde{M},\tilde{N}](r,\omega) = \operatorname{S-lim}_{t\downarrow r}[\delta M,\delta N](t,\omega)$ a.s. and use the polarization identities (7.2.3) and (7.2.11) to see that this is independent of the choice of the lifted pair up to indistinguishability. This shows how to extend (7.2.11) to pairs and the following extends (7.2.6) to pairs.

(7.2.12) LEMMA:

Let $(\tilde{M},\tilde{N})$ be a 2d-dimensional local hyper-martingale with the δt-local martingale lifting (M,N). If σ is a δt-stopping time satisfying $\tilde{\sigma} < \infty$ a.s. and $\tilde{M}(\tilde{\sigma}) = \operatorname{st} M(\sigma)$ a.s., then $[\tilde{M},\tilde{N}](\tilde{\sigma}) = \operatorname{st}[\delta M,\delta N](\sigma)$ a.s.

PROOF:

By the preceding remarks and (7.2.8) we know that

$$[\tilde{M},\tilde{N}](\tilde{\sigma}) = \underset{\Delta t \downarrow 0}{S\text{-}\lim}[\delta M,\delta N](\sigma+\Delta t) \quad \text{a.s.}$$

and that $[\tilde{M},\tilde{M}](\tilde{\sigma}) = st[\delta M,\delta M](\sigma) = st[\delta M,\delta M](\sigma+\Delta t)$ for $\Delta t \approx 0$ a.s. Hence it suffices to show that for every infinitesimal $\Delta t \geq 0$ in $\mathbb{T}_\delta$,

$$[\delta M,\delta N](\sigma+\Delta t) \approx [\delta M,\delta N](\sigma) \quad \text{a.s.}$$

We simply apply the *transform of Cauchy's inequality to the *finite dimensional vectors with components $\delta M_i(t),\delta N_i(t)$.

$$\begin{aligned}
&|[\delta M,\delta N](\sigma+\Delta t)-[\delta M,\delta N](\sigma)| = \\
&\quad = |\textstyle\sum[\delta M_i(t)\delta N_i(t) : 1 \leq i \leq d,\ \sigma \leq t < \sigma+\Delta t,\ t \in \mathbb{T}_\delta]| \\
&\quad \leq \sqrt{\Sigma\delta M_i^2(t)}\ \sqrt{\Sigma\delta N_i^2(t)} \\
&\quad = \sqrt{[\delta M,\delta M](\sigma+\Delta t)-[\delta M,\delta M](\sigma)}\ \sqrt{[\delta N,\delta N](\sigma+\Delta t)-[\delta N,\delta N](\sigma)}.
\end{aligned}$$

We know that $[\delta M,\delta M](\sigma+\Delta t) \approx [\delta M,\delta M](\sigma)$ a.s. and local S-integrability of $[\delta N,\delta N]$ makes $[\delta N,\delta N](\sigma+\Delta t) - [\delta N,\delta N](\sigma)$ finite a.s. This proves the lemma.

(7.3) Square Martingale Integrals

Let $\tilde{M} : [0,1] \times \Omega \to \mathbb{R}$ be a hypermartingale with $\tilde{M}^2(1)$ integrable and $\tilde{M}(0) = 0$. We know from the Martingale Lifting Theorem (6.5.13) that there exists a δt-martingale M with $M^2(1)$ S-integrable and $M(\delta t) \approx 0$ a. s. whose δt-decent path projection is indistinguishable from $\tilde{M}$. By Theorem (7.2.5), we know that $[\delta M, \delta M](t)$ is S-integrable for all $t \leq 1$. In this section we use estimates on the quadratic variation to show that the martingale integral

$$I(r,\omega) = \int_0^r H(q,\omega) d\tilde{M}(q,\omega)$$

is well-defined as the δt-decent path projection of the Stieltjes sums

$$S(t,\omega) = \sum_{s=\delta t}^{t-\delta t} G(s,\omega) \delta M(s,\omega),$$

for a pathwise lifting G of H. "Well-defined" means this a. s. does not depend on the choice of the martingale lifting, M, or the path lifting, G, once M is chosen. Our first construction is the analog of (7.1.4). The development runs parallel to section 7.1, except that we use martingale maximal inequalities (instead of the triangle inequality) and this requires that our summands be predictable.

(7.3.1) DEFINITION:

Let M be a δt-martingale with $M(\delta t) \approx 0$ a. s. and $M^2(1)$ S-integrable. For each $\omega \in \Omega$ define a quadratic path variation measure λ_ω on $\mathbb{T}$ by the weight function

$$\delta\lambda_\omega(t) = \begin{cases} |\delta M(t,\omega)|^2, & t \in \mathbb{T}_\delta \\ 0, & t \in \mathbb{T} \setminus \mathbb{T}_\delta \end{cases}.$$

Define a total quadratic variation measure υ *on* $\mathbb{T} \times \Omega$ *as the hyperfinite extension of the measure with weight function*

$$\delta\upsilon(t,\omega) = \delta\lambda_\omega(t)\cdot\delta P(\omega).$$

Since $E\{[\delta M,\delta M](1)\}$ is limited, υ extends to a bounded hyperfinite measure, $\boldsymbol{\upsilon}$. Since $[\delta M,\delta M](1)$ is S-integrable, υ is P-continuous, i. e., if $P[A] = 0$, then $\boldsymbol{\upsilon}[\mathbb{T} \times A] = 0$. Also, Iterated Integration (7.1.5) applies to $\boldsymbol{\upsilon}$. Since $\tilde{M}$ is right continuous at zero, $\boldsymbol{\upsilon}$ is continuous at zero, $\upsilon[st^{-1}(0) \times \Omega] = 0$.

(7.3.2) **DEFINITION:**

Let $H : [0,1] \times \Omega \to \mathbb{R}$ *be any function. An internal* $G : \mathbb{T} \times \Omega \to {}^*\mathbb{R}$ *such that for almost all* ω

$$\boldsymbol{\lambda}_\omega\{t : st[G(t,\omega)] \neq H(st[t],\omega)\} = 0$$

is called a δM^2*-path lifting of* H.

We can prove a path lifting theorem like (7.1.8) for the quadratic path variation measure, but unpredictable integrands give the "wrong" answer, as shown in the following exercise.

(7.3.3) **EXERCISE:**

Let $B(t,\omega)$ *be Anderson's infinitesimal random walk associated with* δt *as above in* (5.2.3). *Define*

$$\Sigma^t 2B\delta B(\omega) = \sum[2B(s,\omega)[B(s+\delta t,\omega)-B(s,\omega)] : 0 \le s < t,\ s \in \mathbb{T}]$$

$$\beta^t 2B\delta B(\omega) = \sum[2B(s+\delta t,\omega)[B(s+\delta t,\omega)-B(s,\omega)] : 0 \le s < t,\ s \in \mathbb{T}]$$

and

$$S^t 2B\delta B(\omega) = \sum[(B(s+\delta t,\omega)+B(s,\omega))\delta B(s,\omega) : 0 \le s < t,\ s \in \mathbb{T}].$$

Show that

$$\Sigma^t 2B\delta B = B^2(t)-t$$

$$\beta^t 2B\delta B = B^2(t)+t$$

and

$$S^t 2B\delta B = B^2(t).$$

(HINT: Write $B^2(t)$ as a double sum and compare.) *Show that* β^t *and* S^t *are not* **martingales, but* Σ^t *is. Show that when* $H(r) = 2\tilde{B}(r)$, *then all the functions* $K(t,\omega) = 2B(t,\omega)$, $K(t,\omega) = 2B(t+\delta t,\omega)$ *and* $K(t,\omega) = B(t,\omega)+B(t+\delta t,\omega)$ *are* δB^2*-path liftings of* H.

The exercise above shows that δM^2-path lifting alone is not enough to make infinitesimal Stieltjes sums independent of the infinitesimal differences in liftings. Moreover, the internal sum

$$\sum^{t} \operatorname{sgn}[\delta B(s)]\delta B(s)$$

is infinite a. s. for all noninfinitesimal t. The function $\operatorname{sgn}[\delta B(t)]$ depends precisely on $\omega_{t+\delta t}$, but is internal and bounded.

(7.3.4) **DEFINITION:**

An internal process $G : \mathbb{T} \times \Omega \to {}^*\mathbb{R}$ *is* δM^2*-summable if* G *is O-predictable and the function* $|G(t,\omega)|^2$ *is S-integrable with respect to the hyperfinite measure* υ *generated by the weight function* $\delta\upsilon = |\delta M|^2 \cdot \delta P$.

This summability condition is equivalent to the condition

$$\operatorname{st} E[\sum |G(s,\omega)|^2 |\delta M(s,\omega)|^2] = E[\int \operatorname{st}|G(s,\omega)|^2 d\lambda_\omega(s)],$$

by the Iterated Integration Lemma (7.1.5).

Our next result is part of a closure law for stochastic Stieltjes sums. (It lacks the decent path property.) It is understood that the martingale M is as above.

(7.3.5) **PROPOSITION:**

Suppose G *is* δM^2*-summable (where* $M^2(1)$ *is S-integrable). Then* $N(t) = \sum_{s=\delta t}^{t-\delta t} G(s)\delta M(s)$ *is a* **martingale after* δt *with* $N^2(1)$ *S-integrable.*

PROOF:

Since G is nonanticipating after δt,

$$E[\delta N|\omega^t] = G(t)E[\delta M|\omega^t] = 0.$$

Moving st inside always produces the inequalities:

$$\begin{aligned}
\mathrm{st}[\sum_\omega \sum_t G^2\delta\nu] &= \mathrm{st}[E\{[\delta N,\delta N](1)\}] \\
&\geq E\{\mathrm{st}[\delta N,\delta N](1)\} \\
&\geq E\{\int \mathrm{st}\ G^2 d(\lambda_\omega)\} \\
&= \int \mathrm{st}\ G^2 d\nu \qquad \text{by (7.1.5).}
\end{aligned}$$

The two extremes of these inequalities agree because G^2 is ν-S-integrable. Hence $\mathrm{st}\ E\{[\delta N,\delta N](1)\} = E\{\mathrm{st}[\delta N,\delta N](1)\}$, so $[\delta N,\delta N]$ is S-integrable and (7.2.5) completes the proof.

Our next result says ν-equivalent summands pathwise give nearly the same sums. Again, M is as above.

(7.3.6) **PROPOSITION:**

Suppose G_1 *and* G_2 *are* δM^2*-summable and*

$$\nu\{(t,\omega) : \mathrm{st}\ G_1(t,\omega) \neq \mathrm{st}\ G_2(t,\omega)\} = 0.$$

Then the **martingale* $N_1(t) = \Sigma^t\ G_1(s)\delta M(s)$ *is infinitely close to* $N_2(t) = \Sigma^t\ G_2(s)\delta M(s)$, *in fact,*

$$E\{|\max_{\delta t\leq t\leq 1} \Sigma^t[G_1(s)-G_2(s)]\delta M(s)|^2\} \approx 0.$$

PROOF:

Since $|G_1|^2$ and $|G_2|^2$ are ν-S-integrable, so is $G_1G_2 \leq \max[G_1^2, G_\nu^2]$, while $st\ G_1G_2 = st\ G_1^2$ ν-a.e. Therefore,

$$E[\sum_{\delta t}^{1} |G_1(t)-G_2(t)|^2 |\delta M(t)|]$$
$$= E[\Sigma\ G_1^2|\delta M|^2]-2E[\Sigma\ G_1G_2|\delta M|^2]+[\Sigma\ G_2^2|\delta M|^2]$$
$$\approx E[\int st\ G_1^2 d\lambda_\omega]-2E[\int st\ G_1^2 d\lambda_\omega]+E[\int st\ G_1^2 d\lambda_\omega] = 0.$$

Applying the BDG Inequality (7.2.6) to the martingale $N = N_1-N_2$ yields

$$E\{[\max_{\delta t\leq t\leq 1} |\Sigma^t[G_1-G_2]\delta M|]^2\} \leq CE\{\Sigma|G_1-G_2|^2|\delta M|^2\} \approx 0.$$

In particular, the whole path of N_1 is infinitely close to the path of N_2 a.s.

(7.3.7) PROPOSITION:

Suppose M *is a* δt*-martingale with* $M^2(1)$ *S-integrable. If* F,G *and the sequence* $\{F_m\}$ *are all* δM^2*-summable,* $st\ F_m$ *tends to* $st\ F$ *in* ν*-measure and* $|F_m| \leq |G|$*, then*

$$st[\max_{\delta t\leq t\leq 1} |\Sigma^t F_m\delta M - \Sigma^t F\delta M|]$$

tends to zero in probability.

PROOF:

Use comprehension from section (0.4) to extend the sequence $\{F_m\}$ to an internal sequence satisfying $|F_m| \leq |G|$ for all

$m \in {}^*\mathbb{N}$. Use the Internal Definition Principle (in the style of Chapter 1) to pick an infinite n_1 such that whenever $n \leq n_1$ is infinite, then $F_n \approx F$ ν-a.e. Now use the preceding proposition to show that

$$E\{[\max_{\delta t \leq t \leq 1} |\Sigma^t F_n \delta M - \Sigma^t F \delta M|]^2\} \approx 0.$$

This means that for every standard positive ϵ the internal set

$$I(\epsilon) = \{m \in {}^*\mathbb{N} \mid m \leq k \leq n_1 \Rightarrow P\{\max|\Sigma^t[F-F_k]\delta M| > \epsilon\} < \epsilon\}$$

contains a finite $m(\epsilon)$. This proves the claim.

Up to this point we have not assumed that G is a lifting of a standard function of any kind. However, we have not been able to make any conclusions about the paths of $N(t) = \Sigma^t G \cdot \delta M$. The weakest pathwise result is taken up next. In section (7.6) we describe Hoover & Perkins' [1983] better martingale lifting that makes N have δt-decent paths when G is any δM^2-summable internal function—not just the lifting of some standard H.

(7.3.8) **THEOREM:**

Let $\tilde{M} : [0,1] \times \Omega \longrightarrow \mathbb{R}^d$ *be the projection of the* δt*-martingale* M *with* $M^2(1)$ *S-integrable and* $M(\delta t) \approx 0$ a. s. *Suppose that* $H : [0,1] \times \Omega \longrightarrow \mathbb{R}$ *is* ν*-almost previsible or even only previsible with respect to the usual agumentation of* $\mathcal{F}$ *and that the pathwise Stieltjes* *integrals satisfy*

$$E[\int_0^1 H^2(r,\omega)d[\tilde{M},\tilde{M}](r,\omega)] < \infty.$$

Then H *has a* δM^2*-summable* δM^2*-path lifting* $G : \mathbb{T} \times \Omega \longrightarrow {}^*\mathbb{R}$. *Such a* G *makes* $\Sigma^t\ G(s)\delta M(s)$ *have a* δt*-decent path sample, so the square summable martingale*

$$\int_0^r H(q,\omega)d\tilde{M}(q,\omega)$$

may be defined independent of the particular lifting as the δt*-decent path projection of*

$$\sum_{s=\delta t}^{t-\delta t} G(s,\omega)\delta M(s,\omega).$$

PROOF:

First, as long as $N(t) = \Sigma^t G\cdot\delta M$ has a δt-decent path sample, Proposition (7.3.6) shows that any two δM^2-summable G's have the same decent path projection. Hence the two remaining claims are that such a G exists and that its δM-sums have a δt-decent path sample.

PROOF OF LIFTING:

Change of variables gives us

$$\int H^2(r,\omega)d[\tilde{M},\tilde{M}](r,\omega) = \int H^2(st(t),\omega)d\lambda_\omega(t)$$

for all ω such that $[\tilde{M},\tilde{M}](1)$ is finite. Iterated Integration gives

$$E[\int H^2 d[\tilde{M},\tilde{M}]\} = E\{\int H^2(st(t),\omega)d\lambda_\omega(t)\}$$
$$= \int H^2(st(t),\omega)d\nu < \infty.$$

We know by (6.6.8) or (6.6.11) that there is a O-predictable F such that

$$\nu\{(t,\omega) \mid st[F(t,\omega)] \neq H(st[t],\omega)\} = 0.$$

For each $m \in \mathbb{N}$, the truncations of F and H at m satisfy

$$\Sigma\Sigma\ F_m^2 \delta\nu \approx \int st \circ F_m^2 d\nu$$
$$= \int H_m^2 \circ st\ d\nu = h_m$$
$$\to \int H^2 \circ st\ d\nu.$$

Thus by Robinson's Sequential Lemma, for sufficiently small infinite n,

$$\Sigma\Sigma\ F_n^2 \delta\nu \approx \int st\ F_n^2 d\nu$$
$$= \int H^2\ st\ d\nu.$$

The function $G = F_n$ is our square-summable predictable path lifting.

PROOF OF OF δt-DECENT PATHS:

We say that $N(t)$ a. s. only jumps at the same times as $M(t)$ if there is a null set $\Lambda \subseteq \Omega$, $P[\Lambda] = 0$, such that if $\omega \notin \Lambda$, then whenever $t, t+\Delta t \in \mathbb{T}_\delta$ and $0 < \Delta t \approx 0$, then

$$N(t+\Delta t) \not\approx N(t) \quad \text{implies} \quad M(t+\Delta t) \not\approx M(t).$$

If $N(t)$ a. s. only jumps where $M(t)$ does and $M(t)$ has a δt-decent path sample, then $N(t)$ also has a δt-decent path sample. We begin by showing that when H is bounded and G is a bounded δM^2-path lifting of H, then $N(t) = \Sigma^t G \cdot \delta M$ a. s. only jumps where $M(t)$ does.

First suppose that H is bounded and ν-equivalent to a basic almost previsible process of the form in (6.6.13) and that

$$G(t,\omega) = g_0(\omega) I_{[0,t_1]}(t) + \sum_{j=1}^{m} g_j(\omega) I_{(t_j, t_{j+1}]}(t)$$

is the 0-predictable δM^2-path lifting guaranteed by Lemma (6.6.13). Since ν is continuous at zero, we may assume that $g_0 = 0$. The functions g_j are bounded, $|g_j(\omega)| \leq b$. To show that $N(t) = \Sigma^t G \cdot \delta M$ only jumps where M does we assume that $M(t)$ has a δt-decent path sample at ω and consider two cases.

Case 1: $t \leq t_j < t+\Delta t$, for t_j as above in G.

$$N(t+\Delta t) - N(t) =$$

$$= [N(t+\Delta t) - N(t_j+\delta t)] + [N(t_j+\delta t) - N(t)]$$

$$= g_j(\omega)[M(t+\Delta t) - M(t_j+\delta t)] + g_{j-1}(\omega)[M(t_j+\delta t) - M(t)]$$

At most one of the terms with a difference in M can be noninfinitesimal because M has a decent path at ω and g_j and g_{j-1} are bounded. Thus, if $N(t+\Delta t) \not\approx N(t)$, exactly one of the terms $g_j[M(t+\Delta t) - M(t_j)]$ and $g_{j-1}[M(t_j) - M(t)]$ is noninfinitesimal. Therefore, $M(t+\Delta t) \not\approx M(t)$.

Case 2: $t < t+\Delta t \leq t_j$, for t_j as in G above.

$$N(t+\Delta t) - N(t) = g_j(\omega)[M(t+\Delta t) - M(t)],$$

so if $N(t+\Delta t) \not\approx N(t)$ then $M(t+\Delta t) \not\approx M(t)$.

These two cases show that N a.s. only jumps where M does in the case that H is a bounded basic almost previsible process and G is the particular lifting obtained in (6.6.13). Proposition (7.3.6) shows that any other bounded δM^2-path lifting G makes $N(t) = \Sigma^t G \cdot \delta M$ also a. s. only jump where M does. The next step of the proof uses Lemma (6.6.14) to show that every bounded almost previsible process produces nice path sums.

Let $\mathbb{V}$ be the set of all bounded almost previsible processes, H, that have a bounded δM^2-path lifting, G, such that $N(t) = \Sigma^t G \cdot \delta M$ a. s. only jumps where $M(t)$ does. We have shown that $\mathbb{V}$ contains all basic almost previsible processes in the paragraphs above. The set $\mathbb{V}$ is a vector space, because if two stochastic sums only jump where M does, so does a linear combination. Proposition (7.3.7) and Lemma (6.6.14) show that $\mathbb{V}$ contains all bounded almost previsible processes.

Finally, let H_m be the finite truncations of an integrable unbounded H. If G is a δM^2-summable δM^2-path lifting of H, then the finite truncations G_m lift H_m and satisfy $|G_m| \le |G|$ and $G_m \to G$ in ν-measure. The $N_m(t) = \Sigma^t G_m \delta M$ a. s. only jump M does. In particular, N_m a. s. has a δt-decent path sample for the same δt as M. Proposition (7.3.7) says that a subsequence of the N_m a. s. tends uniformly to N in the space of internal functions on $\mathbb{T}_\delta$. Let Λ be the countable union of null sets where some N_m does not have a δt-decent path or the subsequence of N_m does not tend uniformly to N. If $\omega \notin \Lambda$ and $\Sigma^{t+\Delta t} G \cdot \delta M \not\approx \Sigma^t G \cdot \delta M$, then for sufficiently large finite m, $\Sigma^{t+\Delta t} G_m \cdot \delta M \not\approx \Sigma^t G_m \cdot \delta M$, since we may make the uniform error between G_m-sums and G-sums less than one third of the difference. Since each $\Sigma^t G_m \cdot \delta M$ a. s. only jumps where M does, so does $\Sigma^t G \cdot \delta M$. This proves half of the main result of the section.

Since we have actually shown more than that $\Sigma^t G \cdot \delta M$ has a δt-decent path sample, we have the following corollary to our proof.

(7.3.9) **THEOREM:**

If M is an S-continuous δt-martingale with $M^2(1)$ S-integrable and $M(\delta t) \approx 0$ and if G is a δM^2-summable δM^2-path lifting of a standard previsible process, then $\Sigma^t G \cdot \delta M$ is also S-continuous.

PROOF:

Since the proof of (7.3.8) shows that $\Sigma^t G \cdot \delta M$ only jumps where M does and M does not jump by S-continuity, the stochastic sum is S-continuous.

The remaining half of the well-definedness question is: "What if we take a different martingale lifting?"

(7.3.10) THEOREM:

Suppose that $\tilde{M}$ is a hypermartingale with $\tilde{M}^2(1)$ integrable and $\tilde{M}(0) = 0$. Suppose that M is a δt-martingale lifting of $\tilde{M}$ with $M^2(1)$ S-integrable and N is a Δt-martingale lifting of $\tilde{M}$ with $N^2(1)$ S-integrable. Suppose that F is a δM^2-summable δM^2-path lifting of a standard almost previsible process H, while G is a δN^2-summable δN^2-path lifting of H. Then the Δt-decent path projection of $\Sigma^t G \cdot \delta N$ is indistinguishable from the δt-decent path projection of $\Sigma^s F \cdot \delta M$.

PROOF:

First suppose that H is a bounded basic almost previsible process and F and G have the form of the lifting of (6.6.12). In both cases the decent path projections are indistinguishable from

$$\sum_{j=1}^{m} h_j(\omega) \cdot [\tilde{M}(r_{j+1}) - \tilde{M}(r_j)].$$

Proposition (7.3.6) shows that F and G can be any bounded path-liftings of H and still have the above projection.

Next we apply (6.6.13). Let the subset of almost previsible processes $\mathbb{V}$ consist of all bounded almost previsible H such that if F is a bounded δM^2-lifting of H and G is a bounded δN^2-lifting of H, then the decent path projections of $\Sigma^s F \cdot \delta M$ and $\Sigma^t G \cdot \delta N$ are indistinguishable. The set $\mathbb{V}$ is a vector space containing the basic almost previsible processes by the remarks above. The space $\mathbb{V}$ is closed under bounded pointwise convergence by Proposition (7.3.7). Thus $\mathbb{V}$ contains all bounded almost previsible processes.

Finally, a truncation argument similar to the last part of the proof of Theorem (7.3.8) shows that every square summable integrand produces indistinguishable stochastic sums.

(7.3.11) **EXERCISE:**

Show that all the proofs in this section actually apply to the case of d-dimensional linear functionals. That is, we may take $G : \mathbb{T} \times \Omega \to \mathbb{R}^d$ *and interpret our stochastic Stieltjes sums as sums of inner products,*

$$\Sigma_{\delta t}^{t-\delta t} G(s,\omega)\delta M(s,\omega) = \Sigma_{\delta t}^{t-\delta t}(G(s,\omega),\delta M(s,\omega)).$$

HINTS:

The meaning of lifting is clear. A δM^2-summable vector valued function is one for which the square of the vector norm, $|G(s,\omega)|^2$, is υ-S-integrable.

(7.4) Toward Local Martingale Integrals

To extend the treatment of martingale integrals from the square summable martingales of the last section to standard local martingales, we may use the Local Martingale Lifting Theorem (6.7.5) and the special δt-reducing sequence $\{\tau_m\}$ of (6.7.3) together with Theorem (7.2.5). This gives us the basic infinitesimal stability results without use of Iterated Integration (7.1.5). This section proves these stability results, but does not give a general standard lifting theorem. (Sections (7.6) and (7.7) sketch the proofs of the powerful lifting theorems of Hoover & Perkins [1983].)

We do show how these simple extensions of section (7.3) may be applied to prove that paths of the quadratic variation are decent. This is the main application of the section.

For this section we work in the evolution scheme of (5.5) and (6.7).

(7.4.1) DEFINITION:

Let M be a d-dimensional δt-local martingale with δt-reducing sequence $\{\tau_m\}$ and $M(\delta t) \approx 0$ a.s. The quadratic path variation measure is defined by the weight function

$$\delta\lambda_\omega(t) = \begin{cases} |\delta M(t,\omega)|^2 & , \quad t \in \mathbb{T}_\delta \\ 0 & , \quad t \in \mathbb{T}\setminus\mathbb{T}_\delta \end{cases}.$$

The hyperfinite extension λ_ω may now be an unbounded measure, in fact, $[\delta M,\delta M](k)$ may not be S-integrable even for $k \in \mathbb{N}$. Proof of lifting theorems require that we control the growth of $\delta\lambda_\omega$'s so that the total measure $\delta\upsilon$ is a bounded version of $\delta\lambda_\omega \cdot \delta P$. The simple procedure that we used in Definition (7.1.4) will not work, so we postpone this problem to section (7.7).

We *do* know from the definition of δt-reducing sequence that

$$M^\delta(\tau_m) \text{ is S-integrable for each } m.$$

By (7.2.5) we also know that

$$\sqrt{[\delta M, \delta M](\tau_m)} \text{ is S-integrable for each } m.$$

Despite the technical problem with boundedness of the total quadratic variation measure, the notion of path-lifting remains the same.

(7.4.2) **DEFINITION:**

Let M *be a* δt*-local martingale as above. An internal fucntion* G *is a* δM^2*-path lifting of a standard function* H *provided*

$$\lambda_\omega\{t \mid st[G(t,\omega)] \neq H(st[t],\omega)\} = 0 \quad a.s. \quad \omega.$$

Since we do not yet have an analog of the measure υ from the last section, we define local summability with the standard part.

(7.4.3) DEFINITION:

Let M *be a* δt*-local martingale with* δt*-reducing sequence* $\{\tau_m\}$ *and* $M(\delta t) \approx 0$ *a.s. An internal 0-predictable process* G *is called locally* δM^2*-summable provided that there is an increasing sequence of* δt*-stopping times* $\{\sigma_m\}$ *satisfying:*

(a) $\sigma_m \le \tau_m$

(b) $st[M(\sigma_m)] = \tilde{M}(st[\sigma_m])$ *a.s.*

(c) $S\text{-}\lim \sigma_m = \infty$ *a.s.*

(d) $$st\ E\{\sqrt{\sum[\,|G(s,\omega)|^2|\delta M(s,\omega)|^2 : 0 < s < \sigma_m(\omega)]}\}$$

$$= E\{\sqrt{\int_{\{s<\sigma_m(\omega)\}} st|G(s,\omega)|^2\, d\lambda_\omega(s)}\} < \infty.$$

Our first result should be a "closure law" for stochastic sums. Unfortunately, we cannot conclude that internal sums have δt-decent paths without the stronger martingale lifting of section (7.6). However, we will now show that $N(t) = \Sigma^t\, G\cdot\delta M$ has all the other properties of a δt-local martingale.

The maximal function

$$N^\delta(\sigma_m) \quad \text{is S-integrable}$$

by (7.2.5) because we have assumed by (d) that

$$\sqrt{[\delta N, \delta N](\sigma_m)} \quad \text{is S-integrable.}$$

Without knowing that N has a δt-decent path sample, we can still show that

$$st[N(\sigma_m)] = S\text{-}\lim_{t \downarrow \sigma_m} N(t) \quad \text{a.s.}$$

First we show that

$$st[\delta N, \delta N](\sigma_m) = [\tilde{N}, \tilde{N}](st\ \sigma_m) \quad \text{a.s.}$$

We have the inequality:

$$\begin{aligned}
&[\tilde{N}, \tilde{N}](st\ \sigma_m) - st[\delta N, \delta N](\sigma_m) \\
&\quad = S\text{-}\lim_{\Delta t \downarrow 0}[\delta N, \delta N](\sigma_m + \Delta t) - [\delta N, \delta N](\sigma_m) \\
&\quad \leq S\text{-}\lim_{\Delta t \downarrow 0} \Sigma[\,|G(s)|^2 |\delta M(s)|^2 : \sigma_m \leq s < \sigma_m + \Delta t] \\
&\quad = \lim_{\Delta t \downarrow 0} \int_{\{\sigma_m \leq s < \sigma_m + \Delta t\}} st|G(s)|^2 d\lambda_\omega(s).
\end{aligned}$$

The final integral in this inequality tends to zero because we have assumed that $\tilde{M}(st\ \sigma_m) = st\ M(\sigma_m)$ a.s.

(7.4.4) **EXERCISE:**

If N *is a* **martingale after* δt *and* σ *is a* δt*-stopping time such that* $N^\delta(\sigma)$ *is* S*-integrable and*

$$[\tilde{N}, \tilde{N}](st\ \sigma) = S\text{-}\lim_{\Delta t \downarrow 0}[\delta N, \delta N](\sigma + \Delta t), \quad a.s.,$$

then

$$st[N(\sigma)] = S\text{-}\lim_{\Delta t \downarrow 0} N(\sigma + \Delta t) \quad a.s.$$

HINT:

This is a "converse" to Lemma (7.2.8). The ideas in that proof can be used here.

This exercise together with the preceding remarks mean that if we show that N has a δt-decent path sample, then N is a δt-local martingale.

The next result implies that stochastic sums are independent of the choice of lifting, but applies to more general internal summands.

(7.4.5) **PROPOSITION:**

Let M *be a* δt*-local martingale as above and suppose that* G_1 *and* G_2 *are locally* δM^2*-summable and that* $\lambda_\omega\{t \in \mathcal{O} \mid st\, G_1(t,\omega) \neq st\, G_2(t,\omega)] = 0$ *a.s.* ω. *Then except for a single null set of* ω*'s, for all finite* $t \in \mathbb{T}_\delta$,

$$\Sigma^t\, G_1(s,\omega)\delta M(s,\omega) \approx \Sigma^t\, G_2(s,\omega)\delta M(s,\omega).$$

PROOF:

If σ_m^1 and σ_m^2 are the stopping times in the summability conditions for G_1 and G_2, let $\sigma_m = \sigma_m^1 \wedge \sigma_m^2$. The σ_m satisfy the summability conditions for both G_1 and G_2.

Let $N_j(t) = \Sigma^t\, G_j(s,\omega)\delta M(s,\omega)$, for $j = 1,2$, define *martingales. The BDG Inequality (7.2.6) says that

$$E[\max_{\delta t \leq t \leq \sigma_m} |N_1(t)-N_2(t)] \leq CE[\sqrt{\sum^{\sigma_m} |G_1-G_2|^2 |\delta M|^2}].$$

We can see that the right hand side of this inequality is

infinitesimal by showing that the internal function $\sqrt{\sum^{\sigma_m} |G_1 - G_2|^2 |\delta M|^2}$ is P-S-integrable.

Summability of G_j means that $[\delta N_j, \delta N_j](\sigma_m)$ is S-integrable. By (7.2.5) this means that the maximal functions $N_j^\delta(\sigma_m)$ are P-S-integrable. Therefore, $\max_{\delta t \leq t \leq \sigma_m} |\Sigma(G_1 - G_2)\delta M| \leq N_1^\delta(\sigma_m) + N_2^\delta(\sigma_m)$ is S-integrable. Applying (7.2.5) again, we see that $\sum^{\sigma_m} |G_1 - G_2|^2 |\delta M|^2$ is P-S-integrable, so

$$\text{st } E\left[\sqrt{\sum^{\sigma_m} |G_1 - G_2|^2 |\delta M|^2}\right] = E\left[\text{st } \sqrt{\sum^{\sigma_m} |G_1 - G_2|^2 |\delta M|^2}\right].$$

Finally, condition (d) of the summability hypothesis says that a.s. ω,

$$\begin{aligned}
\text{st } \sum^{\sigma_m} |G_1 - G_2|^2 |\delta M|^2 &= \text{st}\left[\sum |G_1|^2 |\delta M|^2 - 2\sum |G_1 G_2| |\delta M^2| + \sum |G_2|^2 |\delta M|^2\right] \\
&= \int \text{st} |G_1|^2 d\lambda_\omega - 2\int \text{st} |G_1 G_2| d\lambda_\omega + \int \text{st} |G_2|^2 d\lambda_\omega \\
&= 0.
\end{aligned}$$

Hence the right side of the BDG Inequality is infinitesimal. This concludes the proof, since $\sigma_m \to \infty$ a.s.

We also have a convergence in measure result similar to the last section.

(7.4.6) PROPOSITION:

Let M *be a* δt*-local martingale as above. Suppose that the functions* F, G *and the sequence* $\{F_k\}$ *are locally* δM^2*-summable all with respect to the same stopping times* $\{\sigma_m\}$*. If* $|F_m| \leq |G|$ *and for each* m *and each standard* $\epsilon > 0$,

$$\text{st } P[\lambda_\omega\{t \leq \sigma_m : |F_k - F| > \epsilon\} > \epsilon] \longrightarrow 0 \quad \textit{as} \quad k \longrightarrow \infty,$$

then for each finite u,

$$\text{st} \max_{\delta t \leq t \leq u} [\Sigma^t F_k \delta M - \Sigma^t F \delta M] \longrightarrow 0$$

in probability.

PROOF:

Extend $\{F_k\}$ to an internal sequence satisfying $|F_k| \leq |G|$. For each $m \in \mathbb{N}$ and $\epsilon \in \mathbb{Q}^+$ there is a finite n_1 and an infinite n_2 such that

$$P[\lambda_\omega\{t < \sigma_m : |F_k - F| > \epsilon\} > \epsilon] < \epsilon$$

for every k between n_1 and n_2. By saturation, all sufficinetly small infinite k satisfy these inequalities for all standard m and ϵ. Hence, we may apply (7.4.5) to prove our claim (along the lines of the analogous result from the last section).

We shall not prove a general lifting theorem, but the following result may be proved along the same lines as the decent path part of (7.3.8). This in turn has an interesting application.

(7.4.7) PROPOSITION:

Let M be a δt-local martingale as above. Suppose that G is a locally δM^2-summable process which lifts a standard previsible process H, that is,

$$\lambda_\omega\{t \in \mathcal{O} \mid st[G(t,\omega)] \neq H(st[t],\omega)\} = 0 \quad a.s. \quad \omega.$$

Then the stochastic Stieltjes sum

$$\Sigma^t G \cdot \delta M$$

a.s. has a δt-decent path sample.

This "half" of the standard stochastic integration theorem for local martingales can be applied to the internal quadratic variation process as follows.

(7.4.8) LEMMA

If the pair $L = (M,N)$ is a 2d-dimensional δt-local martingale, then N is a locally δM^2-summable δM^2-path lifting of the left limit process

$$\tilde{N}(r^-) = \underset{t \uparrow r}{S\text{-lim}}\, N(t) = \underset{\sim}{N}(r).$$

PROOF:

Let τ_m be a δt-reducing sequence for (M,N). Since $N^{\delta}(\tau_m-\delta t) \lesssim m$ on $\{\tau_m > \delta t\}$ and $\sqrt{[\delta M,\delta M](\tau_m)}$ is S-integrable,

$$E\left[\sqrt{\sum(|N(t)|^2|\delta M(t)|^2 : 0 < t < \tau_m)}\right]$$
$$\approx E\left[\sqrt{\int_{\{0<t<\tau_m\}} st|N(t)|^2 d\lambda_{\omega}(t)}\right],$$

or N is δM^2-summable up to τ_m.

Since $L = (M,N)$ has a δt-decent path sample a.s. if $\delta L(t-\delta t) \not\approx 0$, then $st\ M(t) = \tilde{M}(st(t))$ a.s. Similarly, if we define stopping times σ_j^i = "$i^{\underline{th}}$ time t in $\mathbb{T}_\delta$ that $|\delta L(t-\delta t)| > \frac{1}{j}$," for finite $i,j \in \mathbb{N}$, then if $\sigma_j^i << \infty$, $st\ M(\sigma_j^i) = \tilde{M}(\tilde{\sigma}_j^i)$ a.s. so by (7.2.8), $st[\delta M,\delta M](\sigma_j^i) = \operatorname{S-lim}_{\Delta t\downarrow 0} [\delta M,\delta M](\sigma_j^i+\Delta t)$. Also, if $t \in \mathbb{T}_\delta$ is finite and $\delta N(t-\delta t) \not\approx 0$ on a decent path, then $t = \sigma_j^i$ for some finite i,j. Therefore, we a.s. have

$$\lambda_{\omega}\{t \in \mathcal{O} \mid st\ N(t) \neq \underset{\sim}{N}(st(t))\}$$
$$\leq \lambda_{\omega}\{t \in \mathcal{O} \mid t \approx \sigma_j^i(\omega)\ \&\ t \geq \sigma_j^i(\omega) \text{ for some } i,j \in \mathbb{N}\}.$$

We know that

$$\lambda_\omega\{t \in \mathcal{O} \mid t \approx \sigma^i_j(\omega) \ \& \ t \geq \sigma^i_j(\omega)\}$$

$$= \underset{\Delta t \downarrow 0}{\text{S-lim}} \ [\delta M, \delta M](\sigma^i_j(\omega)+\Delta t) - [\delta M, \delta M](\sigma^i_j(\omega))$$

$$= 0 \quad \text{a.s.}$$

by the remarks above. Hence $N(t)$ is a δM^2-path lifting of $\underset{\sim}{N}(r)$.

(7.4.9) **THEOREM:**

If $L = (M,N)$ *is a 2d-dimensional* δt*-local martingale, then* $[\delta M, \delta N]$ *has a* δt*-decent path sample.*

PROOF:

This follows easily from (7.4.8) and (7.4.7) together with the (*transform of the finite) formula for summation by parts:

$$[\delta M, \delta N](t) = (M(t), N(t)) - \Sigma^t(N, \delta M) - \Sigma^t(M, \delta N).$$

(7.5) Notes on Continuous Martingales

In this section we present a few basic facts about local hypermartingales with continuous paths and the corresponding internal objects. The basic references for this material are Keisler [1984], Panetta [1978], Lindstrom [1980a] and especially Hoover & Perkins [1983], section 8, part II, which has the strongest results.

The first result says that sampling in order to obtain nice path properties of a lifting is not necessary.

(7.5.1) THEOREM:

Let $\delta t = \min \mathbb{T}^+$. *If* M *is a local hypermartingale with a.s. continuous paths and* $M(0) = 0$, *then there is a* δt*-local martingale* N *with a.s. S-continuous paths such that the continuous path projection* $\tilde{N}$ *is indistinguishable from* M.

The next result of Hoover & Perkins [1983] says continuity of M can be measured by the quadratic variation [in contrast to the decent path case, see (7.6.2)]. It extends results in Keisler [1984] and Lindstrom [1980a].

(7.5.2) THEOREM:

Let M *be a* **martingale and* $\delta t = \min \mathbb{T}^+$. *Then* M *is S-continuous and locally S-integrable if and only if* $\sqrt{[\delta M, \delta M]}$ *is S-continuous and locally S-integrable.*

This has the following important consequence.

(7.5.3) **COROLLARY:**

Suppose that M *is an* S-*continuous* δt-*local martingale and* G *is a locally* δM^2-*summable internal process. Then* $\Sigma^t G \cdot \delta M$ *is an* S-*continuous* δt-*local martingale.*

We add that G need not be the lifting of a standard process [compare to (7.3.9).] This idea plays a key role in Keisler's existence theorem for stochastic differential equations mentioned below. Since G may be internal, solutions of internal difference equations have standard parts which satisfy an associated stochastic differential equation.

The criterion in (7.5.6) for continuity plays a role in constructing strong martingale liftings which satisfy the decent path version of Corollary (7.5.3).

(7.5.4) **NOTATION:**

Suppose $\delta t \in \mathbb{T}$ *is a positive infinitesimal and* X *is nonanticipating after* δt. *We define*

$$X|_\delta(t) = X(\delta t) + \sum_{\substack{s=\delta t \\ \text{step } \delta t}}^{t-\delta t} E[\delta X(s)\,|\,\omega^s], \quad \textit{for} \quad t \in \mathbb{T}_\delta^+.$$

(7.5.4) **EXAMPLE:**

Let $Z : W \longrightarrow {}^*\mathbb{R}$ be an internal function with zero mean, $\Sigma[Z(w) : w \in W] = 0$, and limited variance, $\sigma^2 = \Sigma[Z^2(w) : w \in W]/{}^\#[W]$. Define a *martingale by

$$M(t) = \sum_{s=\delta t}^{t-\delta t} Z(\omega_{s+\delta t})\sqrt{\delta t},$$

where $\delta t = \min \mathbb{T}^+$. In this case $|\delta M(t)|^2 = Z^2(\omega_{t+\delta t})\delta t$, but $E[|\delta M(t)|^2 \mid \omega^t] = \sigma^2 \cdot \delta t$, so

$$[\delta M, \delta M]|_\delta(t) = \sigma^2 \cdot t,$$

while $[\delta M, \delta M](t)$ itself is usually not so easily computed.

With some extra integrability, Hoover & Perkins [1983] estimate $[\delta M, \delta M]$ as follows.

(7.5.5) **THEOREM:**

If M *is a* δt*-martingale,* M^2 *is locally S-integrable and* $\sup_{t \in \mathcal{O}}[\mathrm{st}|\delta M(t)|] = 0$, *a.s., then*

$$[\delta M, \delta M](t) \approx [\delta M, \delta M]|_\delta(t) \quad \text{for} \quad t \in \mathcal{O} \quad \text{a.s.}$$

The following result uses $[\delta M, \delta M]|_\delta(t)$ to check continuity.

(7.5.6) **THEOREM:**

Let M *be a* δt*-local martingale. If* $[\delta M, \delta M]|_\delta(t)$ *is a.s. S-continuous and if* $\sup_{t \in \mathcal{O}}[\mathrm{st}|\delta M(t)|] = 0$ *a.s., then* M *is S-continuous.*

(7.6) Stable Martingale Liftings

Hoover & Perkins [1983] show that coarse enough δt-martingales have the property that all their Stieltjes sums also have a δt-decent path sample. This section outlines the ingredients of their result in the hyperfinite framework. Example (7.6.2) helps explain why coarser time sampling is necessary. Roughly speaking, the idea is to decompose an internal martingale into continuous and discontinuous martingales. Since the internal time line is discrete this means we "take a limit."

Let M be a δt-local martingale and let m be a natural number. We define

$$M_m(t) = \sum_{s=\delta t}^{t-\delta t} \delta M(s) I_{\{|\delta M(s)| \leq \frac{1}{m}\}}$$

and

$$M^m(t) = \sum_{s=\delta t}^{t-\delta t} \delta M(s) I_{\{|\delta M(s)| > \frac{1}{m}\}}.$$

The conditioned processes [see (7.5.4)]

$$M_m(t) - M_{m|\delta}(t)$$

and

$$M^m(t) - M^m{}_{|\delta}(t)$$

are each *martingales and

$$M(t) = M(\delta t) + [M_m(t) - M_{m|\delta}(t)] + [M^m(t) - M^m{}_{|\delta}(t)].$$

In the limit as $m \longrightarrow \infty$ we expect $M^m(t)$ to contain all the jumps of the martingale, so $M^m{}_{|\delta}(t)$ should "tend toward a continuous process." Hoover & Perkins [1983] show that when m is infinite, $[M_m(t)-M_{m|\delta}(t)]$ is an S-continuous process by using Theorem (7.5.6) above. This means that $[M_m(t)-M_{m|\delta}(t)]$ tends to a continuous process as $m \longrightarrow \infty$. We shall formulate the limit condition on $M^m{}_{|\delta}(t)$ more precisely.

If X is any internal process we denote the infinitesimal oscillation of X up to t by

$$^{o}X(t) = \sup[st|X(u)-X(v)| : u \approx v < t].$$

We want to have

$$^{o}[\delta \text{Var } M^m{}_{|\delta}](t) \longrightarrow 0 \text{ in probability, as } m \rightarrow \infty,$$

for each limited t. Unfortunately, this does not always happen. However, there is always a coarser infinitesimal $\Delta t \in \mathbb{T}_\delta$ which makes this happen. A sample of M along the coarser time line $\mathbb{T}_\Delta$ produces only decent path sums. Recall the special form of a coarser sample of a local martingale from (6.7.6).

(7.6.1) **DEFINITION:**

Let M *be a* δt*-martingale. If* $\Delta t \in \mathbb{T}_\delta$ *is an infinitesimal such that*

$$^{o}[\Delta \text{Var } M^m{}_{|\delta}](t) \longrightarrow 0$$

in probability for all limited t, *then we say that a* Δt*-sample of* M *is a stable* Δt*-local martingale.*

Our next example should help to clarify the situation. Here is the outline of what the example contains: M is a δt-martingale, G is an internal bounded predictable summand, yet $N(t) = \sum^{t} G \cdot \delta M$ has indecent paths. Moreover, quadratic variation does not detect the problem because $[\delta M, \delta M] = [\delta N, \delta N]$. A coarser sample of M is stable.

(7.6.2) EXAMPLE:

Let $\upsilon : W \longrightarrow \{0,1\}$ be an internal function such that

$$\frac{\upsilon^{-1}(1)}{\#_W} = \sqrt{\delta t} + \iota \cdot \delta t = p \ , \quad \text{for} \quad \iota \approx 0.$$

We may think of υ as Bernoulli trials with an infinite success rate starting at $t = 1$ by summing "successes" as

$$\sum [\upsilon(\omega_s) \ : \ 1 < s].$$

Recall that our infinitesimal approximation to the Poisson process in (0.3.7), (4.3.3) and (5.3.8) had a jump rate a when

$$\frac{\pi^{-1}(1)}{\#_W} = b \cdot \delta t \ , \quad \text{with} \quad b \approx a.$$

Now we have

$$\frac{\upsilon^{-1}(1)}{\#_W} = b \cdot \delta t \ , \quad \text{with} \quad b \approx \frac{1}{\sqrt{\delta t}}.$$

Define a δt-stopping time for the first "success,"

$$\tau(\omega) = \min\{t : \sum[\upsilon(\omega_s) : 1 < s \leq t] \geq 1\} \wedge n.$$

Then the probability of beginning with at least k "failures" is

$$P[\tau-1 > k\cdot\delta t] = (1-p)^k$$

or

$$P[\tau-1 > t] = (1-p)^{t/\delta t}.$$

This means that "success" happens within a noninfinitesimal limited multiple of $\sqrt{\delta t}$,

$$P[0 \not\approx \frac{\tau-1}{\sqrt{\delta t}} \in \mathcal{O}] = 1.$$

To see this, observe that if $t = r\sqrt{\delta t}$, for r limited, then

$$P[\tau-1 > r\sqrt{\delta t}] = [1-(\sqrt{\delta t} + 2\cdot\delta t)]^{r/\sqrt{\delta t}} \approx e^{-r}.$$

Now we define a δt-martingale using τ. Let $M(t,\omega) = 0$, for $t \leq 1$, and $M(t) = \sum^t \delta M(s)$, for $t > 1$, where

$$\delta M(t-\delta t,\omega) = \begin{cases} (-1)^{k+1}\, p \ , & \text{if } \ t = 1+k\delta t < \tau(\omega) \\ (-1)^k(1-p) \ , & \text{if } \ t = 1+k\delta t = \tau(\omega) \\ 0 \ , & \text{if } \ t > \tau(\omega). \end{cases}$$

Notice that $\delta M = \pm\sqrt{\delta t} + o(\delta t)$ until the first success, so $[\delta M,\delta M](1+t) \approx t$ until the first success, $1+t = \tau$. However, we have seen that $\tau \approx 1$ a.s., since τ is a.s. a finite multiple of $\sqrt{\delta t}$. Hence

$$[\tilde{M},\tilde{M}](r) = \begin{cases} 0, & r < 1 \\ 1, & r \geq 1 \end{cases}.$$

Next, we define a *martingale

$$N(t) = \sum^{t} G \cdot \delta M,$$

where G is deterministic and bounded,

$$G(1+k\delta t) = (-1)^k.$$

This makes

$$\delta N(t-\delta t) = \begin{cases} (-1)^{k-1}(-1)^{k+1}p = p &, \text{ if } t = 1+k\delta t < \tau(\omega) \\ (-1)^{k-1}(-1)^{k}(1-p) = p-1 &, \text{ if } t = \tau(\omega) = 1+k\delta t \end{cases}.$$

We know that $p = \sqrt{\delta t} + o(\delta t)$ and $\tau(\omega) = 1+r\sqrt{\delta t}$ for some noninfinitesimal limited r a.s., hence $\tau \approx 1$ and

$$N(1) = 0$$

$$N(\tau-\delta t) \approx r$$

$$N(\tau) \approx r-1.$$

Therefore, N does not have a δt-decent path sample because it first builds up to r and then jumps down by 1 for times infinitely near $t = 1$.

The quadratic variation of N is the same as that of M,

$$[\delta M, \delta M](t) \equiv [\delta N, \delta N](t),$$

because $|\delta M| = |\delta N|$ for all t.

Finally, we compute the decomposition

$$M(t) = [M_m(t) - M_{m|\delta}(t)] + [M^m(t) - M^m{}_{|\delta}(t)].$$

This decomposition is the same for all values of $m \in {}^*\mathbb{N}$ such that $2 \leq m < \frac{1}{p}$, so we simply let $m = 2$. First,

$$\delta M(s) I_{\{|\delta M(s)| \leq \frac{1}{2}\}} = \begin{cases} (-1)^h p, & \text{if } s = 1+h \cdot \delta t < \tau - \delta t \\ 0, & \text{otherwise} \end{cases}$$

so that

$$E[\delta M(s) I_{\{|\delta M(s)| \leq \frac{1}{2}\}} | \omega^s] = (-1)^{(s-1)/\delta t} p(1-p), \quad \text{if } 1 < s < \tau.$$

If we denote the martingale

$$X(t) = M_m(t) - M_{m|\delta}(t) \qquad (m \geq 2)$$

then

$$\delta X(s) = \begin{cases} (-1)^{(s-1)/\delta t} p^2, & \text{for } 1 < s < \tau \\ 0, & \text{otherwise,} \end{cases}$$

hence

$$X(t) \approx 0 \quad \text{for all} \quad t \quad \text{a.s.}$$

The other part of the decomposition of M satisfies

$$\delta M(s) I_{\{|\delta M(s)|> \frac{1}{2}\}} = \begin{cases} (-1)^{(s-1)/\delta t}(p-1), & \text{if} \quad s = \tau - \delta t \\ 0, & \text{otherwise} \end{cases}$$

while

$$E\left[\delta M(s) I_{\{|\delta M(s)|> \frac{1}{2}\}} \middle| \omega^s\right] = \begin{cases} (-1)^{(s-1)/\delta t}(p-1)p, & \text{if} \quad s < \tau \\ 0, & \text{otherwise.} \end{cases}$$

This means that the variation in steps of δt satisfies

$$\delta \text{Var } M^m|_\delta(\tau) = \sum_{1+\delta t}^{\tau-\delta t} (1-p)p \approx \frac{\tau - 1}{\sqrt{\delta t}} \not\approx 0 \quad \text{a.s.}$$

On the other hand, if we sample $M^m|_\delta$ on a time axis $\mathbb{T}_\Delta$, where $\Delta t \in \mathbb{T}_\delta$ satisfies $\frac{\sqrt{\delta t}}{\Delta t} \approx 0$, then

$$\Delta \text{Var } M^m|_\delta(\tau) \approx 0 \quad \text{a.s.}$$

and a Δt-sample of M makes it a stable Δt-martingale. In general, such coarser time sampling always works.

(7.6.3) **LEMMA:**

Let M be a δt-local martingale. There is an infinitesimal $\Delta t \in \mathbb{T}_\delta$ so that a Δt-sample of M is a stable Δt-local martingale.

A value of Δt that makes the standard part of the variation equal the variation of the standard part always exists by results in section 7.1. Such a Δt satisfies the lemma above.

The point of this lemma together with the Local Martingale Lifting Theorem is that every local hypermartingale has a stable Δt-local martingale lifting. These are the liftings that Hoover & Perkins [1983] show make internal Stieltjes sums have decent paths. (This is why we called them "stable.")

(7.6.4) THE HOOVER-PERKINS THEOREM:

If M *is a stable* Δt*-local martingale and* G *is a locally* ΔM^2*-summable process, then* $N(t) = \sum^t G\cdot\Delta N$ *is a* Δt*-local martingale.*

This is the main technical result of the Hoover & Perkins [1983] article. It is applied to give a new existence theorem for semimartingale equations. The decent path property is needed to show that the internal sum arising from the solution of a *finite difference equation has a standard part.

(7.7) Semimartingale Integrals

We want to define integrals

$$\int_0^r H(s,\omega)dZ(s,\omega)$$

for a wide class of integrands and 'differentials'. About the most general kind of process Z that we can use for an Itô-calculus is defined in (7.7.3). The term "semimartingale" is not completely standardized in the literature and our use of it is relative to our own evolution scheme. (We used "hypermartingale" to warn you of the latter above.) Our usage is close to a predominant custom and even *our* sense of humor couldn't bear 'local-semi-hyper'...

A semimartingale Z is a process which may be written as

$$Z(r) = Z(0) + N(r) + W(r)$$

where N is a martingale, with $N(0) = 0$ and W has bounded variation with $W(0) = 0$. For the time being let us assume that we work on $r \in [0,1]$ where $N^2(1)$ and $\text{var } W(1)$ are both integrable. The complete carefully developed theory of sections (7.1) and (7.3) applies to the Z-integration of a bounded previsible process. Let M be a lifting of N as in (7.2-3) and let U be a lifting of W as in (7.1). By first choosing a Δt-martingale lifting M and then choosing U with δt-bounded variation for $\delta t \in \mathbb{T}_\Delta$, we may suppose that M is a δt-martingale with $[\delta M, \delta M](1)$ S-integrable and U has δt-bounded variation with $\delta \text{Var } U(1)$ S-integrable.

We may combine the path variation measure for M and U. Define the weight functions

$$\delta\kappa_\omega(t) = \begin{cases} |\delta U(t,\omega)| , & t \in \mathbb{T}_\delta \\ 0 , & t \in \mathbb{T}\setminus\mathbb{T}_\delta \end{cases} \qquad \delta\lambda_\omega(t) = \begin{cases} |\delta M(t,\omega)|^2 , & t \in \mathbb{T}_\delta \\ 0 , & t \in \mathbb{T}\setminus\mathbb{T}_\delta \end{cases}$$

and

$$\delta\mu_\omega(t) = \delta\kappa_\omega(t) + \delta\lambda_\omega(t).$$

In this case $\mu_\omega[\mathbb{T}]$ is P-S-integrable, so Iterated Integration (7.1.5) applies to the hyperfinite measure υ given by the internal weight function

$$\delta\upsilon(t,\omega) = \delta\mu_\omega \cdot \delta P(\omega).$$

The Predictable Lifting Theorems (6.6.8) and (6.6.11) may be applied to υ and a bounded previsible process H, yielding a bounded 0-predictable internal G satisfying

$$\upsilon\{st[G(t,\omega)] \neq H(st[t],\omega)\} = 0.$$

Since this joint total variation measure υ dominates the path measures of sections (7.1) and (7.3), G is both δM^2-summable and δU-summable and

$$\int_0^r HdZ$$

is well-defined as $H(0)\cdot Z(0)$ plus the decent path projection of

$$\sum_{\delta t}^{t} G\cdot(\delta M+\delta U).$$

There is something extra to prove in this definition. The theory of sections (7.1) and (7.3) only shows that the infinitesimal Stieltjes sums give the same answer for different liftings of N, W and H.

The decomposition of a semimartingale Z into martingale plus bounded variation terms is *not unique*. If Z is a martingale of bounded variation [such as $Z(r) = \tilde{J}(r)-\lambda r$ for a Poisson process as in (5.3.8), (4.3.3), (0.3.7)] then we may view either N or W as zero in the decomposition above. [If Z is continuous, then N and W may also be chosen continuous and then are unique.] This means that Stieltjes and martingale integrals must agree when both are defined since classically one defines integrals against dZ by $\int dN + \int dW$. This causes some trouble in the classical approach which we at least avoid conceptually since we use infinitesimal Stieltjes sums for both parts of the lifting, $\Sigma\delta X = \Sigma\delta M + \Sigma\delta U$. (We do still use separate estimates for the two terms.)

The nonuniqueness in the decomposition $Z = Z(0) + N + W$ causes a far more irritating problem when it comes to trying to identify a space of dZ-integrable processes.

(7.7.1) **EXAMPLE**:

Let $\beta : W \longrightarrow \{-1,+1\}$ be internal and equal -1 on one half of W and $+1$ on the other half (see 4.3.2). We know that $\frac{1}{2}, \frac{2}{3}, \frac{3}{4},\cdots,\frac{m-1}{m} \in \mathbb{T}$ for each finite m, so let n be an infinite natural number such that $\{\frac{m-1}{m} : m \leq n\} \subseteq \mathbb{T}$. Define a

δt-martingale by $M(0) = 0$ and $M(t) = \Sigma^t \delta M$, where

$$\delta M(t,\omega) = \begin{cases} \beta(\omega_{t+\delta t})\frac{1}{m^2}, & \text{if } t = \frac{m-1}{m}, \quad m \leq n \\ 0, & \text{otherwise.} \end{cases}$$

Since $\Sigma^1 |\delta M| \approx \Sigma^\infty \frac{1}{m^2} < \infty$, we see that M *has* bounded variation. The semimartingale $X(t) = \tilde{M}(t)$ may be decomposed as $X(t) = 0 + \tilde{M}(t) + 0$ or as $X(t) = 0 + 0 + \tilde{M}(t)$. The deterministic internal function

$$G(t) = \begin{cases} m, & \text{if } t = \frac{m-1}{m}, \quad m \leq n \\ 0, & \text{otherwise} \end{cases}$$

is not S-integrable with respect to the first variation $|\delta M|$, because $\Sigma^1 G|\delta M| = \Sigma^n \frac{1}{m}$ is infinite. However, G is S-integrable with respect to the quadratic variation,

$$\sum^1 |G|^2 |\delta M|^2 = \sum^n \frac{1}{m^2} \text{ for all } \omega$$

and

$$\int^1 st|G|^2 d\lambda_\omega = \sum^\infty \frac{1}{m^2}.$$

It turns out that semimartingale integrability is well-defined by saying that a process H is integrable if it is integrable with respect to *some* decomposition. For another decomposition H may not be integrable, but if it is integrable, then it produces an indistinguishable integral. In other words, given another decomposition of Z, either we get the same answer or "infinity." This is the point of (7.7.10). Strange, but at least consistent.... A simple special case of this may be proved as follows.

Suppose that $Z(r) = 0 + N_1(r) + W_1(r) = 0 + N_2(r) + W_2(r)$ and that (M_j, U_j) lifts (N_j, W_j), for $j = 1,2$, as above. We associate total variation measures υ^1 and υ^2 to each lifting for infinitesimal increments δt_1 and δt_2. If H is an almost previsible process satisfying

$$E[\int_0^1 |H(s,\omega)|^2 d[N_j,N_j](s,\omega) + \int_0^1 |H(s,\omega)|\cdot|dW_j|(s,\omega)] < \infty,$$
$$j = 1,2,$$

Then we can find 0-predictable internal processes G_j such that

$$\begin{aligned} \text{st } E[\sum^1 |G_j|^2|\delta M_j|^2 + \sum^1 |G_j|\cdot|\delta U_j|] \\ = E[\int^1 \text{st}|G_j|^2 d\lambda_{\omega}^j + \int^1 \text{st}|G_j|\cdot d\kappa_{\omega}^j], \quad \text{for } j = 1,2. \end{aligned}$$

This just requires a slightly different truncation argument and the bounded Predictable Lifting Theorem (6.6.8). This shows that H has liftings for both decompositions.

The decent path property of the infinitesimal Stieltjes sums can be proved in the same manner as the decent path part of the proof of (7.3.8). Finally, the indistinguishability can be proved in the style of the proof of (7.3.10). In fact, these two parts can be combined in a single proof using (6.6.12 - 6.6.14) where we let $\mathbb{V}$ be the set of bounded almost previsible processes which have liftings for each decomposition producing δt_j-decent path sums and indistinguishable projections. The details are left as an exercise.

We hope that the separate treatments of sections (7.1) and (7.3) are clearer even than a combined treatment of "square summable plus integrable variation" semimartingales. For the remainder of this section we simply state the full-blown general results needed to define semimartingale integrals by lifting to infinitesimal Stieltjes sums. Further details must be found in Hoover & Perkins [1983].

(7.7.2) **SEMIMARTINGALE INTEGRALS:**

Here is a summary of the effect of this section on classical semimartingale integrals. If Z with $Z(0) = 0$ is an $\mathcal{F}$-semimartingale and H is an almost-previsible Z-integrable process, then the decomposition that makes H integrable has a δt-semimartingale lifting $X(t) = 0 + M(t) + U(t)$ and H has a $(\delta M^2, \delta U)$-summable lifting G. The decent path projection of the δt-semimartingale

$$Y(t) = \sum^t G\delta X$$

is what we take as our definition of the process

$$\int_0^r H(s)dZ(s) = \tilde{Y}(r).$$

A slight variation on Theorem (7.3.8) shows that the decent path projection exists and moreover that Y a.s. only jumps where X does on $\mathbb{T}_\delta$. Theorem (7.7.10) shows that this definition is well-defined up to indistinguishability, that is, the decent path projection is the same no matter which decomposition lifting, which integrand lifting G, or which infinitesimal time sample (subject to all the "lifting" requirements) we take.

We want to study processes $W : [0,\infty) \times \Omega \longrightarrow {}^*\mathbb{R}^d$ with *locally bounded variation*. This means that except for a single null set of ω's, the paths of W have finite classical variation up to any $r \in [0,\infty)$, or that the process $\text{var}\, W(r,\omega) : [0,\infty) \times \Omega \longrightarrow \mathbb{R}$ is defined for a.a. paths.

We extend the notation (7.1.1) to the setting (5.5.4) with the only formal change that restrictions at $r = 1$ are dropped. We need a definition of bounded variation on each interval $[0,r]$, $r < \infty$.

(7.7.3) DEFINITIONS:

We say that the internal process $U : \mathbb{T} \times \Omega \longrightarrow {}^*\mathbb{R}^d$ *has* δt*-locally bounded variaiton if* $(U, \delta \times \text{Var}\, U)$ *has a* δt*-decent path sample and its projection is indistinguishable from* $(\tilde{U}, \text{var}\, \tilde{U})$.

A process $Z : [0,\infty) \times \Omega \longrightarrow \mathbb{R}^d$ *is called a d-dimensional semimartingale if there exist progressively measurable decent path processes* N *and* W *with* $N(0) = W(0) = 0$ *such that* N *is a local hypermartingale,* W *has locally bounded variation, and*

$$Z(r) - Z(0) = N(r) + W(r).$$

An internal process $X : \mathbb{T} \times \Omega \longrightarrow {}^*\mathbb{R}^d$ *is called a d-dimensional* δt*-semimartingale for an infinitesimal* δt *if there exist internal processes* M *and* U *with* $M(\delta t) \approx U(\delta t) \approx 0$ *such that* M *is a stable* δt*-local martingale,* U *is nonanticipating after* δt *and has* δt*-locally bounded variation, the pair* (M,U) *has a* δt*-decent path sample a.s., and*

$$X(t) - X(\delta t) = M(t) + U(t)$$

for $t \in \mathbb{T}_\delta^+$.

An internal process X *is called a semimartingale lifting of a measurable process* Z *if* X *is a* δt*-semimartingale for some infinitesimal* δt *and if the* δt*-decent path projection* $\tilde{X}$ *is indistinguishable from* Z.

If X is a δt-semimartingale, the projection $\tilde{X}$ is clearly a semimartingale because the projected pair $(\tilde{M},\tilde{U})$ is the required decomposition.

If Z is a semimartingale, then there is a semimartingale lifting X for Z.

(7.7.4) THE SEMIMARTINGALE LIFTING AND PROJECTION THEOREM:

A process Z *is a semimartingale if and only if it has a semimartingale lifting* X.

Semimartingales are a "good" class of 'stochastic differentials' because they contain a wide class of traditionally important processes, are closed under integration by a wide class of integrands, and because they are also closed under change of variables in the sense of "Itô's formula." Moreover, there are several technical ways to say that semimartingales are the widest possible class of 'differentials', for example, see Metivier and Pellaumail [1980], 12.12, or Williams [1981], p. 68.

(7.7.5) **EXAMPLES**

We would like to indicate how one goes about verifying that the classical stationary independent increment processes are semimartingales. Suppose Z is such an adapted decent path process, that is, $Z(0) = 0$ and $Z(s) - Z(r)$ is independent of $\mathcal{F}(r)$ and its distribution is only a function of $(s-r)$. For example, we may have $Z(r) = \tilde{X}(r)$ where

$$X(t) = \sum^{t} \delta X$$

for a *independent family $\{\delta X(t) : t \in \mathbb{T}\}$ with $\delta X(t)$ a function of $\omega_{t+\delta t}$ as in (5.3.21) or (4.3.6). The process Z need not be integrable. There is an extensive classical theory of the characteristic functions of these processes which (4.3.4) hints at. By breaking Z up as follows (either measurably with Z or internally with X)

$$Z^b(r) = \text{sum of the jumps of } Z \text{ bigger than } b$$

and

$$Z_b(r) = Z(r) - Z^b(r)$$

we can show that the characteristic function of Z_b is smooth so that Z_b is integrable. We can also show that Z^b has bounded variation. If we let $c = E[Z_b(1)]$, then $[Z_b(r)-cr]$ is a hypermartingale. (We may even say Z_b has bounded variaiton if and only if $\int_{|x|\leq b} |x|\nu(dx) < \infty$ for the classical ν associated with the characteristic function.) The details would show why Z decomposes, $Z = N+W$ with $N(r) = Z_b(r)-cr$ and $W(r) = Z^b(r)+cr$; the discussion is only intended as background to justify the definition.

A more modern justification to single out the semimartingales is the form of the Doob-Meyer decomposition theorem that says a decent path submartingale decomposes, $Z(r) = Z(0) + N(r) + W(r)$, where N is a local martingale and W is increasing and previsible.

We need not use the two decomposition results we just have stated in our development. Now we will state another which we clearly don't use. If M is a local hypermartingale, we may write $M = N+W$ where N is a locally square integrable hypermartingale and W has locally bounded variation. The point of stating this result is to indicate why our seemingly more general local martingale integrals are no better than a square integrable theory once we combine each with Stieltjes integrals.

Let X be a δt-semimartingale lifting of Z. We shall define path liftings of integrands relative to a measure analogous to the one in (7.1.4).

(7.7.6) **NOTATION FOR SEMIMARTINGALE PATH MEASURES:**

Suppose that $X(t) = X(\delta t) + M(t) + U(t)$ is a decomposed δt-semimartingale and $\{\tau_m\}$ is the δt-reducing sequence for M as above. The *joint variation process* of the decomposition pair (M,U),

$$V_M^U(t,\omega) = [\delta M,\delta M](t,\omega) + \delta \mathrm{Var}\ U(t,\omega)\ , \quad \text{for} \quad t \in \mathbb{T}_\delta,$$

is finite a.s., but is not necessarily S-integrable for any t. However, the internal increasing family of δt-*stopping times*,

$$\sigma_n(\omega) = \min[t \in \mathbb{T}_\delta : V_M^U(t,\omega) > n \text{ or } t > n], \quad \text{for} \quad n \in {}^*\mathbb{N},$$

do tend to infinity, $\lim \tilde{\sigma}_n = \infty$ a.s. (as we see by examining $[\delta M,\delta M](\tau_m)$), and make $V_M^U(\sigma_n - \delta t) \le n$. The *unbounded joint variation path measure* η_ω of the pair (M,U) is given by the weight functions

$$\delta\kappa_\omega(t) = \begin{cases} |\delta U(t,\omega)|, & \text{for} \quad t \in \mathbb{T}_\delta \\ 0 \quad , & \text{for} \quad t \in \mathbb{T}\setminus\mathbb{T}_\delta \end{cases}$$

$$\delta\lambda_\omega(t) = \begin{cases} |\delta M(t,\omega)|^2, & \text{for} \quad t \in \mathbb{T}_\delta \\ 0 \quad , & \text{for} \quad t \in \mathbb{T}\setminus\mathbb{T}_\delta \end{cases},$$

and

$$\delta\eta_\omega(t) = \delta\kappa_\omega(t) + \delta\lambda_\omega(t).$$

The internal measures η_ω may very well be unbounded, or make $\eta_\omega[\mathbb{T}]$ infinite with noninfinitesimal probability. This complicates our use of lifting theorems, but may be technically remedied by a *series of truncations with the stopping times σ_n. The *bounded joint variation path measure* of the decomposition pair (M,U) is given by the internal series

$$\mu_\omega(S) = \sum [\frac{1}{n2^n} \eta_\omega\{S \cap [0,\sigma_n(\omega))\} : n \in {}^*\mathbb{N}].$$

We know $\mu_\omega(S) \leq \mu_\omega(\mathbb{T}) = \sum [\frac{1}{2^n} \{\frac{1}{n} V_M^U(\sigma_n(\omega))\} : n \in {}^*\mathbb{N}]$.

This measure is analogous to the path measure μ_ω of section (7.1), but the procedure for bounding η_ω into μ_ω is more complicated. We also define a *bounded total variation measure* of the pair (M,U) by

$$\nu(\mathcal{S}) = E[\mu_\omega(\mathcal{S}_\omega)],$$

where $\mathcal{S}_\omega$ is the section of an internal $\mathcal{S} \subseteq \mathbb{T} \times \Omega$. The measure ν is carried on $\mathbb{T}_\delta \times \Omega$ although we may consider it defined on $\mathbb{T} \times \Omega$, that is,

$$\nu(\mathcal{S}) = E[\mu_\omega(\{t \in \mathbb{T}_\delta : (t,\omega) \in \mathcal{S}\})].$$

Just as ν-liftings in section (7.1) ultimately told us something about almost surely-κ_ω-almost everywhere, ν-liftings now translate into a.s. properties of η_ω.

The next proposition shows how the truncation procedure taking η_ω to μ_ω works.

(7.7.7) **PROPOSITION:**

Let $X = M+U$, *etc. be as above. If* $\mathcal{S} \in \text{Loeb}(\mathbb{T}\times\Omega)$, *then* $\upsilon[\mathcal{S}] = 0$ *if and only if for almost all* ω

$$\eta_\omega\{\mathcal{S} \cap [\mathcal{O} \times \Omega]\} = 0.$$

The proof of the converse part of this proposition is fairly technical. The importance of the result is seen from the partial development of local martingale integrals in section (7.4). We may take a bounded υ-lifting of a standard process H and then apply the condition

$$\eta_\omega[\text{st}[G(t,\omega)] \neq H(\text{st}[t],\omega)] = 0 \quad \text{a.s.}$$

(7.7.8) **DEFINITION:**

Let $X(t) = 0 + M(t) + U(t)$ *be a* δt*-semimartingale (decomposed as above with* δt*-reducing sequence* $\{\tau_m\}$ *for* M). *An internal process* G *is* $(\delta M^2, \delta U)$*-summable if*

(a) G *is nonanticipating after* δt,

(b) *except for a single null set of* ω*'s, whenever* t *is finite (the variables* s *and* t *range over* $\mathbb{T}_\delta$),

$$\text{st} \sum[\,|G(s,\omega)|\,|\delta U(s,\omega)| \;:\; 0 < s < t]$$

$$= \int_{\{0<s<t\}} \text{st}|G(s,\omega)|\,d\kappa_\omega(s,\omega) < \infty$$

and there is an increasing sequence of δt*-stopping times* $\{\sigma_m\}$ *such that*

(c) $\sigma_m \leq \tau_m$,

(d) $\text{st } M(\sigma_m) = \tilde{M}(\tilde{\sigma}_m)$ a.s.,

(e) $\lim \tilde{\sigma}_m = \infty$ a.s.

and

$$\text{(f)} \quad \text{st } E\left[\sqrt{\Sigma(|G(s,\omega)|^2|\delta M(s,\omega)|^2 : 0 < s < \sigma_m(\omega))}\right]$$

$$= E\left[\sqrt{\int_{[s<\sigma_m(\omega)\}} \text{st}|G(s,\omega)|^2 d\lambda_\omega(s,\omega)}\right] < \infty.$$

Because of Proposition (7.7.7), the proof of (7.4.2) carries over almost intact to show the next result.

(7.7.9) PROPOSITION:

Let X *be a* d*-dimensional* δt*-semimartingale with* $X(\delta t) \approx 0$ *decomposed as above. If* F *and* G *are internal* $(\delta M^2, \delta U)$*-summable processes and*

$$\nu\{(t,\omega) : F(t,\omega) \not\approx G(t,\omega)\} = 0$$

then except for a single null set of ω*'s, for all infinite* t,

$$\sum^t F(t)\delta X(t) \approx \sum^t G(t)\delta X(t).$$

A convergence in ν-measure results like (7.3.7) and (7.4.6) can now be proved for semimartingale sums.

Since we have chosen M to be a stable δt-martingale the infinitesimal Stieltjes sums

$$\sum^{t} G \cdot \delta X$$

have δt-decent paths.

The peculiar "integrability" question described after Example (7.7.1) is taken care of by the next result.

(7.7.10) **THEOREM:**

Suppose that $X_1 = 0 + M_1 + U_1$ *and* $X_2 = 0 + M_2 + U_2$ *are both decomposed semimartingale liftings of* $\tilde{X}_1$. *Let* X_1 *be a* δt*-semimartingale while* X_2 *is a* Δt*-semimartingale. If* H *is an almost-previsible process which has a* $(\delta M^2, \delta U)$*-summable* υ^1*-lifting* G_1 *and a* $(\Delta M^2, \Delta U)$*-summable* υ^2*-lifting* G_2, *then the decent path projections of* $Y_1(s) = \Sigma^s\, G_1 \delta X$ *and* $Y_2(t) = \Sigma^t\, G_2 \Delta X$ *are indistinguishable.*

This result is proved using (6.6.12) - (6.6.14) as described in the special case following Example (7.7.1).

The whole standard theory of semimartingale integrals is summarized by

(7.7.11) **THE STOCHASTICALLY INTEGRABLE LIFTING THEOREM:**

Let $X(t)$ *be a d-dimensional* δt*-semimartingale which is a lifting of the* $\mathcal{F}$*-semimartingale* $Z(r)$. *An almost previsible process* $H : [0,\infty) \times \Omega \longrightarrow \mathbb{R}^{k\times d}$ *has a*

O-predictable $(\delta M^2, \delta M)$-*summable lifting* G *for the decomposition* $X(t) = 0 + M(t) + U(t)$ *if and only if the pathwise Stiletjes integrals below with respect to the terms of the projected decomposition* $Z(r) = 0 + N(r) + W(r)$, $N = \widetilde{M}$ *and* $W = \widetilde{U}$ *satisfy:*

$$\int_0^r |H(s,\omega)||dW(s,\omega)| < \infty \quad \textit{for all} \quad r \quad \textit{a.s.}$$

and there is an increasing sequence of $\mathcal{F}$*-stopping times* $\{\rho_m\}$ *such that* $\rho_m \longrightarrow \infty$ *a.s. while*

$$E\left[\sqrt{\int_0^{\rho_m} |H(s,\omega)|^2 \; d[N,N](s,\omega)}\right] < \infty.$$

An *almost-previsible* H *is called a* Z-*integrable process provided these conditions hold for some decomposition* $Z(r) - Z(0) = N(r) + W(r)$.

In this case the stochastic integral

$$\int_0^r H \cdot dZ$$

is well-defined (up to indistinguishability) by the decent path projection of

$$\sum^t G \cdot \delta X.$$

AFTERWORD

This book might very well have been written in two volumes. Had we chosen that format, volume two would have contained various fundamental applications of infinitesimal analysis to processes arising in the study of stochastic integrals. Rather than doing this, we have only given the 'measure-theoretic' foundations of this branch of stochastic analysis. It is fair to criticize our 'definition' of foundations taken by itself, so this afterword points out some topics that the reader might pursue in the literature. We especially recommend the book by Albevario, Fenstad, Höegh-Krohn, and Lindström [1985] and the Memoir by Keisler [1984].

(A.1) Stochastic Calculus

The rules for manipulating stochastic integrals are different from the rules for manipulating classical integrals. Many of these rules can be proved by the three step procedure:

1) Lift the terms to be manipulated.

2) Manipulate the *finite liftings.

3) Project the rearranged terms.

For example, if M and N are d-dimensional local δt-martingales we may apply transfer to ordinary summation to rearrange the inner product (M(t),N(t)) into sums of the form $\sum_{u<v} + \sum_{v<u} + \sum_{u=v}$. This gives:

$$(M(t),N(t)) = \sum_{i=1}^{d}\left[\sum^{t}\delta M_i(u)\right]\left[\sum^{t}\delta N_i(v)\right]$$

$$= \sum_{v=\delta t}^{t-\delta t}\sum_{u=\delta t}^{v-\delta t}\sum_{i=1}^{d} \delta M_i(u)\delta N_i(v)$$

$$+ \sum_{u=\delta t}^{t-\delta t}\sum_{v=\delta t}^{u-\delta t}\sum_{i=1}^{d} \delta N_i(v)\delta M_i(u)$$

$$+ \sum_{s=\delta t}^{t-\delta t} \delta M_i(s)\delta N_i(s)$$

$$= \sum^{t}(M(s),\delta N(s)) + \sum^{t}(N(s),\delta M(s)) + \sum^{t}\delta M(s)\delta N(s).$$

Decent path projection yields the formula for

(A.1.1) **INTEGRATION BY PARTS:**

$$(\widetilde{M}(r),\widetilde{N}(r)) = \int_0^r \widetilde{M}(s^-)d\widetilde{N}(s) + \int_0^r \widetilde{N}(s^-)d\widetilde{M}(s) + [\widetilde{M},\widetilde{N}](r).$$

We know that every 2d-dimensional local hypermartingale $(\widetilde{M},\widetilde{N})$ has a local δt-martingale lifting (M,N). We have also seen that $M(t)$ is a locally δN^2-summable lifting of the left-limit process

$$\widetilde{M}(r^-) = \underset{t\uparrow r}{\text{S-lim}}\ M(t).$$

This means that the decent path projections of both sides of the internal calculation yield the standard formula shown above.

A similar argument can be applied to prove that if $\widetilde{M}$ and $\widetilde{N}$ are one-dimensional local hypermartingales with $M(0) = N(0) = 0$ and if $\widetilde{H}$ is a locally δM^2-summable process, then the quadratic variation "commutes" with integration.

(A.1.2) **QUADRATIC VARIATION OF INTEGRALS:**

$$[\int \tilde{H} d\tilde{M}, \tilde{N}] = \int \tilde{H} d[\tilde{M}, \tilde{N}].$$

The first integral is a stochastic integral yielding a martingale. The second integral is a pathwise Stieltjes integral. The internal rearrangement of sums is clear and the Stieltjes summability condition needed for H is proved in Hoover & Perkins [1983], Theorem 7.17].

The most famous fundamental theorem of Stochastic Calculus is known as "Itô's Lemma." It has various generalizations up to the local result for discontinuous semimartingales. Let us consider a special case of the classical formula.

Let $B(t,\omega)$ denote Andersons's infinitesimal random walk and let $f : \mathbb{R} \longrightarrow \mathbb{R}$ be a twice continuously differentialbe function. Whenever $x \in {}^*\mathbb{R}$ is finite and δx is infinitesimal, the uniform Taylor "small oh" formula says

$$f(x+\delta x) - f(x) = f'(x)\delta x + \frac{1}{2} f''(x)(\delta x)^2 + \iota(x,\delta x)(\delta x)^2,$$

where $\iota(x,\delta x) \approx 0$. This means that if the random sample ω makes $B(t)$ S-continuous, so $B(t)$ is finite for finite t, then

$$f(B(t+\delta t)) - f(B(t)) = f'(B(t))\delta B(t) + \frac{1}{2}f''(B(t))\delta t + \iota(B(t))\delta t$$

for an infinitesimal $\iota(B(t))$, since $[\delta B(t)]^2 = \delta t$. We sum the increments of the process $X(t,\omega) = f(B(t,\omega))$ to obtain the approximation:

$$X(t) = \sum^{t} \delta X$$

$$\approx f(B(0)) + \sum^{t} f'(B)\delta B + \frac{1}{2}\sum^{t} f''(B)\delta t$$

for all finite t. The term $\frac{1}{2}\sum f''(B)\delta t$ is a lifting of the pathwise integral $\frac{1}{2}\int f''(\widetilde{B}(t))dt$ whenever B is S-continuous. We wish to show

(A.1.3) **ITÔ'S LEMMA:**

$$f(\widetilde{B}(r)) = f(\widetilde{B}(0)) + \int_{0}^{r} f'(B)dB + \frac{1}{2}\int_{0}^{r} f''(B)\delta t$$

or

$$df(\widetilde{B}) = f'(\widetilde{B})d\widetilde{B} + \frac{1}{2}f''(\widetilde{B})dt.$$

In order to prove this we need to know that the $\sum^{t} f'(B)\delta B$ term lifts the corresponding standard stochastic integral. Since f' is S-continuous at finite points and B is finite and S-continuous a.s., we only need to show that $f'(B)$ is locally δB^2-summable. The reducing sequence

$$\tau_m(\omega) = m \wedge \min[t : |B(t)| > m]$$

does the trick.

In general if M is an S-continuous δt-martingale, the same sort of argument produces the Itô formula for M. The reader may find additional details of the classical case in Anderson [1976].

(A.2) Stochastic Integral Equations

A classical stochastic differential equation is something of the form

$$dX(t) = f(t,X(t))dt + g(t,X(t))dB$$

in the unknown process X, where B is a d-dimensional Brownian motion, $f : [0,\infty) \times \mathbb{R}^d \longrightarrow \mathbb{R}^d$ and $g : [0,\infty) \times \mathbb{R}^d \longrightarrow \mathbb{R}^d \otimes \mathbb{R}^d$ (so $g(t,b)$ is a $d \times d$ matrix acting on the vector dB). Since no classical meaning is given to the unintegrated differentials, an integral equation with an initial condition $X_0(\omega)$ is actually intended. Itô showed that these equations have solutions under Lipschitz assumptions on f and g. "Generalizations" of Itô's theorem sometimes assert existence of "weak solutions," meaning that there is a new probability space and a new Brownian motion so that a process X exists on the new space. Keisler [1984] proves the following very general "strong" existence theorem. The fact that X exists as a process on the original hyperfinite evolution scheme is a nice feature of this scheme (also see Hoover & Perkins [1983] and Hoover & Keisler [1982]).

(A.2.1) THEOREM:

Suppose

$$f : [0,1] \times \mathbb{R}^d \longrightarrow \mathbb{R}^d, \quad g : [0,1] \times \mathbb{R}^d \longrightarrow \mathbb{R}^d \otimes \mathbb{R}^d$$

are bounded Lebesgue measurable functions such that

$[\det g]^{-2}$ is bounded.

Let $\tilde{B}(r,\omega)$ be a d-dimensional Brownian motion adapted to $\mathcal{F}$. Then for each $\mathcal{F}(0)$-measurable initial condition $X_0(\omega)$, the equation

$$X(r,\omega) = X_0(\omega) + \int_0^r f(s,X(s,\omega))ds + \int_0^r g(s,X(s,\omega))d\tilde{B}(s,\omega)$$

has a solution on our hyperfinite scheme.

PROOF:

See Keisler [1984], Theorem 5.5.

Keisler [1984], Theorem 5.2, also shows very simply that if f and g are bounded measurable functions, continuous in the X-variable (but not needing $[\det g]^{-2}$ bounded), then the equation above has a solution on the original hyperfinite scheme. The $[0,\infty)$-version of this result is given in Perkins [1982] as Theorem 9. Cutland [1980] allows g to depend on the past of X and further explores applications to control theory in Cutland [1982, 1983a].

Hoover & Perkins [1983], Theorem 10.3, show that the stochastic equation

$$X(r,\omega) = H(r,\omega) + \int_0^r f(s,\omega,X(\cdot,\omega))dZ(s,\omega)$$

has a strong solution where Z is a semimartingale with $Z(0) = 0$, H is an $\mathcal{F}$-adapted decent path process and f is a previsibly measurable function of the past of the paths in the third variable.

The starting point for all these existence proofs is to lift the functions and known processes, replace the integrals by *finite sums against infinitesimal differences such as δB. This "lifted" equation has an inductively defined internal solution, e.g., $X(t+\delta t) = X(t) + f(t,X(t))\delta t + g(t,X(t))\delta B(t)$, once $X(t)$ is known. The difficulties involve showing that $X(t)$, $f(t,X(t))$, $g(t,X(t))$ are each liftings of some classical process...

(A.3) Markov Processes

One omission from this book closely related to the last section is the study of probabilistic properties of solutions of integral equations. Keisler [1984], Theorem 6.11, shows that when f and g of (A.2.1) are bounded and continuous, then the associated integral equation has a path continuous strongly-Markov solution. The internal solution of the "lifted" infinitesimal difference equation must satisfy a property that is not the transform of the strong-Markov property, but rather a property that makes its standard part have that property. Keisler gives the S-notions of the Markov property, the strong-Markov property and the Feller property. He also shows when various integral equations have solutions with these properties.

Markovian properties of stochastic processes are certainly fundamental, but we refer our reader to Keisler's excellent memoir for the corresponding internal notions.

(A.4) **Reshuffling**

Keisler [1984] proves his existence theorems for Brownian motions on a hyperfinite evolution scheme by first proving them for Anderson's infinitesimal random walk and then appealing to the following "homogeniety property" of this scheme. If X and Y are continuous Markov processes with the same finite dimensional distributions, then there is an internal bijection which reshuffles the sample space, respects the progressively measurable filtration and almost surely maps X onto Y. (We saw a faint shadow if this result in Chapter 4. The result is easily extended to decent path Markov processes, but Hoover & Keisler [1983] have gone far beyond this.)

(A.5) **Universality**

Keisler [1984] also proves a "universal" property of the hyperfinite evolution scheme. If a stochastic integral equation has a solution on any space then we may replace the Brownian motion with one on our hyperfinite space and obtain a solution with respect to that Brownian motion on our same space, moreover, with the same finite dimensional distributions as the original solution. He proves that for any continuous process on any space there is a continuous progressively measurable process on the hyperfinite scheme with the same finite dimensional distributions as the original.

Hoover & Keisler [1982] contains an important generalization of this result. They give a logic of "adapted distribution" which has common finite dimensional distributions as its zeroth step. Two Markov processes have the same adapted distribution if and only if they have the same finite dimensional distributions, but adapted distributions measure properties of the progressive filtration for more general processes. The hyperfinite evolution scheme is universal (and "saturated") for the full theory of adapted distribution. Sections 6 and 7 of their paper give a precise meaning to our earlier claim that there is no loss of generality in studying semimartigale integration on our hyperfinite scheme.

Works of Keisler, Hoover, Rodenhausen and others study "probability logic" in greater detail.

(A.6) Local Time

Perkins [1981b] gives impressive results on Brownian local time. This is also partly explained in Perkins [1983]. All of Perkins' articles in the references are worthy of study. Most of them involve key uses of infinitesimal analysis. They are beyond the scope of this book.

(A.7) Applications

The works of Helms [1982], Arkeryd [1981-] and Cutland [1982,3] are nice uses of infinitesimal analysis in problems aimed at scientific applications (rather than applications of mathematics to mathematics). Lawler [1980] has an interesting treatment of loop-erased random walks that may turn out to have applications. Kosciuk [1982] gives another potentially applicable use of infinitesimals. The most ambitious discussion of applied infinitesimal stochastic analysis is contained in the forthcoming book of Albeverio, Fenstad, Höegh-Krohn and

Lindström [1985]. Their book treats a broad selection of physically oriented applications of infinitesimal stochastic analysis. It is essentially self-contained, but overlaps remarkably little with this book. Since it summarizes the research of its authors, we have not mentioned that separately.

(A.8) Other Infinitesimal Methods

The recommended reading above does not adhere closely to the hyperfinite framework we have chosen. However, most of it is fairly close to our treatment in spirit at least and should be easy to move into from here. Lawler [1980] is written in the framework of Nelson's [1977] Internal Set Theory, so it must be read a little more carefully because of that restriction. Henson and Wattenberg [1981] solve an interesting technical problem of infinitesimal measure theory raised in Robinson's [1966] book, but their methods essentially fall outside the framework we consider. Our framework is not the only possible one. We do think that it is a coherent highly promising one for applciation to mathematics and science. Naturally, we also think it is interesting in its own right.

So now, dear reader, we wish you farewell and good hunting. We hope these methods help you find solutions to problems that interest you.

Sergio Albeverio, Jens Erik Fenstad, Raphael Höegh-Krohn and Tom Lindström

[1985] "Nonstandard Methods in Stochastic Analysis and Mathematical Physics, Academic Press, to appear.

Robert M. Anderson

[1976] A Non-standard Representation for Brownian Motion and Itô Integration, *Israel J. Math.*, **25**, PP. 15-46.

[1982] Star-Finite Representation of Measure Spaces, *Trans. Amer. Math. Soc.*, **271**, pp. 667-687.

Robert M. Anderson & Salim Rashid

[1978] A Nonstandard Characterization of Weak Convergence, *Proc. Amer. Math. Soc.* **69**, pp. 327-332.

L. Arkeryd

[1981a] A Nonstandard Approach to the Boltzmann Equation, *Arch. Rational Mech. Anal.*, **77**, pp. 1-10.

[1981b] A Time-wise Approximated Boltzman Equation, *I M A Journal of Applied Math.*, **27**, pp. 373-383.

[1981c] Intermolecular Forces of Infinite Range and the Boltzman Equation, *Arch. Rational Mech. Anal.*, **77**, pp. 11-21.

[1982] Asymptotic Behavior of the Boltzman Equation with Infinite Range Forces, *Comm. Math. Phys.*, **86**, pp. 475-484.

[1984] Loeb Sobolev Spaces with Applications to Variational Integrals and Differential Equations, preprint.

Bernd J. Arnold

[1982] "*Die Verwendung von Infinitesimalien in der Theorie Der Brownschen Bewegung*," Diplomarbeit, Technische Hochschule Darmstadt.

Jon Barwise (editor)

[1977] "*Handbook of Mathematical Logic*," North-Holland, Studies in Logic, vol. **90**, Amsterdam.

Patrick Billingsley

[1968] "*Convergence of Probability Measures*," John Wiley & Sons, New York.

Leo Brieman

[1968] "*Probability*," Addison-Wesley, Reading.

Rafael V. Chacon, Yves Le Jan and S. James Taylor

[1982] Generalized Arc Length for Brownian Motion and Levy Processes, preprint with appendix by Edwin Perkins.

C. C. Chang & H. Jerome Keisler

[1977] "*Model Theory*," North-Holland Studies in Logic & Found. of Math., **73**, Amsterdam.

Nigel J. Cutland

[1980] On the Existence of Strong Solutions to Stochastic Differential Equations on Loeb Spaces, *Z. Wahrsch. Verw. Gebiete*, **60**, pp. 335-357.

[1981] Internal Controls and Relaxed Controls, *J. London Math. Soc.*, **27**, pp. 130-140.

[1982] Optimal Controls for Partially Observed Stochastic Systems: An Infinitesimal Approach, *Stochastics*, **8**, pp.239-257.

[1983] Nonstandard Measure Theory and Its Applications, *Bull. of London Math. Soc.*, **15**, pp.529-589.

Martin Davis

[1977] "*Applied Nonstandard Anaylsis*," Wiley-Interscience, Pure and Appl. Math. Series, New York.

Claude Dellacherie and Paul-André Meyer

[1978] "*Probabilities and Potential*," North-Holland, Math. Studies, Amsterdam.

J. L. Doob

[1953] "*Stochastic Processes*," John Wiley & Sons, Inc., New York.

Nelson Dunford and Jacob T. Schwartz

[1958] "*Linear Operators* Part I," Interscience, New York.

Stewart N. Ethier and Thomas G. Kurtz

[1980] "*Markov Process: Existence and Approximation*," Preliminary version of a book.

William Feller

[1968] "An *Introduction to Probability and Its Applications*, Vol. I," Third Ed., John Wiley & Sons, Series in Prob. & Math. Stat., New York.

I. I. Gihman and A. V. Skorohod

[1974] "*The Theory of Stochastic Processes*," Springer-Verlag, Grundlehren band 210, New York.

B. V. Gnedenko and A. N. Kolmogorov

[1968] "*Limit Distributions for Sums of Independent Random Variables*," revised ed., Addison-Wesley, Reading.

P. Greenwood and R. Hersh

[1975] Stochastic Differentials and Quasi-Standard Random Variables, in the vol. "*Probabilistic methods in differential Equations*, "Springer-Verlag, Notes in Math. #451, Berlin,

Priscilla Greenwood and Edwin Perkins

[1983] Conditional Limit Theorem for Random Walk, and Brownian Local Time on Square Root Boundaries, *Ann. Prob. II*, pp. 227-261.

L. L. Helms

[1983] A Nonstandard Approach to the Martingale Problem for Spin Models, in the vol. *Nonstandard Analysis - Recent Developments*, see Hurd [1983].

L. L. Helms and P. A. Loeb

[1979] Applications of Nonstandard Analysis to Spin Models, *J. Math. Anal. Appl.* **69**, pp. 341-352.

[1982] A Nonstandard Proof of the Martingale Convergence Theorem, *Rocky Mountain J. Math.* **12**, pp. 165-170.

[1982b] Bounds on the Oscillation of Spin Systems, J. Math. Anal. Appl., **86**, pp. 493-502.

C. Ward. Henson

[1979a] Unbounded Loeb Measures, *Proc. Amer. Math. Soc.* **74**, pp. 143-150.

[1979b] Analytic Sets, Baire Sets and the Standard Part Map, *Canad. J. Math.* **XXXI**, pp. 663-672.

C. Ward Henson & L. C. Moore, Jr.

[1974] Nonstandard Hulls of the Classical Banach Spaces, *Duke Math. J.* **41**, pp. 277-284.

C. Ward Henson & Frank Wattenberg

[1981] Egoroff's Theorem and the Distribution of Standard Points in a Nonstandard Model, *Proc. Amer. Math. Soc.* **81**, pp. 455-461.

Edwin Hewitt & Karl Stromberg

[1965] "*Real and Abstract Analysis*," Springer-Verlag, New York.

Douglas N. Hoover

[1978] Probability Logic, *Ann. Math. Logic* **14**, pp. 287-313.

[1981] A Normal Form Theorem for $L_{\omega,p}$, with Applications, preprint.

Douglas N. Hoover & H. Jerome Keisler

[1982] Adapted Probability Distributions, preprint.

Douglas N. Hoover & Edwin Perkins

[1983] Nonstandard Construction of the Stochastic Integral and Applications to Stochastic Differential Equations I and II, *Trans. Amer. Math. Soc.*, **275**, pp. 1-36 & 37-58.

A. E. Hurd

[1981] Nonstandard Analysis and Lattice Statistical Mechanics: A Variational Principle, *Trans. Amer. Math. Soc.*, 263, pp. 89-110.

[1983] "*Nonstandard Analysis - Recent Developments*," Springer Verlag Lecture Notes in Math., **983**, New York.

Nobuyuki Ikeda & Shinzo Watanabe

[1981] "*Stochastic Differential Equations and Diffusion Processes*," North-Holland Math. Library, vol. **24**, Amsterdam.

H. Jerome Keisler

[1976] "*Foundations of Infinitesimal Calculus*," Prindle, Weber & Schmidt, Boston.

[1977] Hyperfinite model theory, in the volume "*Logic Colloquium* **76**, pp. 5-110, North-Holland, Amsterdam.

[1984] An Infinitesimal Approach to Stochastic Analysis, *Mem. Amer. Math. Soc.*, vol. **48**, nr. 297.

S. A. Kosciuk

[1982] *Nonstandard Methods in Diffusion Theory*, Ph. D. Thesis, University of Wisc.-Madison.

[1983] Stochastic Solutions to Partial Differential Equations, in the vol., *Nonstandard Analysis - Recent Developments*, Hurd [1983] above.

K. Kunen

[1979] Personal communication.

K. Kuratowski

[1966] "*Topology I & II*," Academic Press, New York.

A. U. Kussmaul

[1977] "*Stochastic Integration and Generalized Martingales*," Pitman Publishing, Research Notes in Math., London.

Gregory F. Lawler

[1980] A Self-Avoiding Random Walk, *Duke Math. J.* **47(3)**, pp. 655-693.

Tom L. Lindström

[1980a] Hyperfinite Stochastic Integration I, II, III, and Addendum, *Math. Scand.* 46, pp. 265-292, pp. 293-314, pp. 315-331, pp. 332-333.

[1982] A Loeb-Measure Approach to Theorems by Prohorov, Sazonov and Gross, *Trans. Amer. Math. Soc.*, 269, pp. 521-534.

[1983] Stochastic Integration in Hyperfinite Dimensional Linear Spaces, in the vol., *Nonstandard Analysis - Recent Developments*, Hurd [1983] above.

[1980d] The Structure of Hyperfinite Stochastic Integrals, preprint series, Univ. of Oslo.

Peter A. Loeb

[1972] A Non-Standard Representation of Measurable Spaces, L_∞, and L_∞^*, in the vol. "*Contributions to Non-Standard Analysis*" (Luxemburg & Robinson, eds.), North-Holland.

[1975] Conversion from Nonstandard to Standard Measure Spaces and Applications in Probability Theory, *Trans. Amer. Math. Soc.* 211, pp. 113-122.

[1979a] Weak Limits of Measures and the Standard Part Map, *Proc. Amer. Math. Soc.* 77, pp. 128-135.

[1979b] An Introduction to Nonstandard Analysis and Hyperfinite Probability Theory, in the vol. "*Probabilistic. Analysis and Related Topics* 2," Bharucha-Reid (editor), Academic Press, New York.

M. Loève

[1977] "*Probability Theory I & II*," Springer-Verlag, New York.

W. A. J. Luxemburg

[1973] What is Nonstandard Analysis?, in a supplement to *Amer. Math. Monthly* 80, pp. 38-67.

Michel Metivier & J. Pellaumail

[1980] "*Stochastic Integration*," Academic Press series on Probability and Mathematical Statistics, New York.

P. A. Meyer

[1976] "*Un Cours sur les Integrales Stochastiques*," Seminaire de Probabilités X, Springer-Verlag Lecture Notes in Math. vol. 511, Berlin.

Edward Nelson

[1977] Internal Set Theory, *Bull. Amer. Math. Soc.* 83, pp. 1165-1198.

[1980] The Syntax of Nonstandard Analysis, preprint.

Richard Lee Panetta

[1978] "*Hyperreal Probability Spaces: Some Applications of the Loeb Construction*," Ph.D. Thesis, Univ. of Wisconsin-Madison.

Edwin Perkins

[1979] "*A Nonstandard Approach to Brownian Local Time*," Ph.D. Thesis, Univ. of Illinois at Urbana-Champaign.

[1982a] On the Construction and Distribution of a Local Martingale of Given Absolute Value, *Trans. Amer. Math. Soc.*, 271, pp.261-281.

[1981a] On the Uniqueness of a Local Martingale of Given Absolute Value, *Z. Wahrsch. Verw. Gebiete* 56, pp. 255-281.

[1981b] A Global Intrinsic Characterization of Local Time, *Ann. Probability* 9, pp. 800-817.

[1981c] The Exact Hausdorff Measure of the Level Sets of Brownian Motion, *Z. Wahrsch. Verw. Gebiete*, 58, pp. 373-388.

[1982b] Weak Invariance Principles for Local Time, *Z. Wahrsch. Verw. Gebiete*, 60, pp.437-451.

[1983] Stochastic Processes and Nonstandard Analysis (expository), in the vol., *Nonstandard Analysis - Recent Developments*, Hurd [1983]

Also see Greenwood & Perkins, Hoover & Perkins and Chacon, Le Jan & Taylor.

Michael M. Richter

[1982] "*Ideale Punkte Monaden und Nichtstandardmethoden*," preprint of book to appear.

Abraham Robinson

[1966] "*Non-Standard Analysis*," North-Holland Studies in Logic and the Found. of Math., Amsterdam.

Hermann Rodenhausen

[1982] "*The Completeness Theorem for Adapted Probability Logic*," Inaugural-Dissertation, Ruprecht-Karl, Universität Heidelberg.

David A. Ross

[1983] *Measurable Transformations in Saturated Models of Analysis*, Ph. D. Thesis, Univ. of Wisc. - Madison.

[1984] Automorphisms of the Loeb Algebra, preprint.

H. L. Royden

[1968] "*Real Analysis*," second edition, Macmillan, New York.

Andreas Stoll

[1982] "A *Nonstandard Construction of Levy Brownian Motion with Applications to Invariance Principles*," Diplomarbeit, Universität Freiburg.

K. D. Stroyan & W. A. J. Luxemburg

[1976] "*Introduction to the Theory of Infinitesimals*," Academic Press, Series on Pure and Appl. Math., New York.

K. D. Stroyan

[1984] Previsible Sets for Hyperfinite Filtrations, preprint.

Richard L. Wheeden & Antoni Zygmund

[1977] "*Measure and Integral*," Marcel Dekker, Pure and Appl. Math., New York.

David Williams

[1979] "*Diffusions, Markov Processes and Martingales*, vol. 1: *Foundations*," John Wiley & Sons series in Prob. and Math. Stat., Chichester.

[1981] "*Stochastic Integrals, Proceedings of the LMS Durham Symposium*," Springer-Verlag Lecture Notes in Math., vol. 851, Berlin.

APPENDIX 1: A PRIMER OF INFINITESIMAL ANALYSIS

This Appendix is a collection of results that are useful in various parts of the book. They are given here in detail to serve as a further help at learning to apply the logical principles to Analysis.

(APP.1.1) PROPOSITION:

A *natural number $n \in {}^*N$ is either standard, $n \in {}^\sigma N$, or unlimited, $n > m$ for every $m \in {}^\sigma N$.

PROOF:

We offer a direct proof as an application of Leibniz' Principle; below, (APP.1.2)(4), we suggest a different proof. Take any limited $n \in {}^*N$, i.e., a *natural number n such that $n < m$ for some (standard) $m \in {}^\sigma N$. Apply Leibniz' Principle to the bounded sentence

$$\forall x \in N, \ (x < m \ \text{ imply } \ x = 0 \ \text{ or } \ \ldots \ \text{ or } \ x = m)$$

and conclude that n equals one of the standard numbers $0, 1, \ldots, m$.

(APP.1.2) REMARK:

$^*N$ "looks like" N followed by a dense ordinal Θ times copies of Z, and Θ is uncountable.

Specifically:

(1) If m, n belong to $^*N$ and $|m - n| < 1$, then $m = n$ (apply Leibniz' Principle to the 'same' statement in N).

(2) If $n \in {}^*N$ is unlimited and $m \in {}^\sigma N$, then $n + m$ and $n - m$ are also in $^*N$ and unlimited. Therefore, around each

unlimited $n \in {}^*N$ there is a whole copy of Z embedded into the unlimited part of ${}^*N$.

(3) If $m,n \in {}^*N$ are unlimited, then the *integer part of $(m+n)/2$ is also unlimited. Hence, between two disjoint copies of Z there is also another (disjoint) copy of Z.

(4) Once we have shown (1), you can give another proof of (APP.1.1) using the standard part map: if $n \in {}^*N$ and n is limited, call $r = \text{st}\, n$; then

$$|n - [r]| \leqq |n - r| + |r - [r]| ,$$

which is strictly smaller than 1, because it is the sum of an infinitesimal and a standard real number smaller than 1 ; therefore, $n = [r] \in {}^\sigma N$.

(5) If $n \in {}^*N$ is unlimited and $r,s \in {}^\sigma(0,1)$ are standard real numbers, $r < s$, then the hyperreal numbers n^r, n^s are unlimited (why?), and at an infinite distance apart of each other, because $n^r - n^s = n^r(n^{s-r} - 1)$. Hence, the hypernatural numbers $[n^r]$, $[n^s]$ lie on two different copies of Z. This proves that there are uncountably such copies.

(APP.1.3) PROPOSITION:

(a) The set of standard *natural numbers ${}^\sigma N$ is external.

(b) The set of unlimited *natural numbers ${}^*N \setminus {}^\sigma N$ is external.

(c) For any set $X \in \mathcal{X}$, either X is a finite set and ${}^\sigma X = {}^*X$, or ${}^\sigma X$ is external and the inclusion ${}^\sigma X \subset {}^*X$ is strict.

(d) The sets of limited scalars $\mathcal{O}$, of unlimited scalars, of infinitesimals o, the map st, and the relation $\approx$, are all external entities.

PROOF:

First, we claim that for a subset of an internal set V , the property of being internal is equivalent to belonging to ${}^*P(V)$: assume T , V are internal and $T \subset V$; then $T, V \in {}^*X_p$ for some p ; apply Leibniz' Principle to the sentence

$$\forall\ v \in X_p,\ \forall\ t \in X_p,\ t \in P(V) \text{ iff } (\forall\ x \in t,\ x \in v),$$

to obtain

$$\forall\ v \in {}^*X_p,\ \forall\ t \in {}^*X_p,\ t \in {}^*P(V) \text{ iff } (\forall\ x \in t,\ x \in v),$$

hence if T is internal, then $T \in {}^*P(V)$ (the converse is obvious).

Proofs of 'externality' are best handled by contradiction: it is usually convenient to show that a set does not have some property which is known to hold (by an appropiate application of Leibniz' Principle) for internal sets.

(a) Consider the following true statement in N :

$$\forall\ T \in P(N),\ (T \text{ is bounded in } N) \text{ implies } (T \text{ has a maximum}).$$

Its *transform says that if T is an internal subset of ${}^*N$ and is *bounded, i.e., bounded by some member of ${}^*N$, then it has a *maximum, i.e., a maximum (writte down the whole sentences in detail if you do not feel sure about the last assertions).

Now regard ${}^\sigma N$ as a subset of ${}^*N$; if it were internal, then it would have a maximum m , because it is certainly *bounded. But then $m+1$ would also be limited, and hence a member of ${}^\sigma N$, by (APP.1.1).

(b) The reader can work out a proof similar to the last one, by considering the statement

$$\forall\ T \in P(N),\ T \neq \emptyset \text{ implies } T \text{ has a mimimum.}$$

Alternately, one can show, by *transform of appropiate obvious sentences, that finite boolean operations with internal sets always give internal sets; then $^*N \setminus {}^\sigma N$ has to be external, otherwise $^\sigma N$ would be internal.

(c) Assume $X \in \mathcal{X}$ is a finite set. The following proof of $^*X = {}^\sigma X$ is a generalization of the proof of (APP.1.1) given above. If $X = \{x_1,..,x_n\} \in X_p$, for some p, then $p \geqq 1$ and $X \subseteq X_{p-1}$, so $^*X \subseteq {}^*X_{p-1}$, and by Leibniz' Principle, $\forall x \in {}^*X_{p-1}$, $x \in {}^*X$ iff ($x = {}^*x_1$ or ... or $x = {}^*x_n$) .

Now, let $X \in \mathcal{X}$ be infinite, and assume that $^\sigma X$ is internal; then every finite subset $A \subset {}^\sigma X$ is also internal, so by the saturation property included as a part of Leibniz' Principle (see (0.2.3)(b) and (0.4.2)), the set

$$\bigcap \; [A : A \subset {}^\sigma X \;\&\; {}^\sigma X \setminus A \text{ is finite}]$$

is not empty; and this is absurd: if x is a member of this intersection, then x must belong to $^\sigma X \setminus \{x\}$.

Hence $^\sigma X$ is external, and therefore the inclusion $^\sigma X \subset {}^*X$ of an external set into an internal set has to be strict. (See a hint for a more direct proof of this fact in Exercise (0.4.5).)

(d) If $\mathcal{O}$ were internal, then $^\sigma N = {}^*N \cap \mathcal{O}$ would also be internal; a similar argument works for $^*R \setminus \mathcal{O}$. Since the map $x \longrightarrow x^{-1}$ is internal (indeed, even standard), if o were internal ($o \setminus \{0\}$ would be internal and) then $^*R \setminus \mathcal{O}$ would be internal as well.

It is not difficult to see that for a map to be internal it is necessary that both its domain and its range be internal (apply the Internal Definition Principle), so the map st : $\mathcal{O} \dashrightarrow R$ cannot be internal. Again by the same principle, if $\approx$ were internal, the set of infinitesimals would be internal too:

$$o = \{x \in {}^*R : x \approx 0\} .$$

In the rest of this Appendix, we are to introduce some elementary properties of ${}^*R$ that appear very often throughout the book and have in common that they are translations to this setting of analogous standard topological properties, but keeping the standard tolerances, whence the 'S-' prefix.

(APP.1.4) DEFINITION:

A hyperreal function f is said to be S-continuous at a in ${}^*R$ if $x \approx a$ implies $f(x) \approx f(a)$ (and $x, a \in \text{dom}(f)$ in particular). We say f is S-continuous at a relative to $D \subseteq {}^*R$ if $x \in D$ and $x \approx a$ imply $f(x) \approx f(a)$ (and $f(x)$ is defined for such x in particular). We say f is S-continuous relative to D if f is S-continuous at a relative to D for each a in D ($D \subseteq \text{dom}(f)$ in particular).

Similar definitions apply for hyperreal functions defined on subsets of arbitrary metric spaces (in this case, $x \approx a$ means that the *distance $d(x,a)$ is infinitesimal).

(APP.1.5) PROPOSITION:

Let f be an internal function and $D \subseteq \text{dom}(f)$ be an internal set. Then f is S-continuous relative to D if and only if for every standard positive ε in ${}^\sigma R^+$, there exists a standard positive θ in ${}^\sigma R^+$, such that for all x, y in D,

$$|x-y| < \theta \quad \text{implies} \quad |f(x)-f(y)| < \varepsilon .$$

PROOF:

If f is S-continuous and ε in ${}^{\sigma}R^{+}$ is fixed, the set

$T(\varepsilon) = \{\theta \in {}^{*}R^{+} : x,y \in D \;\&\; |x-y| < \theta \quad \text{imply } |f(x)-f(y)| < \varepsilon\}$

is internal by the Internal Definition Principle. By hypothesis, $T(\varepsilon) \supseteq o \cap {}^{*}R^{+} = o^{+}$. Since o^{+} is external, $T(\varepsilon)$ contains a noninfinitesimal δ. Any $\theta < \delta$ is in $T(\varepsilon)$, so there is a standard θ in $T(\varepsilon)$. This proves the ε-θ condition.

Conversely, if $x,y \in D$ and $x \approx y$, then $|x-y| < \theta$ for every standard positive θ. Hence, $|f(x)-f(y)| < \varepsilon$ for every standard positive ε, and $f(x) \approx f(y)$.

(APP.1.6) COROLLARY:

Let f be an internal function and $D \subset O$ be an internal set of limited points. If $f(x)$ is limited for each x in D and if f is S-continuous relative to D, then $\hat{f}: D/\approx \dashrightarrow O/\approx$ is uniformly continuous, where $\hat{f}(\hat{x}) = \operatorname{st} f(x)$ for $x \in \operatorname{st}^{-1}(\hat{x})$ is the <u>infinitesimal</u> <u>hull</u> of f.

PROOF:

The map $\hat{f}$ is well-defined on equivalence classes, $\operatorname{st}^{-1}(\hat{x})$, by S-continuity. The $\varepsilon - \theta$ condition is at most disturbed by changing $<$ to $\leqq$ by taking standard parts. That proves that $\hat{f}$ is ε-θ continuous uniformly on D.

(APP.1.7) REMARK:

The function $\hat{f}$ may not be uniformly continuous if D is external: consider $f(x) = 1/x$ on $O \setminus o$.

(APP.1.8) DEFINITION:

(a) Let f be a hyperreal function and let $D \subseteq {}^{*}R$

be a subset of its domain that contains a real interval ${}^{\sigma}(r,s)$, with $r,s \in {}^{\sigma}R$. Then we say that $b \in {}^{\sigma}R$ is the right S-limit of f at r within D ,

$$\operatorname*{S-lim}_{\substack{x \downarrow r \\ x \in D}} f(x) = b ,$$

if for every standard positive ε there is a standard positive θ , such that

$$x \in D \text{ and } r \ll x < r+\theta \text{ implies } |f(x) - b| < \varepsilon .$$

Similarly, we define the left S-limit $\operatorname*{S-lim}_{\substack{x \uparrow s \\ x \in D}} f(x)$ and the (bilateral) S-limit $\operatorname*{S-lim}_{\substack{x \to t \\ x \in D}} f(x)$ (for $t \in {}^{\sigma}(r,s)$) .

Finally, we will drop any reference to D whenever it is the whole domain of f .

(b) If $(x_n : n \in {}^{\sigma}N)$ is an external sequence of hyperreal numbers and $b \in {}^{\sigma}R$, we sasy that b is the S-limit of x_n ,

$$b = \operatorname*{S-lim}_{n \to \infty} x_n ,$$

if for every standard positive ε there is a standard n_ε such that

$$n \in {}^{\sigma}N \text{ and } n > n_\varepsilon \quad \text{imply} \quad |x_n - b| < \varepsilon .$$

And we say that (x_n) is S-Cauchy if for every $\varepsilon \in {}^{\sigma}R$ there is an $n_\varepsilon \in {}^{\sigma}N$ such that

$$m,n \in {}^{\sigma}N \text{ and } m,n > n_\varepsilon \quad \text{imply} \quad |x_m - x_n| < \varepsilon .$$

(APP.1.9) PROPOSITION:

Let f be an internal function and D an internal subset of its domain of definition such that $D \supset {}^{\sigma}(r,s)$.

Then in order that the standard real number b be the right S-limit of f at r within D, it is necessary and sufficient the following: there is an $x_1 \in D$, $x_1 \approx r$, such that for all $x \in D$,

$$x \approx r \text{ and } x > x_1 \text{ imply } f(x) \approx b .$$

PROOF:

Assume that $b = \underset{\substack{x \downarrow r \\ x \in D}}{\text{S-lim}} f(x)$. Then for each standard positive ε there is another $\theta_\varepsilon \in {}^\sigma R^+$, that we can take smaller than ε, such that

$$x \in D \text{ and } r << x < r+\theta_\varepsilon \text{ imply } |f(x) - b| < \varepsilon .$$

Call $D_\varepsilon = \{y \in D : r < y < r+\theta_\varepsilon\}$ and

$$F_\varepsilon = \{y \in D_\varepsilon : x \in D_\varepsilon \text{ and } x > y \text{ imply } |f(x)-b| < \varepsilon\},$$

(notice that the relation $<<$ is external, that is why in order to build a family F of internal sets to apply the Saturation Principle, we include in D_ε all points y where the inequality $r < y$ holds, regardless of whether they are finitely apart from r or not). Then the family

$$F = [F_\varepsilon : \varepsilon \in {}^\sigma R^+]$$

has the finite intersection property and cardinal strictly smaller than that of X, therefore $\cap[F_\varepsilon : \varepsilon \in {}^\sigma R^+]$ is not empty. It is easy to see that any member of this intersection is one of those x_1 that we were looking for.

Conversely, suppose there is an x_1 satisfying the condition in the statement. Then, given any $\varepsilon \in {}^\sigma R^+$, the set

$$\{\theta \in {}^*R^+ : x \in D \text{ and } x_1 < x < r+\theta \text{ imply } |f(x)-b| < \varepsilon\}$$

is internal and, by hypothesis, contains the set of infinitesimals, which is external; take as θ_ε the standard part of one of the noninfinitesimal members of this set.

(APP.1.10) DEFINITION:

A set $D \subseteq {}^*A$, with $A \subseteq R^d$, is called S-dense in ${}^*A$ if $st(D) \supseteq A$, i.e., if for every standard a in A there exists an x in D which is infinitely close to a : $x \approx a$. When A is the whole space R^d, we say that D is S-dense.

The same definitions make sense for subsets of arbitrary metric spaces.

(APP.1.11) REMARK:

The Lemma (2.1.1) might have gone here in our development. The standard part of an internal set is always closed, so "S-dense" means roughly that the closure is everything. Its role is so important in section (2.1), that we didn't want experts who are not reading this Appendix to miss it.

(APP.1.12) EXAMPLE:

An example of an internal S-dense subset of ${}^*R^2$ is

$T^2 = \{(x,y) \in R^2 : \exists\, h,k \in {}^*Z,\ -n^2 \leq h,k \leq n^2\ \&\ x=h/n, y=k/n\}$

where n is an infinite *natural number. In fact, T^2 is *finite, it has $(2n^2+1)^2$ elements.

If S is an S-dense, *finite subset of ${}^*R$, we may use Leibniz' Principle to write $S = \{s_k : 1 \leq k \leq n\}$ for some internal sequence $s(.)$ with $s_1 < s_k < s_{k+1} < s_n$. If we let $\alpha(s_k) = s_k - s_{k-1}$ for $k>1$ and $\alpha(s_1) = 0$, then α is an internal function (verify this). Since D is S-dense, s_1 is negative infinite. If I is a standard finite interval, $(a,b]$, for example, then

$$\text{length}(I) \approx \sum[\alpha(s) : s \in S \cap {}^*I] .$$

(APP.1.13) EXERCISE:

Use the telescoping sum property to verify the approximate length formula above. We can use α to build Lebesgue measure, see Chapter 2.

Verify the comments above about an S-dense *finite set S and prove the converse remark as follows. If

$$S = \{ s_k : k \in {}^*N, 1 \leq k \leq n \}$$

is a *finite set represented by the internal increasing sequence $\{s_k\}$ and if s_1 is negative infinite, $s_k - s_{k-1}$ is positive infinitesimal when s_k is finite and s_n is positive infinite, then S is S-dense in $^*R$.

Give similar necessary and sufficient conditions for a set $S = \{s_k\}$ to be S-dense in $^*[0,1]$.

Our final basic example to illustrate the interplay between internal and external concepts can be used to construct Lebesgue measure on R^d with a *finite measure (see Chapter 2).

(APP.1.14) PROPOSITION:

If T is an S-dense *finite subset of $^*R^d$, then there is an internal function $\alpha : T \text{----}> {}^*[0,1]$ such that if $I \subset R^d$ is a bounded d-rectangle (for example, $I = \{x \in R^d : a^j \leq x^j \leq b^j, 1 \leq j \leq d\}$), then

$$\text{d-vol}(I) \approx \sum [\alpha(t) : t \in T \cap {}^*I] .$$

PROOF:

For finite m in $^\sigma N$ consider the partial paving P(m) of $^*R^d$ given by the rectangles

$$I(k) = \left[\frac{k^1}{m},\frac{k^1+1}{m}\right) \times \ldots \times \left[\frac{k^d}{m},\frac{k^d+1}{m}\right)$$

for $k \in {}^*Z^d$ such that $-m^2 \leq k^j \leq m^2$, $1 \leq j \leq d$. Since T is S-dense, for each $I(k) \in P(m)$ we may select a single point $s(k,m) \in T \cap {}^*I(k)$. Let $S(m) = \{s(k,m) : |k^j| \leq m^2\}$ be the set of these selections. When m is finite, so is $S(m)$ and therefore it is internal (it is described by the statement "$x=s_1$ or $x=s_2$ or ... or $x=s_n$").

The process of forming the paving $P(m)$ is internal, since its description can be internally formalized. The statement $\varphi(m)$ "there is an internal subset $S \subset T$, so that S contains exactly one point of each rectangle of $P(m)$" is also internally formalizable: $\varphi(m) =$

$$(\exists\ S \in {}^*P(T))(\forall\ I \in P(m))(\ \exists|s \in S \cap I)\ .$$

Thus $\{m \in {}^*N : \varphi(m)\}$ is internal and contains ${}^\sigma N$. Since ${}^\sigma N$ is external, there is an infinite n and an internal selection $S \subseteq T$ such that for every $k \in {}^*Z^d$ with $k^j \leq n^2$, there is a unique s in S such that $k^j/n \leq s^j < (k^j+1)/n$, for $1 \leq j \leq d$. Fix such an S for the rest of the argument.

The indicator function of S, $I_S(x) = 0$ if $x \notin S$, $I_S(x) = 1$ if $x \in S$, is internal provided we restrict its domain to an internal set:

$$I_S = \{(x,y) \in T \times \{0,1\} : y=1 \text{ if } x \in S \ \& \ y=0 \text{ if } x \notin S\}\ .$$

We take our weighting function $\alpha = I_S/n^d$.

Now let I be a bounded d-rectangle in R^d, say $|x^j| \leq B$ for $1 \leq j \leq d$. Let $\pi_j(x) = x^j$ denote projection on the jth coordinate and let

$$p_j = {}^\#[\{k \in {}^*Z : k/n \in \pi_j({}^*I)\}]\ ,$$

the *cardinality of that set. We know that $p_j \leq 2Bn$ by transfer, so p_j/n is finite.

For $1 \leq j \leq d$ we know

$$p_j - 1 \leq {}^{\#}[\{k \in {}^*Z : [k/n,(k+1)/n) \subseteq \pi_j({}^*I)\}]$$

and

$${}^{\#}[\{k \in {}^*Z : [k/n,(k+1)/n) \cap \pi_j({}^*I) \neq \emptyset\}] \leq p_j + 1 ,$$

hence the *number of little boxes inside $^*I$, r , is bounded below by $\prod_{j=1}^{d} (p_j - 1)$ while the *number of little boxes that touch $^*I$, s , is bounded above by $\prod_{j=1}^{d} (p_j + 1)$. The difference between these is a finite sum of products of <u>less than</u> d factors p_j (since the $\prod_{j=1}^{d} p_j$-term cancels out). Each p_j/n is finite and $1/n$ is infinitesimal, therefore the difference $(s-r)/n^d \approx 0$.

By the *finite additivity of volume and monotony, $r/n^d \leq {}^*(d\text{-vol})({}^*I) \leq s/n^d$. Since we also know $r \leq {}^{\#}[S \cap {}^*I] \leq s$, and ${}^*(d\text{-vol})({}^*I) = d\text{-vol}(I)$, we have our result,

$$d\text{-vol}(I) \approx \sum[\alpha(t) : t \in T \cap {}^*I] .$$

APPENDIX 2: ULTRAPOWERS THAT ARE ENLARGEMENTS

The goal of this Appendix is to construct a special kind of nonstandard models, the so-called enlargements, that are needed in Section (0.4) as a first step in the process of defining the superstructure extensions for which the important principles (0.4.2), (0.4.3) and (0.4.4) are shown to hold. In order to do that, we define the ultrapower of a superstructure with respect to an ultrafilter, and then we show that for suitable ultrafilters, this construction produces enlargements.

(APP.2.1) DEFINITION:

Let J be a nonempty set. A family of subsets of J, $\mathcal{U} \subseteq P(J)$ is called an ultrafilter when

(1) if $U, V \in \mathcal{U}$, then $U \cap V \in \mathcal{U}$,

(2) if $U \in \mathcal{U}$ and V satisfies $U \subseteq V \subseteq J$, then $V \in \mathcal{U}$,

(3) for every subset $V \subseteq J$, either $V \in \mathcal{U}$ or its complement $J \setminus V \in \mathcal{U}$, and not both.

(APP.2.2) REMARK:

Nonempty families of nonempty subsets of J that satisfy (1) and (2) above are called filters. It is not difficult to prove that 'ultrafilter' is the same as 'maximal filter' for the relation of inclusion, $\subseteq$. A standard application of Zorn's Lemma shows that each filter is contained in some ultrafilter.

(APP.2.3) DEFINITION:

Let X be a superstructure (see (0.1.1)), J an infinite set, and $\mathcal{U}$ an ultrafilter of subsets of J. If f,g are functions from J into X, $f,g \in X^J$, call

$$f \in_{\mathcal{U}} g \text{ iff } \{j \in J : f(j) \in g(j)\} \in \mathcal{U}$$

$$f =_{\mathcal{U}} g \text{ iff } \{j \in J : f(j) = g(j)\} \in \mathcal{U} .$$

(APP.2.4) EXERCISE:

Here are some simples exercises for the reader to get used to these definitions. Show that for any $f,g \in X^J$,

(1) $=_{\mathcal{U}}$ is an equivalence relation;

(2) $f \neq_{\mathcal{U}} g$ iff $\{j \in J : f(j) \neq g(j)\} \in \mathcal{U}$;

(3) $f \notin_{\mathcal{U}} g$ iff $\{j \in J : f(j) \notin g(j)\} \in \mathcal{U}$;

(4) $f \subseteq_{\mathcal{U}} g$ (i.e., all $\in_{\mathcal{U}}$-members of f are $\in_{\mathcal{U}}$-members of g) iff $\{j \in J : f(j) \subseteq g(j)\} \in \mathcal{U}$;

(5) if the ultrafilter is free, that is, if $\cap[U : U \in \mathcal{U}] = \emptyset$, and $f(j) = g(j)$ except for finitely many j's, then $f =_{\mathcal{U}} g$.

For the rest of this Appendix, X will be a fixed superstructure over a fixed ground set X_0 .

As a consequence of Łos' Theorem (see Stroyan-Luxemburg, [1976, Theorem (3.8.3)], or Barwise [1978, Ch. A.3, Theorem 3.1]), we know that for the embedding of X into X^J given by

$$x \dashrightarrow x' , \text{ with } x'(j) = x \text{ for all } j \in J ,$$

an analog of the Transfer Principle (0.2.3)(a) holds, <u>provided</u> that in *transformed sentences, $\in$ is changed into $\in_{\mathcal{U}}$ and $=$ into $=_{\mathcal{U}}$. When we deal with nonstandard extension, though, we want to replace $\in_{\mathcal{U}}$, $=_{\mathcal{U}}$ by actual 'belongs to' and 'equals to'. That is why we make the next construction.

(APP.2.5) DEFINITION:

Let $\mathcal{U}$ be any ultrafilter over J. The Mostowski collapsing function M is defined as follows.

(i) Take an element f of X^J whose images are all in X_0 and call $M(f)$ the corresponding $=_{\mathcal{U}}$-equivalence class; let Y_0 be the set of those equivalence classes interpreted as new individuals.

(ii) Assume that for some $p \geqq 0$ we have defined $M(f)$ for each $f \in X_p{}^J$; we are going to define it for the elements in $X_{p+1}{}^J \setminus X_p{}^J$: if f is in $X_{p+1}{}^J$ and is not in $X_p{}^J$,

$$M(f) = \{M(g) : g \in X_p{}^J , \; g \in_{\mathcal{U}} f\}$$

(APP.2.6) DEFINITION:

With the same notations as above, call $\mathcal{Y}$ the superstructure on the ground set Y_0, and define $* : \mathcal{X} \dashrightarrow \mathcal{Y}$ by

$$\text{for all } x \in \mathcal{X} , \quad {}^*x = M(x') ,$$

where x' is the constant function $x'(j) = x$ for all $j \in J$. The ultrapower $\mathcal{X}^J/\mathcal{U}$ is the image of $\mathcal{X}$ under $*$.

(APP.2.7) REMARK:

The map $*$ is a superstructure extension. This means (see its definition in the comments before (0.4.6)):

(i) $*$ is injective; the reader is invited to supply a proof of this.

(ii) $*$ satisfies part (a) of Leibniz' Principle; this is

not difficult to prove, but it is outside the scope of this Appendix, since a much more careful treatment of bounded formulas is needed. For a full proof of this and the Internal Definition Principle, we refer to any of the following books: Stroyan-Luxemburg [1976, Chapter 3, Sections 4 & 8], Davis [1977, Chapter 1, Sections 7 & 8], Barwise [1978, Chapters A.6 & A.3]; and for an elementary version, to Keisler [1976, Section 1D*].

(APP.2.8) REMARK:

Moreover, whenever the ultrafilter $\mathcal{U}$ contains a descending chain of sets with empty intersection, the map * is not onto (prove it), so it becomes a meaningful extension. Examples of such ultrafilters are not hard to find; maybe the simplest one is a maximal filter of parts of N that contains the so-called Fréchet filter

$$\{A \subseteq N : N \setminus A \text{ is finite}\}$$

(apply Zorn's Lemma to prove the existence of such a maximal filter).

However, we are interested in superstructure extensions that enjoy the stronger property of being enlargements, in the sense of (0.4.6). The existence of enlargements can be established through the Compactness Theorem ("a set of sentences is consistent if each of its finite subsets is consistent"), see Robinson [1966, Chapter 2] for that approach; instead, we prefer to show that for 'adequate' ultrafilters (with a property stronger than that of the descending chain), the ultrapower is not only a true extension, but also an enlargement of the original superstructure.

(APP.2.9) DEFINITION:

We say the ultrapower $X^J/\mathcal{U}$ is <u>adequate</u> when for every nonempty family $\mathcal{B}$ of subsets of X with the finite intersection property (i.e., such that every finite subfamily of $\mathcal{B}$ has nonempty intersection), there is a map $s \in X^J$ so that

$$\forall B \in \mathcal{B} , \exists U \in \mathcal{U} , B \supseteq s(U) .$$

(APP.2.10) EXAMPLES:

(1) Let J be the set of finite parts of the power set of X , $J = P_f(P(X))$, and $\mathcal{U}$ an ultrafilter that contains the sets

$$\{ j \in J : A \in j \}, \quad A \in X .$$

We claim the ultrapower $X^J/\mathcal{U}$ is adequate. If $\mathcal{B} \in P(X)$ has the finite intersection property, then for each $j \in J$, the set

$$A_j = \cap[B \in \mathcal{B} : B \in j|$$

is not empty. Let $s:J \dashrightarrow X$ be a choice function

$$s(j) \in A_j , \quad j \in J .$$

Then for every $B \in \mathcal{B}$, it is obvious that

$$B \supseteq s(\{j \in J : B \in J\}) .$$

(2) Call a binary relation $r \in X$ <u>concurrent</u> if for every finite subset of its domain, $S \subseteq dom(r)$, there is always an element $b \in X$ such that

$$\forall a \in S , (a,b) \in r .$$

Let J be the cartesian product of the family

$$\{P_f(dom(r)) : r \in X \text{ is a concurrent relation}\}$$

i.e., the set of all maps j defined on the set of concurrent relations of X and such that the image $j(r)$ of r is a finite subset of its domain. (It is obvious that concurrent relations do exist, so J is not empty.) Next, take as $\mathcal{U}$ an ultrafilter that contains the sets

$U(j_0) = \{j \in J : \text{if } r \in X \text{ is a concurrent relation, then } j_0(r) \subseteq j(r)\}$.

Then the ultrapower $X^J/\mathcal{U}$ is adequate: let $\mathcal{B} \in \mathcal{P}(X)$ be a family with the finite intersection property, and define the following binary relation:

$$(B,x) \in r_0 \quad \text{iff} \quad x \in B \in \mathcal{B} ;$$

r_0 is a member of X and is concurrent: for each $j \in J$, the set $\cap j(r_0)$ is not empty. Let $s : J \dashrightarrow X$ be a choice function

$$s(j) \in \cap j(r_0) , \quad j \in J .$$

For every B in $\mathcal{B}$, take a $j_B \in J$ so that $j_B(r_0) = \{B\}$. Then, if $j \in U(j_B)$, the set B is a member of $j(r_0)$; therefore, by definition of s, $s(j) \in B$. We have shown that $B \supseteq s(U(j_B))$.

(APP.2.11) PROPOSITION:

Every adequate ultrapower is an enlargement.

PROOF:

Let A be any member of X. We need to show that there is a *finite entity F in $X^J/\mathcal{U}$ such that $^\sigma A \subseteq F$.

The family

$$\{F \in \mathcal{P}_f(A) : a \in F\} , \quad a \in A ,$$

has the finite intersection property. Since the ultrapower is adequate, there is a map $s : J \dashrightarrow X$ so that for each $a \in A$

$$s(U) \subseteq \{F \in \mathcal{P}_f(A) : a \in F\}$$

for some U in $\mathcal{U}$. Then

$$\{j \in J : a \in s(j)\} \in \mathcal{U} .$$

By the definition of M, it follows that $^*a \in M(s) \in {}^*\mathcal{P}_f(A)$.

Thus, $M(s)$ is *finite and contains all $^\sigma A = \{^*a : a \in A\}$.

More simple ultrafilters, like the one mentioned in Remark (APP.2.8), enjoy weaker forms of the Saturation and Comprehension Principles (0.4.2) and (0.4.3). We refer the interested reader to Stroyan-Luxemburg [1976] for a more detailed discussion of properties of this type.

INDEX